Borel Games

Borel Games are multiplayer games with infinite horizon and general payoff functions. These types of games, first introduced by Gale and Stewart (1953), were studied within descriptive set theory in the context of two-player zero-sum games. Only recently have they attracted attention from the broader game theory community. This book is the first attempt to present a comprehensive exploration of Borel Games in a single volume.

The book can be used as a main text for a graduate-level course on Borel Games, or as a supplementary text for a more general course on game theory. Students are assumed to be familiar with set theory and to have a basic understanding of general game theory.

Features

- Replete with exercises, with solutions available online for course instructors
- Includes a selection of open problems to invite further study
- The first comprehensive coverage of Borel Games in a single volume.

Borel Games

Eilon Solan

CRC Press
Taylor & Francis Group
Boca Raton London New York

CRC Press is an imprint of the
Taylor & Francis Group, an **informa** business

A CHAPMAN & HALL BOOK

Designed cover image: @evilcreatives

First edition published 2026
by CRC Press
2385 NW Executive Center Drive, Suite 320, Boca Raton FL 33431

and by CRC Press
4 Park Square, Milton Park, Abingdon, Oxon, OX14 4RN

CRC Press is an imprint of Taylor & Francis Group, LLC

ISBN: 978-1-032-94865-2 (hbk)
ISBN: 978-1-032-94869-0 (pbk)
ISBN: 978-1-003-58210-6 (ebk)

DOI: 10.1201/9781003582106

Typeset in Latin Modern font
by KnowledgeWorks Global Ltd.

Publisher's note: This book has been prepared from camera-ready copy provided by the author.

Access the Support Material: https://www.routledge.com/9781032948652

To my parents, Chaim and Zafrira.

Contents

Author Biography

Eilon Solan is a professor at the School of Mathematical Sciences at Tel Aviv University, specializing in Game Theory. He earned his Ph.D. in 1998 from the Hebrew University of Jerusalem and spent two years at the Kellogg School of Management at Northwestern University before joining Tel Aviv University.

Solan's research focuses on dynamic games, including stochastic games, stopping games, games with vector payoffs, infinite horizon games with general payoffs, and dynamic decision problems. He has made significant contributions to the study of equilibrium existence in dynamic games and the impact of information in games.

Solan co-authored the textbook *Game Theory* with Michael Maschler and Shmuel Zamir, and authored the graduate-level text *Introduction to Stochastic Game Theory*.

CHAPTER 1

Introduction

This book lies in the area of Game Theory and focuses on multiplayer games with infinite horizon and general payoff functions. We will examine both alternating-move and simultaneous-move games, as well as two-player zero-sum and multiplayer settings. Additionally, we will explore games in which players observe their opponent's moves with a delay.

These types of games, first introduced by Gale and Stewart (1953), were studied within Descriptive Set Theory in the context of two-player zero-sum games. Only recently have they attracted attention from the broader Game Theory community.

1.1 ALTERNATING-MOVE GAMES AND ZERMELO'S THEOREM

The earliest result of modern Game Theory is Zermelo's Theorem (1913), which states that in every game where (a) two players move alternately, (b) the number of rounds in the game is bounded from above, and (c) the game can either end in a win of Player I or a win of Player II, one of the players has a *winning strategy*, namely, a strategy that guarantees a win whatever moves the other player selects.[1]

Many popular games, like Chess, Go, or Checkers, do not satisfy all three properties (a)–(c). Nevertheless, using Zermelo's Theorem, one can show that in these games, exactly one of the following three alternatives holds:

- Player I has a winning strategy;

- or Player II has a winning strategy;

- or both players have strategies that guarantee they do not lose (and so, if the players follow these strategies, the game ends with a draw).

For example, Checkers satisfies the third alternative: a player who plays Checkers perfectly will never lose a match, whether she takes the role of White or Black. For other games, like Chess or Go, we do not know which of the three alternatives applies.

---

[1]In his paper, Zermelo considered games that have three outcomes – win for Player I, win for Player II, and draw. Zermelo also allowed for infinite plays, provided such plays lead to a draw. For simplicity of presentation, we discuss in the introduction games where a draw is not possible.

DOI: 10.1201/9781003582106-1

One can prove Zermelo's Theorem by backward induction. Suppose that the game lasts at most n rounds, so that the game position after round n is *determined*: it is either a winning position for Player I or a winning position for Player II.

Consider now a position that can arise after the $(n-1)$'th round of the game, and suppose without loss of generality that it is Player I's turn to move at round n. Whatever move Player I selects, the new position will be a position after round n, so that it is determined. If Player I has a move that leads to a winning position for herself, then the current position is a winning position for Player I as well: Player I can guarantee a win by selecting a move that leads to a winning position after round n. If all moves of Player I at the current position lead to winning positions of Player II, then the current position is a winning position of Player II. We can continue with this argument by induction to earlier rounds, and ascertain which player can guarantee a win at each position of the game.

A crucial assumption for the validity of the argument we just made is that the number of rounds in the game is bounded. When the number of rounds is not bounded, the base case of the induction is not valid, and the whole proof crumbles. Here is an example of a game with infinitely many rounds. Suppose that $W \subseteq [0,1]$ is a given set. The players alternately select bits 0 or 1, thereby constructing (the binary representation of) a number x in $[0,1]$; Player I selects the bits in even positions $0, 2, 4, \ldots$, and Player II selects the bits in odd positions $1, 3, 5, \ldots$. Player I wins if $x \in W$, and Player II wins otherwise. Is this game necessarily determined? That is, is it true that for every set $W \subseteq [0,1]$, either Player I has a strategy that guarantees that the number x constructed by the players is in W, whatever bits Player II selects, or Player II has a strategy that guarantees that the number x is not in W, whatever bits Player I selects? This is one of the questions we will be addressing in this book.

1.2   CONTENT OF THE BOOK

Games with infinitely many rounds have been studied in the literature mainly when the players obtain stage payoffs, and each player evaluates her infinite stream of stage payoffs by their discounted sum or their long-run average. In this book we will study games with infinitely many rounds, both alternating-move games and simultaneous-move games, when the players' evaluations are general. We will present the classical results in this topic:

- In Chapter 3, we examine alternating-move games in which the outcome is a win for one of the players and explore the extent to which Zermelo's Theorem applies in this setting. A *play* is a sequence of moves of the two players throughout the game, and the *winning set* is the set of all plays that end with a win of Player I. We will present the result of Gale and Stewart (1953), which states that the game is determined as soon as the winning set is closed or open; the result of Martin (1975), which states that the game is determined as soon as the winning set is Borel; and the result of Martin (1970), which states that under a large cardinal axiom, the game is determined as soon as the winning set is analytic.

- In Chapter 4 we study two variations of the model: the outcome of the game is not necessarily a win or a loss, but can be an arbitrary real number (Section 4.1) and the game need not be zero-sum (Section 4.2). This chapter is based on a proof by Mertens and Neyman; see Mertens (1987).

- In Chapter 5 we study yet another variation of the model, where the players move simultaneously rather than alternately, and we present a result by Martin (1998), which states that the game is determined as soon as the winning set is Borel.

- In Chapter 7 we study a different variation of the basic model, where the players no longer observe each other's past moves. We prove that the game is determined as soon as along the play (a) no player forgets information that she knew in the past, and (b) each player eventually learns all the moves selected by the other player. This chapter is based on Shmaya (2011).

Chapter 6 is dedicated to the presentation of results from recent years about multiplayer simultaneous-move games, where the goal of each player is to maximize her own payoff.

- In Section 6.3 we prove the existence of a "good" vector of strategies, that yield to all players a relatively high payoff in all subgames. This result is based on Flesch and Solan (2024).

- In Section 6.4 we show how the players can identify deviations from a given vector of behavior strategies. This section is based on Alon, Gunby, He, Shmaya, and Solan (2024).

- In Section 6.7 we prove that every repeated game admits an ε-equilibrium for every $\varepsilon > 0$ as soon as the payoff functions are bounded, Borel measurable, and tail measurable. This section is based on Ashkenazi-Golan, Flesch, Predtetchinski, and Solan (2022).

- In Section 6.8 we prove that two-player nonzero-sum Big Match games, where the nonabsorbing payoff is tail measurable, admit an ε-equilibrium for every $\varepsilon > 0$. This section is based on Ashkenazi-Golan, Flesch, and Solan (2022).

- In Section 6.9 we prove that in every simultaneous-move game there is a subgame that admits an ε-equilibrium. This section is based on Flesch and Solan (2024).

The model and proofs use different mathematical tools, ranging from topology and measure theory to probability and stochastic processes. For the benefit of the reader, we review these tools in Chapter 2.

1.3 HOW TO USE THE BOOK

The book can be used for a graduate-level course on Borel Games. Students are assumed to be familiar with Set Theory including transfinite induction; topology; and probability on finite and infinite sample spaces, including laws of large numbers, conditional expectation, martingales, and zero-one laws. In Chapter 2 we recall the basic definitions and results that are required from these areas.

Chapters 3 (Sections 3.1–3.12), 4, and 5, present classical material that arguably should appear in any course on the topic. Among the remaining material, the instructor can make any selection that fits his or her taste and time constraints. I taught a 39-hour course that included, in addition to the material above, Sections 6.1–6.8. I also used the book in 13-week-long reading classes, where the students read the book at home and solved exercises, and we discussed the material they read in our frontal meetings.

Some of the material in the book can be taught as short topics in general classes of Game Theory. These include Section 3.4, which requires knowledge of topology in infinite product spaces; Section 6.4, which requires von Neumann's Minmax Theorem; and Section 3.13, which requires familiarity with large cardinals and analytic sets, but no prior knowledge of Game Theory.

Some proofs require theorems or exercises that appear in earlier chapters. The following table details the material from earlier chapters needed to follow each chapter.

Chapter	Required material
Chapter 3	Sections 2.1, 2.2.
Chapter 4	Sections 2.1, 2.2; Theorem 3.37; Example 3.43.
Chapter 5	Sections 2.1, 2.2, 2.4, 2.5, 2.6, 2.7; Theorems 4.11, 3.37.
Chapter 6	Sections 2.1, 2.2, 2.5, 2.7, 2.8; Theorems 4.11, 5.17, 5.46; Remark 5.32; Example 5.44.
Chapter 7	Sections 2.1, 2.2; Theorems 5.46, 5.17; Remark 5.32; Exercise 6.25.

All chapters are accompanied by exercises, that will help students digest the material and deepen their understanding. The solution to some exercises requires earlier exercises; these dependencies are mentioned at the beginning of each exercise section. Some of the exercises are mechanical, others require a deeper understanding of the material, and a few are more challenging. Exercises that require deeper insight are marked with a single star, and the more challenging ones with two stars. Solutions to the exercises are available to instructors through the publisher.

Students who find the topic interesting will find several open problems in Chapter 8. If you solve some of these problems, please let me know.

1.4 ACKNOWLEDGMENT

I thank Ziv Hellman, David Lagziel, John Levy, Ron Peretz, and Arkadi Predtetchinski for commenting on earlier versions of the book. I thank my students Uri Aviad, Vivian Bar Nir, Sahar Bashan, Itamar Bellaïche, Shelly Cohen, Yuval Cornfeld, Evyatar Freimann Barbash, Shachar Herpe, Ilay Hoshen, Gil Bar Koltun, Dean Kraizberg, Uri Kreitner, Nick Kushnir, Sean Landsberg, Roy Maulbogat, Maya Rat, Ofek Raizman-Kedar, Sahar Rofe, Auriel Rosenzweig, Natali Shalom, Gilad Unikoski, and Itamar Zangvil, who studied earlier versions of the book and provided comments that improved it. I thank Alon Amit, Omer Ben Neria, and Moti Gitik for their help in writing

Section 3.13; Ron Peretz for his help in writing Section 2.7 and Chapter 5; Alexander Kechris for proposing the solution to Exercise 6.20; and the anonymous reviewers whose suggestions improved the presentation. Finally, I thank Galit Ashkenazi-Golan, János Flesch, and Arkadi Predtetchinski (once again) for studying with me the fascinating topic of infinite-horizon games with general payoff functions. Our joint research led me to write this book.

CHAPTER 2

Mathematical Preliminaries

In this chapter, we provide definitions and results from topology, measure theory, and probability theory, which we will need throughout the book. All of these results are standard and can be found in textbooks. We will provide the proofs of some of them, but not all.

2.1 NOTATIONS

The set of non-negative integers is $\mathbb{N} = \{0, 1, 2, \ldots\}$. For every set $\mathcal{A}$ we denote by $\mathcal{A}^{\mathbb{N}}$ the set of all sequences of elements in $\mathcal{A}$, and by $\mathcal{A}^{<\mathbb{N}}$ the set of all finite sequences of elements in $\mathcal{A}$. An element of $\mathcal{A}^{\mathbb{N}}$ is denoted $x = \langle a_0, a_1, \ldots \rangle$, and an element of $\mathcal{A}^{<\mathbb{N}}$ is denoted $\langle a_0, a_1, \ldots, a_n \rangle$. The *length* of a finite sequence $p = \langle a_0, a_1, \ldots, a_n \rangle$ is denoted $\operatorname{len}(p) := n+1$. The empty sequence in $\mathcal{A}^{<\mathbb{N}}$ is denoted by $\langle \, \rangle$; by convention, its length is 0.

If $p = \langle a_0, a_1, \ldots, a_n \rangle$ and $p' = \langle a_0', a_1', \ldots, a_m' \rangle$ are two finite sequences in $\mathcal{A}^{<\mathbb{N}}$, we denote their concatenation by

$$p \circ p' := \langle a_0, a_1, \ldots, a_n, a_0', a_1', \ldots, a_m' \rangle.$$

If $\operatorname{len}(p') = 1$, that is, $m = 0$, then we write $\langle p, a_0' \rangle$ rather than $p \circ \langle a_0' \rangle$. If p' is a prefix of p we say that p *extends* p' and we write $p' \preceq p$. The relation $\preceq$ defines a partial order on T.

When $x = \langle a_0, a_1, \ldots \rangle$ and $k \in \mathbb{N}$, we denote the prefix of length $k + 1$ of x by $x_k := \langle a_0, a_1, \ldots, a_k \rangle$. We also denote $x_{-1} = \langle \, \rangle$ and $x_\infty = x$. Similarly, when $p = \langle a_0, a_1, \ldots, a_n \rangle$ and $k \leq n$, we denote the prefix of length $k + 1$ of p by $p_k := \langle a_0, a_1, \ldots, a_k \rangle$.

When the set $\mathcal{A}$ is finite, Tychonoff's Theorem ensures that $\mathcal{A}^{\mathbb{N}}$ is compact in the product topology. In particular, a sequence $(x^n)_{n \in \mathbb{N}}$ of points in $\mathcal{A}^{\mathbb{N}}$ converges to a point $x \in \mathcal{A}^{\mathbb{N}}$ if every prefix of x is a prefix of all but finitely many elements of $(x^n)_{n \in \mathbb{N}}$.

2.2 TOPOLOGY AND MEASURABILITY

In this section we provide definitions in topology and measurability that we will need in the rest of the book.

DOI: 10.1201/9781003582106-2

Definition 2.1 (Topology, open sets, closed sets, clopen sets). *When X is a set, a topology on X is a family $\mathcal{X}$ of subsets of X that satisfy the following properties:*

- *$\emptyset \in \mathcal{X}$ and $X \in \mathcal{X}$.*

- *$\mathcal{X}$ is closed under unions: if $(X_\alpha)_\alpha$ is a (not necessarily finite or countably infinite) collections of sets in $\mathcal{X}$, then $\bigcup_\alpha X_\alpha \in \mathcal{X}$.*

- *$\mathcal{X}$ is closed under finite intersections: if $(X_i)_{i=1}^n$ are sets in $\mathcal{X}$, then $\bigcap_{i=1}^n X_i \in \mathcal{X}$.*

The elements of $\mathcal{X}$ are called open sets. *A* closed set *is a set whose complement is an open set. A* clopen set *is a set that is both open and closed.*

Definition 2.2 (Topological space). *A topological space is a pair $(X, \mathcal{X})$, where X is a set and $\mathcal{X}$ is a topology on X.*

We next define by transfinite induction the concept of the *Borel hierarchy*, which is a hierarchy of families of subsets of a topological space $(X, \mathcal{X})$.

Definition 2.3 (Borel hierarchy). *Let $(X, \mathcal{X})$ be a topological space. For every countable ordinal α define the families $\Sigma_\alpha = \Sigma_\alpha(X)$, $\Pi_\alpha = \Pi_\alpha(X)$, and $\Delta_\alpha = \Delta_\alpha(X)$ of subsets of X as follows:*

- *$\Sigma_1 = \mathcal{X}$.*

- *Π_α consists of all complements of sets in Σ_α.*

- *$\Delta_\alpha := \Sigma_\alpha \cap \Pi_\alpha$.*

- *For $\alpha > 1$, a set $A \subseteq X$ is in Σ_α if $A = \bigcup_{n \in \mathbb{N}} A_n$, where for every $n \in \mathbb{N}$ there is an ordinal $\alpha_n < \alpha$ such that $A_n \in \Pi_{\alpha_n}$.*

The family Σ_1 consists of all open sets. The family Π_1 consists of all closed sets. The family Δ_1 consists of all clopen sets. The family Σ_2 consists of all countable unions of closed sets, that is, all F_σ-sets. The family Π_2 consists of all complements of sets in Σ_2, namely, all complements of countable unions of closed sets. By De Morgan's Law, Π_2 consists of all countable intersections of open sets, namely, all G_δ-sets. The family Δ_2 consists of all sets that are both F_σ and G_δ. The family Σ_3 consists of all countable unions of closed sets and of countable intersections of open sets. Etc.

The reader can verify the following.

Lemma 2.4. *For every topological space $(X, \mathcal{X})$ and every countable ordinal α,*

$$\Sigma_\alpha \subseteq \Pi_{\alpha+1} \subseteq \Sigma_{\alpha+2} \subseteq \Pi_{\alpha+3}.$$

Remark 2.5 (Strict inclusion in Lemma 2.4). *Fix $n \geq 1$. For the topological space $(\mathbb{R}^n, \mathcal{O})$, where $\mathcal{O}$ is the collection of open sets in $\mathbb{R}^n$, all the inclusions in Lemma 2.4 are strict, for each countable ordinal α.*

Definition 2.6 (Borel sets of rank α). *For every ordinal α, the sets in $\Sigma_\alpha \cup \Pi_\alpha$ are called* Borel sets of rank α.

Note that a Borel set of rank α is also a Borel set of rank β, for every ordinal $\beta > \alpha$. Below we will not use the families $(\Delta_\alpha)_\alpha$.

Remark 2.7 (On the maximal rank of Borel sets). *One may wonder why we do not define Borel sets of rank α for uncountable ordinals. The reason is that if $A = \bigcup_{n \in \mathbb{N}} A_n$, where for each $n \in \mathbb{N}$ there is a countable ordinal α_n such that $A_n \in \Pi_{\alpha_n}$, then (assuming the Axiom of Choice) there is a countable ordinal β such that $A \in \Sigma_\beta$. This ordinal is $\beta = \sup_n \alpha_n$. In particular, continuing the definition of the Borel ranks beyond countable ordinals does not add new sets.*

Definition 2.8 (Sigma-algebra). *Let X be a space. A* sigma-algebra *on X is a family $\mathcal{F}$ of subsets of X that contains the empty set, and is closed under complements and under countable unions. That is,*

- $\emptyset \in \mathcal{F}$.

- *If $A \in \mathcal{F}$, then $A^c \in \mathcal{F}$.*

- *If $(A_n)_{n \in \mathbb{N}}$ is a sequence of sets in $\mathcal{F}$, then $\bigcup_{n \in \mathbb{N}} A_n \in \mathcal{F}$.*

Definition 2.9 (Generated sigma-algebra). *A family $\mathcal{Y}$ of subsets of X generates a sigma-algebra $\mathcal{X}$ if $\mathcal{X}$ is the smallest sigma-algebra of X that contains all the subsets in $\mathcal{Y}$. In such a case, we will write $\mathcal{X} = \sigma(\mathcal{Y})$.*

The Borel sigma-algebra is generated by the family of all open sets.

Definition 2.10 (Borel sigma-algebra). *When $(X, \mathcal{X})$ is a topological space, the* Borel sigma-algebra, *denoted $\mathcal{B}(X)$, is the sigma-algebra generated by $\mathcal{X}$; namely, it is the smallest sigma-algebra that contains all open sets.*

Lemma 2.4 implies that $\bigcup_\alpha \Sigma_\alpha(X) = \bigcup_\alpha \Pi_\alpha(X)$. The reader can verify that this union is equal to the Borel sigma-algebra.

Definition 2.11 (Measurable space). *A* measurable space *is a pair $(X, \mathcal{F})$, where X is a set and $\mathcal{F}$ is a sigma-algebra on X. The elements of $\mathcal{F}$ are called* measurable sets.

Definition 2.12 (Continuous function, measurable function). *Let $(X, \mathcal{X})$ and $(Y, \mathcal{Y})$ be two topological spaces. A function $f : X \to Y$ is* continuous *if the inverse image of every open set in Y is open in X. A function $f : X \to Y$ is* Borel measurable *if the inverse image of every set in $\mathcal{B}(Y)$ is in $\mathcal{B}(X)$.*

As the following lemma states, the rank of a set in the Borel hierarchy is preserved under the inverse of a continuous function.

Lemma 2.13. *Let $f : X \to Y$ be a continuous function between two topological spaces $(X, \mathcal{X})$ and $(Y, \mathcal{Y})$. If $A \subseteq Y$ is a Borel set of rank α, then $f^{-1}(A)$ is a Borel set of rank α.*

Proof. We will prove by transfinite induction the following claims for all countable ordinals α:

H.1) If $A \in \Sigma_\alpha(Y)$, then $f^{-1}(A) \in \Sigma_\alpha(X)$.

H.2) If $A \in \Pi_\alpha(Y)$, then $f^{-1}(A) \in \Pi_\alpha(X)$.

Step 1: Base of the induction: Condition (H.1) for $\alpha = 1$.

If $A \in \Sigma_1(Y)$, then A is open in Y. Since f is continuous, $f^{-1}(A)$ is open in X, so that $f^{-1}(A) \in \Sigma_1(X)$.

Step 2: If (H.1) holds for all $A \in \Sigma_\alpha(Y)$, then (H.2) holds for all $A \in \Pi_\alpha(Y)$.

Let α be a countable ordinal, and suppose that (H.1) holds for every $A \in \Sigma_\alpha(Y)$. Let now $A \in \Pi_\alpha(Y)$. Then $A^c \in \Sigma_\alpha(Y)$. By (H.1), $f^{-1}(A^c) \in \Sigma_\alpha(X)$. We conclude that

$$f^{-1}(A) = X \setminus f^{-1}(A^c) \in \Pi_\alpha(X).$$

Step 3: If (H.2) holds for all $A \in \bigcup_{\beta<\alpha} \Pi_\beta(Y)$, then (H.1) holds for all $A \in \Sigma_\alpha(Y)$.

Let α be a countable ordinal, and suppose that (H.2) holds for all $A \in \bigcup_{\beta<\alpha} \Pi_\beta(Y)$. Let now $A \in \Sigma_\alpha(Y)$. Then $A = \bigcup_{n\in\mathbb{N}} A_n$, where for every $n \in \mathbb{N}$ there is an ordinal $\beta_n < \alpha$ such that $A_n \in \Pi_{\beta_n}(Y)$. By the induction hypothesis, $f^{-1}(A_n) \in \Pi_{\beta_n}(X)$. Hence,

$$f^{-1}(A) = \bigcup_{n\in\mathbb{N}} f^{-1}(A_n) \in \Sigma_\alpha(X).$$

$\square$

The sum and product of real-valued measurable functions is a measurable function, and the composition of measurable functions is a measurable function. As the next result states, the limit of uniformly bounded measurable functions is measurable as well.

Lemma 2.14. *Let $(X, \mathcal{F})$ be a measurable space. If $(f_n)_{n\in\mathbb{N}}$ is a sequence of uniformly bounded measurable functions from X to $\mathbb{R}$, where $\mathbb{R}$ is equipped with the Borel sigma-algebra, then $\limsup_{n\to\infty} f_n$ is Borel measurable as well.*

Definition 2.15 (Relative topology). *Let X be a set, let $\mathcal{X}$ be a topology on X, and let $Y \subseteq X$. The relative topology on Y is the family of sets*

$$\mathcal{Y} := \{A \cap Y : A \in \mathcal{X}\}.$$

As the next result states, the relative topology is indeed a topology.

Lemma 2.16. *When $(X, \mathcal{X})$ is a topological space and $Y \subset X$, the pair $(Y, \mathcal{Y})$ is a topological space.*

2.3 REGULAR PROBABILITY DISTRIBUTIONS

In this section we recall the definition of a probability distribution and the notion of regular probability distribution. We will use these notions in Chapters 5, 6, and 7. Theorem 2.20 is used in Sections 6.4, 6.8, and 6.9.

Definition 2.17 (Probability distribution). *A probability distribution over the measurable space* $(X, \mathcal{X})$ *is a function* $\mathbf{P} : \mathcal{X} \to [0, 1]$ *that satisfies the following properties:*

- $\mathbf{P}(\emptyset) = 0$.

- $\mathbf{P}(W^c) = 1 - \mathbf{P}(W)$, *for every* $A \in \mathcal{X}$.

- $\mathbf{P}(\bigcup_{k=0}^{\infty} A_k) = \sum_{k=0}^{\infty} \mathbf{P}(A_k)$ *for every sequence* $(A_k)_{k=0}^{\infty}$ *of pairwise disjoint sets in* $\mathcal{X}$.

A probability distribution is also called *countably additive probability.*

Notation 2.18 (Set of probability distributions). *For every measurable space* X, *denote by* $\Delta(X)$ *the set of all probability distributions on* X. *When* X *is finite or countable,*

$$\Delta(X) = \left\{ x \in \mathbb{R}_+^X : \sum_{a \in X} x(a) = 1 \right\}.$$

Definition 2.19 (Regular probability distribution). *A probability distribution* $\mathbf{P}$ *on a topological space* X *is* regular *if for every* $\varepsilon > 0$ *and every Borel set* $A \subseteq X$ *there is a compact set* $B \subseteq A$ *such that* $\mathbf{P}(B) \geq \mathbf{P}(A) - \varepsilon$.

The following result states that when A is a finite set, every probability measure on $A^{\mathbb{N}}$ is regular.

Theorem 2.20. *Let* A *be a finite set. Equip* $A^{\mathbb{N}}$ *with the Borel sigma-algebra* $\mathcal{B}(\mathcal{O})$, *where* $\mathcal{O}$ *is the product topology on* $A^{\mathbb{N}}$. *Every probability measure on* $A^{\mathbb{N}}$ *is regular.*

2.4 ALGEBRAS AND CARATHÉODORY'S EXTENSION THEOREM

In this section we recall the definition of an algebra, and we state Carathéodory's Extension Theorem. We will use this theorem in Chapters 5, 6, and 7.

Definition 2.21 (Algebra). *Let* X *be a set. A set* $\mathcal{Y}$ *of subsets of* X *is an* algebra *if it contains the empty set, and is closed under complements and under finite unions.*

Every sigma-algebra is an algebra, but the converse is not true. The analogous concept of a probability distribution for algebras is the following.

Definition 2.22 (Finitely additive probability). *Let* X *be a set, and let* $\mathcal{Y}$ *be an algebra on* X. *A* finitely additive probability *on the pair* $(X, \mathcal{Y})$ *is a function* $\mathbf{P} : \mathcal{Y} \to [0, 1]$ *that satisfies the following properties:*

- $\mathbf{P}(\emptyset) = 0$.

- $\mathbf{P}(A^c) = 1 - \mathbf{P}(A)$, *for every* $A \in \mathcal{Y}$.

- $\mathbf{P}(\bigcup_{k=0}^{n} A_k) = \sum_{k=0}^{n} \mathbf{P}(A_k)$ *for every sequence* $(A_k)_{k=0}^{n}$ *of pairwise disjoint sets in* $\mathcal{Y}$.

Let $\mathcal{Y}$ be an algebra on X, and suppose we are given a finitely additive probability $\mathbf{P}$ on $(X, \mathcal{Y})$. Can we extend $\mathbf{P}$ to a countably additive probability on $(X, \mathcal{X})$, where $\mathcal{X}$ is the sigma-algebra generated by $\mathcal{Y}$? A necessary condition that this can be done is that if we happen to have a sequence $(A_k)_{k \in \mathbb{N}}$ of disjoint subsets in $\mathcal{Y}$ such that $\bigcup_{k \in \mathbb{N}} A_k \in \mathcal{Y}$, then necessarily $\sum_{k \in \mathbb{N}} \mathbf{P}_{\mathcal{Y}}(A_k) = \mathbf{P}_{\mathcal{Y}}(\bigcup_{k \in \mathbb{N}} A_k)$. As the following theorem states, this condition is also sufficient. This result, which is called the Carathéodory Extension Theorem, was probably first proved by Fréchet. For a proof, see Billingsley (1995, Theorem 3.1).

Theorem 2.23 (Carathéodory's Extension Theorem). *Let* $(X, \mathcal{X})$ *be a measurable space, let* $\mathcal{Y} \subseteq \mathcal{X}$ *be an algebra that generates* $\mathcal{X}$, *and let* $\mathbf{P}_{\mathcal{Y}} : \mathcal{Y} \to [0, 1]$ *be a finitely additive probability. Suppose that for every sequence* $(A_k)_{k \in \mathbb{N}}$ *of disjoint subsets in* $\mathcal{Y}$ *that satisfy* $\bigcup_{k \in \mathbb{N}} A_k \in \mathcal{Y}$ *we have* $\sum_{k \in \mathbb{N}} \mathbf{P}_{\mathcal{Y}}(A_k) = \mathbf{P}_{\mathcal{Y}}(\bigcup_{k \in \mathbb{N}} A_k)$. *Then there is a unique countably additive probability* $\mathbf{P}_{\mathcal{X}} : \mathcal{X} \to [0, 1]$ *such that*

$$\mathbf{P}_{\mathcal{X}}(A) = \mathbf{P}_{\mathcal{Y}}(A), \quad \forall A \in \mathcal{Y}.$$

2.5 LAWS OF LARGE NUMBERS

In this section we briefly remind the reader about the weak and the strong laws of large numbers. We will use these laws in Chapters 5 and 6.

Definition 2.24 (Integrable random variable). *Let* $(\Omega, \mathcal{F}, \mathbf{P})$ *be a probability space, and let* $X : \Omega \to \mathbb{R}$ *be a random variable. The random variable* X *is* integrable *if* $\int_{\Omega} |X(\omega)| d\mathbf{P}(\omega) < \infty$.

Recall that when Ω is finite,

$$\int_{\Omega} |X(\omega)| d\mathbf{P}(\omega) = \sum_{\omega \in \Omega} |X(\omega)| \mathbf{P}(\omega).$$

Definition 2.25 (Expectation). *Let* $(\Omega, \mathcal{F}, \mathbf{P})$ *be a probability space, and let* $X : \Omega \to \mathbb{R}$ *be an integrable random variable. The* expectation *of* X, *denoted* $\mathbf{E}[X]$, *is given by*

$$\mathbf{E}[X] := \int_{\Omega} X(\omega) d\mathbf{P}(\omega).$$

Theorem 2.26 (Weak law of large numbers). *Let* $(\Omega, \mathcal{F}, \mathbf{P})$ *be a probability space, and let* $(X_n)_{n \in \mathbb{N}}$ *be a sequence of independent and identically distributed (i.i.d.) integrable random variables with mean* μ. *For every* $n \in \mathbb{N}$ *define the random variable* $\overline{X}_n$ *by*

$$\overline{X}_n := \frac{1}{n+1} \sum_{k=0}^{n} X_k.$$

Then

$$\lim_{n\to\infty} \mathbf{P}(|\overline{X}_n - \mu| < \varepsilon) = 1.$$

Theorem 2.27 (Strong law of large numbers). *Let* $(\Omega, \mathcal{F}, \mathbf{P})$ *be a probability space, and let* $(X_n)_{n\in\mathbb{N}}$ *be a sequence of i.i.d. integrable random variables with mean* μ. *For every* $n \in \mathbb{N}$ *denote*

$$\overline{X}_n := \frac{1}{n+1} \sum_{k=0}^{n} X_k.$$

Then

$$\mathbf{P}\left(\lim_{n\to\infty} \overline{X}_n = \mu\right) = 1.$$

The reader can find proofs of the weak and strong laws of large numbers in any textbook on probability, e.g., Feller (1971).

2.6 CONDITIONAL EXPECTATION

In this section we define the concept of conditional expectation. The interested reader is referred to Williams (1991) or Billingsley (1995) for more material on this topic. We will use conditional expectation in Section 2.7 and Chapter 5.

Definition 2.28 (Conditional expectation). *Let* $(\Omega, \mathcal{F}, \mathbf{P})$ *be a probability space, let* $X : \Omega \to \mathbb{R}$ *be an integrable random variable, and let* $\mathcal{G} \subseteq \mathcal{F}$ *be a sub-sigma-algebra. The random variable* $Y : \Omega \to \mathbb{R}$ *is called the* conditional expectation of X given $\mathcal{G}$ *if the following properties hold:*

1. *Y is measurable with respect to $\mathcal{G}$.*

2. *$\mathbf{E}[\mathbf{1}_A \cdot Y] = \mathbf{E}[\mathbf{1}_A \cdot X]$ for every subset $A \in \mathcal{G}$.*

The conditional expectation Y is often denoted by $\mathbf{E}[X \mid \mathcal{G}]$.

Example 2.29 (Conditional expectation on the unit square). *Let* $\Omega = [0,1]^2$, *let* $\mathcal{F}$ *be the Borel sigma-algebra on* $[0,1]^2$, *and let* $\mathbf{P}$ *be the uniform distribution on* Ω. *Let* $\mathcal{G}$ *be the sigma-algebra generated by the first coordinate:*

$$\mathcal{G} := \{A \times [0,1] \colon A \in \mathcal{B}([0,1])\},$$

i.e., a set is in $\mathcal{G}$ if and only if it is a rectangle: the product of a measurable set in $[0,1]$ *times the whole interval* $[0,1]$.
  Define a random variable X by

$$X(\omega_1, \omega_2) = \omega_1 + \omega_2, \quad \forall \omega = (\omega_1, \omega_2) \in [0,1]^2.$$

The conditional expectation of X given $\mathcal{G}$ is the random variable

$$Y(\omega) = \omega_1 + \frac{1}{2}, \quad \forall \omega = (\omega_1, \omega_2) \in \Omega.$$

The following result lists several useful properties of conditional expectations. For a proof, see, e.g., Williams (1991, Chapter 9).

Theorem 2.30 (Properties of conditional expectations). *Let $(\Omega, \mathcal{F}, \mathbf{P})$ be a probability space, let $X : \Omega \to \mathbb{R}$ be an integrable random variable, and let $\mathcal{G} \subseteq \mathcal{F}$ be a sub-sigma-algebra. The following statements hold.*

1. *(Existence) The conditional expectation $\mathbf{E}[X \mid \mathcal{G}]$ exists.*

2. *(Uniqueness) Let Y_1 and Y_2 be two random variables that satisfy properties (1) and (2) in Definition 2.28. Then $Y_1 = Y_2$ a.s., that is $\mathbf{P}(Y_1 = Y_2) = 1$.*

3. *(Positivity) If $X \geq 0$ a.s., then $\mathbf{E}[X \mid \mathcal{G}] \geq 0$ a.s.*

4. *(Identity) If X is measurable with respect to $\mathcal{G}$, then $\mathbf{E}[X \mid \mathcal{G}] = X$ a.s.*

5. *(Linearity) If X_1 and X_2 are integrable random variables, and $a_1, a_2 \in \mathbb{R}$, then*

$$\mathbf{E}[a_1 X_1 + a_2 X_2 \mid \mathcal{G}] = a_1 \mathbf{E}[X_1 \mid \mathcal{G}] + a_2 \mathbf{E}[X_2 \mid \mathcal{G}] \ a.s.$$

6. *(Monotonicity) If $(X_n)_{n \in \mathbb{N}}$ is an increasing sequence of non-negative random variables that converge to X, then*

$$\mathbf{E}[X \mid \mathcal{G}] = \lim_{n \to \infty} \mathbf{E}[X_n \mid \mathcal{G}] \ a.s.$$

7. *(Law of Iterated Expectation) If $\mathcal{H} \subseteq \mathcal{G}$ is a second sub-sigma-algebra, then*

$$\mathbf{E}[\mathbf{E}[X \mid \mathcal{G}] \mid \mathcal{H}] = \mathbf{E}[X \mid \mathcal{H}] \ a.s.$$

8. *(Product) If $Z : \Omega \to \mathbb{R}$ is an integrable random variable that is measurable with respect to $\mathcal{G}$, then*

$$\mathbf{E}[ZX \mid \mathcal{G}] = Z\mathbf{E}[X \mid \mathcal{G}] \ a.s.$$

2.7 MARTINGALES AND THE MARTINGALE CONVERGENCE THEOREM

In this section, we define the concept of martingales and present the Martingale Convergence Theorem. The interested reader is referred to Williams (1991) or Billingsley (1995) for more material on martingales. We will use martingales in Chapter 5 and Section 6.9.

Definition 2.31 (Stochastic process). *A stochastic process on a probability space $(\Omega, \mathcal{F}, \mathbf{P})$ is a sequence of random variables, i.e., a sequence $(X_n)_{n \in \mathbb{N}}$, where $X_n : \Omega \to \mathbb{R}$ is $\mathcal{F}$-measurable for every $n \in \mathbb{N}$.*

Definition 2.32 (Filtration). *Let $(X, \mathcal{F})$ be a measurable space. A filtration on $(X, \mathcal{F})$ is an increasing sequence of sigma-algebras $\mathcal{F}_0 \subseteq \mathcal{F}_1 \subseteq \cdots \subseteq \mathcal{F}$. A filtration will be denoted by $(\mathcal{F}_n)_{n \in \mathbb{N}}$.*

Example 2.33 (A filtration on the unit interval). *Consider* $([0, 1), \mathcal{B})$, *the half-open half-closed unit interval equipped with the Borel sigma-algebra. For every* $n \in \mathbb{N}$, *let* $\mathbb{Q}_n$ *be the set of all rational numbers whose denominator is* n, *i.e.,* $\mathbb{Q}_n := \{\frac{k}{n} : k \in \mathbb{Z}\}$. *Let* $\mathcal{F}_n$ *be the set of all (finite) unions of sub-intervals of* $[0, 1)$ *of the form* $[a, b)$, *where* $a, b \in \mathbb{Q}_n$. *For every* $n \geq 1$, $\mathcal{F}_n$ *is a sub-sigma-algebra of* $\mathcal{B}$. *Furthermore, if* n *divides* m, *then* $\mathcal{F}_n \subset \mathcal{F}_m$. *It follows that* $\{\mathcal{F}_{2^n}\}_{n=1}^{\infty}$ *is a filtration on* $([0, 1), \mathcal{B})$.

Example 2.34 (A filtration on $\{0, 1\}^{\mathbb{N}}$). *Let* $X = \{0, 1\}^{\mathbb{N}}$. *Define a topology on* X *as follows: for every* $n \in \mathbb{N}$ *and every* $p \in \{0, 1\}^n$, *the set of all infinite sequences that extend* p, *i.e.,*

$$T_p := \{x \in X : p \prec x\},$$

is a basic open set. Denote by $\mathcal{B}$ *the smallest sigma-algebra that contains all basic open sets. Let* $\mathcal{F}_n$ *be the collection of all unions of sets from* $(T_p)_{p \in \{0,1\}^n}$. *The sequence* $(\mathcal{F}_n)_{n \in \mathbb{N}}$ *is a filtration on* X.

Example 2.35. *Let* $X_0, X_1, \ldots$ *be random variables on a measurable space* $(\Omega, \mathcal{F})$. *For every* $n \in \mathbb{N}$, *let* $\mathcal{F}_n$ *be the sigma-algebra generated by* $X_0, \ldots, X_n$. *That is,* $\mathcal{F}_n$ *consists of all the events of the form* $\{(X_0, \ldots, X_n) \in A\}$, *where* A *is a Borel subset of* $\mathbb{R}^n$. *Then,* $(\mathcal{F}_n)_{n \in \mathbb{N}}$ *is a filtration on* $(\Omega, \mathcal{F})$.

Definition 2.36 (Adapted stochastic process). *Let* $(X, \mathcal{F}, P)$ *be a probability space, and let* $(\mathcal{F}_n)_{n \in \mathbb{N}}$ *be a filtration. A stochastic process* $(X_n)_{n \in \mathbb{N}}$ *is adapted to* $(\mathcal{F}_n)_{n \in \mathbb{N}}$ *if* X_n *is* $\mathcal{F}_n$-*measurable, for every* $n \in \mathbb{N}$.

Example 2.34, continued. Let $f : \{0, 1\}^{\mathbb{N}} \to [0, 1]$ be a function. For every $n \in \mathbb{N}$ define

$$X_n(x) := f(\langle a_0, a_1, \ldots, a_{n-1}, 0, 0, 0, \ldots \rangle), \quad \forall x = \langle a_0, a_1, \ldots \rangle \in \{0, 1\}^{\mathbb{N}}.$$

The sequence $(X_n)_{n \in \mathbb{N}}$ is a stochastic process adapted to $(\mathcal{F}_n)_{n \in \mathbb{N}}$. ▽

Definition 2.37 (Super-martingale, sub-martingale, martingale). *Let* $(X, \mathcal{F}, P)$ *be a probability space, and let* $(\mathcal{F}_n)_{n \in \mathbb{N}}$ *be a filtration on this space. A stochastic process* $(X_n)_{n \in \mathbb{N}}$ *adapted to* $(\mathcal{F}_n)_{n \in \mathbb{N}}$ *is a* super-martingale *if (a)* X_n *is integrable for every* $n \in \mathbb{N}$, *and (b)* $\mathbf{E}[X_{n+1}|\mathcal{F}_n] \leq X_n$ *for every* $n \in \mathbb{N}$.
    The stochastic process $(X_n)_{n \in \mathbb{N}}$ *is a* sub-martingale, *if it satisfies (a) and (b')* $\mathbf{E}[X_{n+1}|\mathcal{F}_n] \geq X_n$ *for every* $n \in \mathbb{N}$; *that is, the stochastic process* $(-X_n)_{n \in \mathbb{N}}$ *is a super-martingale. The stochastic process* $(X_n)_{n \in \mathbb{N}}$ *is a* martingale *if it is both a super-martingale and a sub-martingale.*

Example 2.38 (Asymmetric random walk). *Let* $(Y_n)_{n \in \mathbb{N}}$ *be a sequence of i.i.d. random variables such that* $\mathbf{P}(Y_n = +1) = 1 - \mathbf{P}(Y_n = -1) = \frac{1}{3}$ *for every* $n \in \mathbb{N}$. *For every* $n \in \mathbb{N}$ *let* $\mathcal{F}_n$ *be the sigma-algebra generated by* $Y_0, \ldots, Y_n$. *Define* $X_n = \sum_{i=0}^{n} Y_i$, *for every* $n \in \mathbb{N}$. *Then,* $X_0, X_1, \ldots$ *is a super-martingale adapted to* $(\mathcal{F}_n)_{n \in \mathbb{N}}$, *since*

$$\mathbf{E}[X_{n+1} \mid \mathcal{F}_n] = \frac{1}{3}(X_n + 1) + \frac{2}{3}(X_n - 1) = X_n - \frac{1}{3} < X_n, \quad \forall n \in \mathbb{N}.$$

Example 2.39 (Random walk with barrier). *Define random variables $(X_n)_{n\in\mathbb{N}}$ recursively as follows:*

$$X_0 = 1, \quad \text{with probability 1,}$$

and conditional on X_n,

$$X_{n+1} = \begin{cases} X_n + 1, & \text{with probability } \frac{1}{2}, \text{ if } X_n > 0, \\ X_n - 1, & \text{with probability } \frac{1}{2}, \text{ if } X_n > 0, \\ 0, & \text{with probability 1, if } X_n = 0. \end{cases}$$

Then $\{X_n\}_{n\in\mathbb{N}}$ is a martingale (adapted to the filtration $\{\mathcal{F}_n\}_{n\in\mathbb{N}}$ of sigma-algebras generated by $X_1, \ldots, X_n$).

The Law of Iterated Expectation (see Theorem 2.30(7)) implies the chain rule of conditional expectations: given a probability space $(\Omega, \mathcal{F}, \mathbf{P})$ and an integrable random variable X, we have $\mathbf{E}[\mathbf{E}[X|\mathcal{G}]] = \mathbf{E}[X]$, for any sub-sigma-algebra $\mathcal{G}$ of $\mathcal{F}$. It follows that if $(X_n)_{n\in\mathbb{N}}$ is a super-martingale, then $\mathbf{E}[X_0], \mathbf{E}[X_1], \ldots$ is a non-increasing sequence of numbers. In fact, the notion of a super-martingale is a generalization of the notion of a non-increasing sequence of real numbers. The following theorem generalizes the result that bounded non-increasing sequences converge.

Theorem 2.40 (Doob's Martingale Convergence Theorem). *Let $(X_n)_{n\in\mathbb{N}}$ be a super-martingale such that $X_n \geq 0$ a.s., for every $n \in \mathbb{N}$. Then there exists a random variable X such that the limit $X := \lim_{n\to\infty} X_n$ exists a.s.*

In the proof of Theorem 2.40 we utilize Fatou's lemma which we state now.

Theorem 2.41 (Fatou's Lemma). *If $(X_n)_{n\in\mathbb{N}}$ are non-negative random variables, then*

$$\mathbf{E}\left[\liminf_{n\to\infty} X_n\right] \leq \liminf_{n\to\infty} \mathbf{E}[X_n]. \tag{2.1}$$

Remark 2.42. *Equation (2.1) may be strict. Indeed, let $E \subseteq [0,1]$ be a measurable set with Lebesgue measure μ in $(0,1)$. Define*

$$X_n := \begin{cases} \mathbf{1}_E, & \text{if } n \text{ is even,} \\ \mathbf{1}_{E^c}, & \text{if } n \text{ is odd.} \end{cases}$$

Then $\liminf_{n\to\infty} X_n = 0$ always, whereas $\mathbf{E}[X_n]$ is either $\mu(E)$ or $1 - \mu(E)$, depending on whether n is even or odd, and hence

$$\liminf_{n\to\infty} \mathbf{E}[X_n] = \min\{\mu(E), 1 - \mu(E)\}.$$

Remark 2.43. *The convergence in Theorem 2.40 may not hold in L_1. Indeed, if the convergence $X := \lim_{n\to\infty} X_n$ were in L_1, then we would have $\mathbf{E}[X] = \lim_{n\to\infty} \mathbf{E}[X_n]$. To prove that the convergence is not in L_1, we will show that there is a non-negative martingale $(X_n)_{n\in\mathbb{N}}$ that converges to a limit X yet $\mathbf{E}[X] \neq \lim_{n\to\infty} \mathbf{E}[X_n]$.*

Take, for example, the non-negative martingale $(X_n)_{n\in\mathbb{N}}$ of Example 2.39. Theorem 2.40 implies that this martingale converges a.s. Since for almost every $\omega \in \Omega$, the *sequence* $(X_n(\omega))_{n \in \mathbb{N}}$ *is a convergent sequence of integers, it is eventually constant: there is* $n_0 = n_0(\omega) \in \mathbb{N}$ *such that* $X_n(\omega) = X_{n_0}(\omega)$ *for every* $n \geq n_0$. *The definition of* $(X_n)_{n \in \mathbb{N}}$, *together with the Borel-Cantelli Lemma, imply that the limit is 0 a.s. However, by the chain rule of conditional expectation,* $1 = \mathbf{E}[X_0] = \mathbf{E}[X_1] = \mathbf{E}[X_2] = \cdots$.

To prove Theorem 2.40, we will need the concept of a stopping time.

Definition 2.44 (Stopping time). *Let* $(X, \mathcal{F})$ *be a measurable space, and let* $\mathcal{F}_0 \subseteq \mathcal{F}_1 \subseteq \cdots \subseteq \mathcal{F}$ *be a filtration on* $(X, \mathcal{F})$. *A stopping time* for *the filtration* $(\mathcal{F}_n)_{n \in \mathbb{N}}$ *is a random variable* $T \colon X \to \mathbb{N} \cup \{\infty\}$ *satisfying*

$$\{x \in X : T(x) \leq n\} \in \mathcal{F}_n, \quad \forall n \in \mathbb{N}.$$

The definition captures the following idea. Suppose that an element $x \in X$ is determined at the outset, yet we do not know its identity. As time goes on, we gain information about x, which is captured by the sigma-algebra $\mathcal{F}_n$: as stage n, we know for each set in $\mathcal{F}_n$ whether it contains x. If $\mathcal{F}_n$ is derived from a partition of X, then our information is the element of the partition that contains x. Based on this information, we have to decide when to "stop". The random variable T is a stopping time if the information at stage n is sufficient to determine[1] whether $T = n$. Note that a stopping time can be equal to infinity,[2] which corresponds to the situation that we never "stop".

Proof of Theorem 2.40. Let $(X_n)_{n \in \mathbb{N}}$ be a non-negative super-martingale adapted to a filtration $(\mathcal{F}_n)_{n \in \mathbb{N}}$. Suppose to the contrary that $P\left(\limsup_{n \to \infty} X_n > \liminf_{n \to \infty} X_n\right) > 0$. We shall construct a non-negative super-martingale $(Y_n)_{n \in \mathbb{N}}$ adapted to $(\mathcal{F}_n)_{n \in \mathbb{N}}$ such that $\mathbf{P}(\lim_{n \to \infty} Y_n = \infty) > 0$. This will lead to a contradiction, since by Fatou's Lemma (Theorem 2.41),

$$\mathbf{E}\left[\liminf_{n \to \infty} Y_n\right] \leq \liminf_{n \to \infty} \mathbf{E}[Y_n] \leq \mathbf{E}[Y_1] < \infty.$$

As it is assumed that $\mathbf{P}\left(\limsup_{n \to \infty} X_n > \liminf_{n \to \infty} X_n\right) > 0$, there are (rational) numbers $0 \leq \alpha < \beta$ such that

$$\mathbf{P}\left(\limsup_{n \to \infty} X_n > \beta > \alpha > \liminf_{n \to \infty} X_n\right) > 0. \qquad (2.2)$$

Informally, we will define the sequence $(Y_n)_{n \in \mathbb{N}}$ as follows. At time 0, Y_0 is set equal to X_0. Then, Y_n remains constant until the first time, denoted T_1, at which X_{T_1} drops below α. After time T_1, Y_n follows the same increments as X_n until X_n hits above β; the first time this happens is denoted T_2. After time T_2, Y_n remains constant until X_n drops below a; the first time this happens is denoted T_3. And so on.

---

[1] Since the sequence $(\mathcal{F}_n)_{n \in \mathbb{N}}$ is a filtration, the information at stage n is sufficient to know whether $T = n$ if and only if the information at stage n is sufficient to know whether $T \leq n$.

[2] In the literature, stopping times are sometimes required by definition to be finite.

Formally, define a non-decreasing sequence of stopping times $(T_i)_{i \in \mathbb{N}}$ by

$$
\begin{aligned}
T_0 &:= 0, \\
T_1 &:= \inf\{n \geq 0 \colon X_n < \alpha\}, \\
T_{2i} &:= \inf\{n > T_{2i-1} : X_n > \beta\}, \quad \forall i \geq 1, \\
T_{2i+1} &:= \inf\{n > T_{2i} : X_n < \alpha\}, \quad \forall i \geq 1.
\end{aligned}
$$

with the standard convention that $\inf \emptyset = \infty$. Note that $T_i < T_{i+1}$ for every $i \in \mathbb{N}$, unless $i = 0$ and $X_0 < \alpha$, in which case $T_0 = T_1 = 0$.

Define the sequence $(Y_n)_{n \in \mathbb{N}}$ of random variables as follows (see Figure 1):

$$
\begin{aligned}
Y_n &:= X_0, & &\text{if } n = 0, \ldots, T_1, \\
Y_n &:= Y_{T_{2i+1}} + X_n - X_{T_{2i+1}}, & &\text{if } T_{2i+1} < n \leq T_{2i+2} \text{ for some } i \in \mathbb{N}, \\
Y_n &= Y_{T_{2i}}, & &\text{if } T_{2i} < n \leq T_{2i+1} \text{ for some } i \in \mathbb{N}.
\end{aligned}
$$

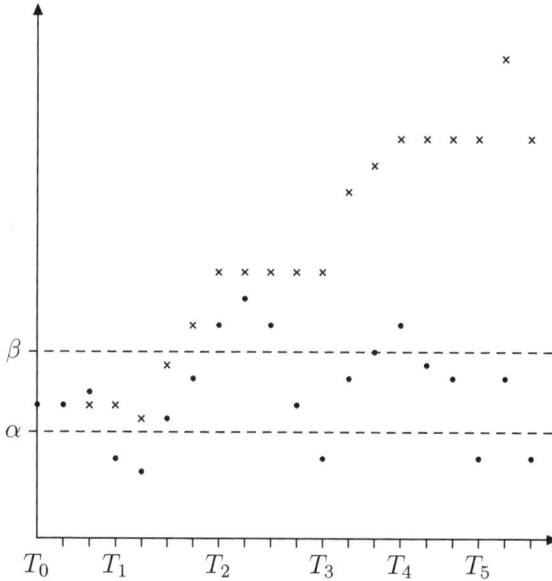

Figure 1: The sequences $(X_n)_{n \in \mathbb{N}}$ (in dots) and $(Y_n)_{n \in \mathbb{N}}$ (in x's).

The definition implies that $(Y_n)_{n \in \mathbb{N}}$ is a super-martingale, since both staying put and following the increments of X_n ensure non-increasing conditional expectations. We claim that Y_n is non-negative for every $n \in \mathbb{N}$. Since $Y_{n+1} = Y_n$ whenever $T_{2i} \leq n < T_{2i+1}$ for some $i \in \mathbb{N}$, it is sufficient to verify that $Y_n \geq 0$ whenever $T_{2i+1} \leq n < T_{2i+2}$ for some $i \in \mathbb{N}$. We will show the following stronger assertion:

$$
Y_n \geq (\beta - \alpha)i + X_n, \quad \forall i \in \mathbb{N}, \forall n \text{ such that } T_{2i+1} \leq n < T_{2i+2}. \tag{2.3}
$$

Since $(Y_n)_{n \in \mathbb{N}}$ follows the same increments as $(X_n)_{n \in \mathbb{N}}$ between T_{2i+1} and T_{2i+2}, it is sufficient to verify that

$$
Y_{T_{2i+1}} \geq (\beta - \alpha)i + X_{T_{2i+1}}, \quad \forall i \in \mathbb{N}. \tag{2.4}
$$

We prove this claim by induction on i. For $i = 0$, we have $Y_{T_1} = Y_0 = X_0$ and $X_{T_1} < \alpha$. If $X_0 \geq \alpha$, then Equation (2.4) holds, while if $X_0 < \alpha$, then $T_1 = 0$ and Equation (2.4) holds as well. For $i > 0$, note that $X_{T_{2i+1}} < \alpha$, whereas,

$$Y_{T_{2i+1}} = Y_{T_{2i}} > (\beta - \alpha) + Y_{T_{2i-1}},$$

hence Equation (2.4) holds by the induction hypothesis.

Equation (2.3) proves also that on the event[3]

$$\left\{ \limsup_{n \to \infty} X_n > \beta > \alpha > \liminf_{n \to \infty} X_n \right\},$$

we have $\lim_{n \to \infty} Y_n = \infty$, since on this event the stopping times $(T_i)_{i \in \mathbb{N}}$ are finite. Since $(Y_n)_{n \in \mathbb{N}}$ is a super-martingale, $\mathbf{E}[Y_n] \leq \mathbf{E}[Y_0] = \mathbf{E}[X_0] < \infty$, hence

$$\mathbf{P}\left(\limsup_{n \to \infty} X_n > \beta > \alpha > \liminf_{n \to \infty} X_n \right) = 0, \tag{2.5}$$

where Equation (2.5) holds since by definition a super-martingale is composed of integrable random variables. Equation (2.5) contradicts Equation (2.2). ☐

Theorem 2.40 admits the following consequence.

Theorem 2.45. *Let $(X_n)_{n \in \mathbb{N}}$ be a bounded martingale such that $X_n \geq 0$ a.s., for every $n \in \mathbb{N}$, and let $X := \lim_{n \to \infty} X_n$. For every $n \in \mathbb{N}$ and every set $A \in \mathcal{F}_n$ with positive probability, we have*

$$\mathbf{E}[X_n \mid A] = \mathbf{E}[X \mid A].$$

If we apply Theorem 2.40 to a bounded sub-martingale (or super-martingale), then, by Lebesgue's Dominated Convergence Theorem, the convergence is not just a.s., but in L_1 as well. Hence we obtain the following corollary.

Corollary 2.46. *For every bounded sub-martingale $(X_n)_{n \in \mathbb{N}}$ and every bounded super-martingale $(X_n)_{n \in \mathbb{N}}$, the limit $\lim_{n \to \infty} X_n$ exists a.s. and in L_1. Furthermore, if $(X_n)_{n \in \mathbb{N}}$ is a bounded sub-martingale, then*

$$\sup_{n \in \mathbb{N}} \mathbf{E}[X_n] \leq \mathbf{E}[\lim_{n \to \infty} X_n],$$

while if $(X_n)_{n \in \mathbb{N}}$ is a super-martingale $(X_n)_{n \in \mathbb{N}}$, then

$$\inf_{n \in \mathbb{N}} \mathbf{E}[X_n] \geq \mathbf{E}[\lim_{n \to \infty} X_n].$$

---

[3]We say that an event A *holds* on an event B if $\mathbf{P}(A \cap B) = \mathbf{P}(B)$.

2.8   ZERO-ONE LAWS

In this section we present two consequences of the Martingale Convergence Theorem (Theorem 2.40): Lévy's Zero-One Law and Kolmogorov's Zero-One Law. We will use zero-one laws in Chapter 6: Theorem 2.48 in Section 6.7, Corollary 2.49 in Sections 6.8 and 6.9, and Theorem 2.53 in Section 6.1.

Notation 2.47 $(\sigma((\mathcal{F}_n)_{n\in\mathbb{N}}))$. *When Ω is a set and $(\mathcal{F}_n)_{n\in\mathbb{N}}$ is a collection of sigma-algebras over Ω, we denote by $\sigma((\mathcal{F}_n)_{n\in\mathbb{N}})$ the smallest sigma-algebra on Ω that contains $\mathcal{F}_n$ for every $n \in \mathbb{N}$.*

Theorem 2.48 (Lévy's Zero-One Law). *Let $(\Omega, \mathcal{F}, \mathbf{P})$ be a probability space, let $\mathcal{F}_0 \subseteq \mathcal{F}_1 \subseteq \cdots \subseteq \mathcal{F}$ be a filtration such that $\mathcal{F} = \sigma((\mathcal{F}_n)_{n\in\mathbb{N}})$, and let $f : \Omega \to [0,1]$ be a measurable function. Define $f_n := \mathbf{E}[f \mid \mathcal{F}_n]$. Then $\lim_{n\to\infty} f_n$ exists with probability 1, and*

$$\mathbf{P}\left(\lim_{n\to\infty} f_n = f\right) = 1.$$

Proof. By the law of iterated expectation, for every $n \in \mathbb{N}$ we have

$$f_n = \mathbf{E}[f \mid \mathcal{F}_n] = \mathbf{E}\big[\mathbf{E}[f \mid \mathcal{F}_{n+1}] \mid \mathcal{F}_n\big] = \mathbf{E}[f_{n+1} \mid \mathcal{F}_n],$$

hence the process $(f_n)_{n\in\mathbb{N}}$ is a bounded martingale. By the Martingale Convergence Theorem (Theorem 2.40), the limit $g := \lim_{n\to\infty} f_n$ exists with probability 1. We will prove that $g = f$ with probability 1.

By Theorem 2.45, for every set $A \in \mathcal{F}_n$ with $\mathbf{P}(A) > 0$ we have

$$\mathbf{E}[g \mid A] = \mathbf{E}[f_n \mid A] = \mathbf{E}[f \mid A]. \tag{2.6}$$

Since $\mathcal{F} = \sigma((\mathcal{F}_n)_{n\in\mathbb{N}})$, this implies that Equation (2.6) holds for every Borel set $A \in \mathcal{F}$ that satisfies $\mathbf{P}(A) > 0$. This in turn implies that $\mathbf{P}(g = f) = 1$. Indeed, if, for example, $\mathbf{P}(g > f) > 0$, let $A := \{g > f\}$. Then

$$\mathbf{E}[g \mid A] > \mathbf{E}[f \mid A] = \mathbf{E}[g \mid A], \tag{2.7}$$

where the inequality holds since $g > f$ on A and since $\mathbf{P}(A) > 0$, and the equality holds by Equation (2.6). Since Equation (2.7) cannot hold, this implies that $\mathbf{P}(g = f) = 1$.   □

The following result is a consequence of Lévy's Zero-One Law.

Corollary 2.49. *Let $(\Omega, \mathcal{F}, \mathbf{P})$ be a probability space, let $\mathcal{F}_0 \subseteq \mathcal{F}_1 \subseteq \cdots \subseteq \mathcal{F}$ be a filtration such that $\mathcal{F} = \sigma((\mathcal{F}_n)_{n\in\mathbb{N}})$, and let $A \in \mathcal{F}$ such that $\mathbf{P}(A) > 0$. For every $\varepsilon > 0$ there are $n \in \mathbb{N}$ and $B \in \mathcal{F}_n$ with $\mathbf{P}(B) > 0$ such that*

$$\mathbf{P}(A \mid \mathcal{F}_n)(\omega) > 1 - \varepsilon, \quad \forall \omega \in B.$$

Proof. Applying Lévy's Zero-One Law (Theorem 2.48) to the function $f = \mathbf{1}_A$ we obtain that with probability 1,

$$\lim_{n\to\infty} \mathbf{P}(A \mid \mathcal{F}_n) = \lim_{n\to\infty} \mathbf{E}[A \mid \mathcal{F}_n] = \mathbf{1}_A.$$

Therefore, for almost every $\omega \in A$ we have $\lim_{n\to\infty} \mathbf{P}(A \mid \mathcal{F}_n)(\omega) = 1$. For every $n \in \mathbb{N}$ define

$$B_n := \{\omega \in A \colon \mathbf{P}(A \mid \mathcal{F}_n)(\omega) > 1 - \varepsilon\}.$$

The set B_n is $\mathcal{F}_n$-measurable for every $n \in \mathbb{N}$, and $\lim_{n\to\infty} \mathbf{P}(B_n) = \mathbf{P}(A)$. Hence, for every sufficiently large $n \in \mathbb{N}$, the set $B = B_n$ satisfies the requirements in the corollary. $\qquad\square$

Definition 2.50 (Tail sigma-algebra). *Let $(\Omega, \mathcal{F}, \mathbf{P})$ be a probability space, and let $(\mathcal{F}_n)_{n\in\mathbb{N}}$ be a sequence of sub-sigma-algebras of $\mathcal{F}$. The* tail sigma-algebra *(of $(\mathcal{F}_n)_{n\in\mathbb{N}}$) is the sigma-algebra $\mathcal{F}^{\mathrm{tail}}$ defined by:*

$$\mathcal{F}^{\mathrm{tail}} := \bigcap_{n\in\mathbb{N}} \sigma\left(\bigcup_{m\geq n} \mathcal{F}_m\right).$$

The tail sigma-algebra of $(\mathcal{F}_n)_{n\in\mathbb{N}}$ is a subset of the sigma-algebra $\sigma\left(\bigcup_{n\in\mathbb{N}} \mathcal{F}_n\right)$ that is generated by the sigma-algebras $(\mathcal{F}_n)_{n\in\mathbb{N}}$. We now provide a few examples of sets in the tail sigma-algebra.

Example 2.51. *Let $\Omega = \{0,1\}^{\mathbb{N}}$, and let $\mathcal{F}$ be the product sigma-algebra on Ω. For every $n \in \mathbb{N}$, let $\mathcal{F}_n$ be the sigma-algebra that is determined by the n'th coordinate:*

$$\mathcal{F}_n := \{\emptyset, \Omega, B_{n,0}, B_{n,1}\},$$

where

$$\begin{aligned} B_{n,0} &:= \{\omega = (\omega_0, \omega_1, \ldots) \in \Omega \colon \omega_n = 0\}, \\ B_{n,1} &:= \{\omega = (\omega_0, \omega_1, \ldots) \in \Omega \colon \omega_n = 1\}. \end{aligned}$$

The set

$$A_1 := \{\omega = (\omega_0, \omega_1, \ldots) \in \Omega \colon \omega_k = 1 \text{ for infinitely many } k\text{'s}\}$$

is in the tail sigma-algebra of $(\mathcal{F}_n)_{n\in\mathbb{N}}$, as is the set

$$A_2 := \left\{\omega = (\omega_0, \omega_1, \ldots) \in \Omega \colon \limsup_{n\to\infty} \frac{\sum_{k=0}^{n} \omega_k}{n+1} \geq \frac{1}{3}\right\}.$$

The set

$$A_3 := \{\omega = (\omega_0, \omega_1, \ldots) \in \Omega \colon \omega_0 = 0\}$$

is not in the tail sigma-algebra of $(\mathcal{F}_n)_{n\in\mathbb{N}}$, because it is not in $\sigma\left(\bigcup_{n\geq 1} \mathcal{F}_n\right)$.

A sequence of sigma-algebras $(\mathcal{F}_n)_{n\in\mathbb{N}}$ is *independent* if every *finite* collection of sets, each one lying in a different sigma-algebra in the sequence, is independent.

Definition 2.52 (Independent sigma-algebras). *Let $(\Omega, \mathcal{F}, \mathbf{P})$ be a probability space, and let $(\mathcal{F}_n)_{n\in\mathbb{N}}$ be a sequence of sub-sigma-algebras of $\mathcal{F}$. The sigma-algebras $(\mathcal{F}_n)_{n\in\mathbb{N}}$ are* independent *if for every finite collections of different natural numbers $0 \le n_0 < n_1 < \cdots < n_k$ and every collection of sets $(A_i)_{i=0}^k$ such that $A_i \in \mathcal{F}_{n_i}$ for each $i \in \{0, 1, \ldots, k\}$, we have*

$$\mathbf{P}\left(\bigcap_{i=0}^k A_i\right) = \prod_{i=0}^k \mathbf{P}(A_i).$$

Theorem 2.53 (Kolmogorov's zero-one Law). *Let $(\Omega, \mathcal{F}, \mathbf{P})$ be a probability space, and let $(\mathcal{F}_n)_{n\in\mathbb{N}}$ be an independent sequence of sub-sigma-algebras of $\mathcal{F}$. Let $\mathcal{F}^{\text{tail}} := \bigcap_{n\in\mathbb{N}} \sigma\left(\bigcup_{m\ge n} \mathcal{F}_m\right)$ be the tail sigma-algebra. For every $A \in \mathcal{F}^{\text{tail}}$ we have $\mathbf{P}(A) = 0$ or $\mathbf{P}(A) = 1$.*

Proof. Denote the sigma-algebra generated by $\{\mathcal{F}_{n+1}, \mathcal{F}_{n+2}, \ldots\}$ by

$$\mathcal{G}_n := \sigma\big((\mathcal{F}_k)_{k>n}\big).$$

By assumption, the sequence of sigma-algebras $(\mathcal{F}_n)_{n\in\mathbb{N}}$ is independent. Hence, for every $n \in \mathbb{N}$, the sigma-algebras $\{\mathcal{F}_0, \mathcal{F}_1, \cdots, \mathcal{F}_n, \mathcal{G}_n\}$ are independent. Since $\mathcal{F}^{\text{tail}} \subseteq \mathcal{G}_n$, it follows that the sigma-algebras $\{\mathcal{F}_0, \mathcal{F}_1, \ldots, \mathcal{F}_n, \mathcal{F}^{\text{tail}}\}$ are independent for every $n \in \mathbb{N}$. Since this is true for every $n \in \mathbb{N}$, the sequence of sigma-algebras $\{\mathcal{F}_0, \mathcal{F}_1, \cdots, \mathcal{F}^{\text{tail}}\}$ is independent as well. This implies that the sigma-algebras $\sigma((\mathcal{F}_n)_{n\in\mathbb{N}})$ and $\mathcal{F}^{\text{tail}}$ are independent. Finally, since $\mathcal{F}^{\text{tail}} \subseteq \sigma((\mathcal{F}_n)_{n\in\mathbb{N}})$, the sigma-algebras $\mathcal{F}^{\text{tail}}$ and $\mathcal{F}^{\text{tail}}$ are independent.

Therefore, for every $A \in \mathcal{F}^{\text{tail}}$ we have

$$P(A) = P(A \cap A) = P(A) \cdot P(A),$$

which imply that $\mathbf{P}(A) = 0$ or $\mathbf{P}(A) = 1$. $\qquad\square$

CHAPTER 3

Determinacy of Games

In this chapter we study the determinacy of two-player alternating-move win/lose games; namely, games where the outcome is a win to Player I or a win to Player II. In Section 3.1 we will define the concept of trees, which are used to describe games. In Section 3.2 we will define two-player alternating-move win/lose games and the concept of determinacy: a game is determined if one of the players has a winning strategy. In Section 3.3 we will prove Zermelo's Theorem, which states that games with finitely many stages are determined. In Section 3.4 we will prove that the game is determined as soon as the winning set of Player I is closed or open. In Section 3.5 we formulate the main result of the chapter (Theorem 3.37), stating that every two-player alternating-move win/lose game, where the winning set of Player I is Borel, is determined. The proof of this result appears in Sections 3.6–3.12. In Section 3.13 we will prove an independent result, stating that the game is determined as soon as the winning set is analytic. This result will depend on a large cardinal axiom.

3.1   TREES

In this section we define trees, subtrees, and a topology on trees.

Definition 3.1 (Alphabet). *An* alphabet *is a nonempty set* $\mathcal{A}$.

Definition 3.2 (Tree). *A* tree *over an alphabet* $\mathcal{A}$ *is a nonempty set* $T \subseteq \mathcal{A}^{<\mathbb{N}}$ *of finite sequences in* $\mathcal{A}$ *that satisfies the following properties:*

T.1) *If* $p \in T$ *and* p' *is a prefix of* p, *then* $p' \in T$.

T.2) *For every* $p \in T$ *there is* $p' \in T$ *such that* p *is a strict prefix of* p'.

An element of T *is called a* position.

The empty sequence $\langle \, \rangle$ is the root of the tree. The descendants of a position p in the tree are all positions p' that extend p. Condition (T.2) implies that the tree has no terminal nodes.

Remark 3.3 (Tree and pruned tree). *In Set Theory, the concept introduced in Definition 3.2 is called a* pruned tree, *and a tree is defined solely by Property (T.1). Since*

DOI: 10.1201/9781003582106-3

all our trees satisfy (T.1) *and* (T.2), *to make the writing lighter we omit the adjective* "*pruned*".

Remark 3.4 (Suppresing the alphabet from the notation). *To simplify notation, we will suppress the dependence of the tree on the alphabet whenever possible.*

Definition 3.5 (Subtree). *A subtree of T is a subset of T that forms a tree.*

Every position p defines a subtree, which consists of all its prefixes and all positions that extend it. This subtree is denoted T_p and is formally defined as follows.

Definition 3.6 (The subtree T_p). *Let T be a tree. For every position $p \in T$ define*

$$T_p := \{p' \in T : p' \preceq p \text{ or } p \preceq p'\}.$$

Definition 3.7 (Play). *Let T be a tree over an alphabet $\mathcal{A}$. A* play *in T is an element $x \in \mathcal{A}^{\mathbb{N}}$ with the property that all its prefixes are in T. The set of all plays in T is denoted $[T]$.*

Remark 3.8 (Equivalent definition of a play). *The set of prefixes of a play is a maximal linearly ordered subset of T, with respect to the partial order $\preceq$ defined in Section 2.1. Conversely, every maximal linearly ordered subset of T is the set of prefixes of some play.*

Example 3.9 (Complete binary tree). *When $\mathcal{A} = \{0,1\}$ is the alphabet, the tree $T = \mathcal{A}^{<\mathbb{N}}$ is the* complete binary tree. *In this case, the set $[T]$ is $\mathcal{A}^{\mathbb{N}}$, the set of all infinite sequences of 0's and 1's.*

The set $T_{\langle 0 \rangle}$ is the set of all positions in T that start with 0, and the set $T_{\langle 1,1 \rangle}$ is the set of all positions in T that start with 11.

Example 3.10. *Suppose that the alphabet is $\mathcal{A} = \{0,1,2\}$. Define a tree T by*

$$T := \{\langle a_0, a_1, \ldots, a_n \rangle : n \in \mathbb{N}, a_k = 1 \ \Rightarrow \ a_{k+1} = 1 \ \ \forall k = 0, 1, \ldots, n-1\}.$$

The tree T consists of all finite sequences of 0's, 1's, and 2's such that if the k'th element is 1, then all subsequent elements are 1 as well. The set $[T]$ consists of all infinite sequences of 0's, 1's, and 2's such that if the k'th element is 1, then all subsequent elements are 1 as well.

Definition 3.11 (Tree topology). *We endow the set of plays $[T]$ with the following topology. The collection of* basic open sets *is the collection $\{[T_p] : p \in T\}$. The topology $\mathcal{O}$ over $[T]$ is the family that consists of all unions (not necessarily finite or countable) of basic open sets. That is, a set $A \subseteq [T]$ is in $\mathcal{O}$ if there exists a set of positions $U \subseteq T$ such that $A = \bigcup\{[T_p] : p \in U\}$. The topology $\mathcal{O}$ is called the* tree topology.

The verification that $\mathcal{O}$ is indeed a topology is left to the reader (Exercise 3.2). We have seen a special case of the tree topology in Example 2.34.

Remark 3.12 (The tree topology and the product topology). *When $T = \mathcal{A}^{<\mathbb{N}}$, the tree topology coincides with the product topology on T.*

Lemma 3.13. *For every tree T and every $p \in T$, the set $[T_p]$ is clopen.*

Proof. The set $[T_p]$ is open by definition. The complement of $[T_p]$ is the set

$$([T_p])^c = \bigcup \{[T_{p'}] : p' \in T, \ \mathrm{len}(p') = \mathrm{len}(p), \ p' \neq p\},$$

which is a union of basic open sets. It follows that the set $([T_p])^c$ is open, hence $[T_p]$ is closed. □

Notation 3.14 ($\mathcal{B}(T)$, tree sigma-algebra). *Let T be a tree. Denote by $\mathcal{B}(T)$ the Borel sigma-algebra that is generated by the tree topology of T. This sigma-algebra is called the tree sigma-algebra.*

3.2 GAMES

In this section we define the notion of a game.

Definition 3.15 (Game). *A game (over an alphabet $\mathcal{A}$) is a pair $G = (T, W)$, where T is a tree (over $\mathcal{A}$) and $W \subseteq [T]$ is a winning set.*

The game is played as follows. Player I plays first, and selects a move $a_0 \in \mathcal{A}$ such that $\langle a_0 \rangle \in T$. Then Player II plays, and selects a move $a_1 \in \mathcal{A}$ such that $\langle a_0, a_1 \rangle \in T$. Then Player I plays again, and selects a move $a_2 \in \mathcal{A}$ such that $\langle a_0, a_1, a_2 \rangle \in T$. And so on. In general, at each stage $n \in \mathbb{N}$,

- If n is even, Player I selects a move $a_n \in \mathcal{A}$ such that $\langle a_0, a_1, \dots, a_{n-1}, a_n \rangle \in T$.

- If n is odd, Player II selects a move $a_n \in \mathcal{A}$ such that $\langle a_0, a_1, \dots, a_{n-1}, a_n \rangle \in T$.

The infinite sequence of moves $\langle a_0, a_1, \dots \rangle$ selected by the players throughout the interaction is a play, see Definition 3.7. Player I *wins* the interaction if $\langle a_0, a_1, \dots \rangle \in W$, and Player II *wins* otherwise.

The alphabet $\mathcal{A}$ consists of all possible moves in the game (by both players, in all positions). The tree T describes all possible positions in the game, as well as the set of moves that are available to the players at each position. The set W describes the winning conditions of Player I, namely, all plays that lead to a win of Player I. The complement of W, that is, the set $W^c := [T] \setminus W$, is the winning set of Player II.

Remark 3.16 (Alternating-move game). *In games as defined in Definition 3.15, the two players select their moves alternately. For this reason these games are also called alternating-move games. In Chapters 5 and 6 we will study games where the players select their moves simultaneously.*

Example 3.17. *(Nim) The game Nim is played between two players who move alternately as follows. On the table lie several heaps of matches. The player who must make the next move selects one of the heaps that still contains matches, and removes* *from it some number of matches. The player is allowed to remove any number of matches from that heap, yet she cannot remove matches from more than one heap. The loser is the player who has to remove the last match from the table.*

Suppose that there are two heaps, each containing two matches. We describe the game using the alphabet $\mathcal{A} = \{1A, 2A, 1B, 2B, 0\}$, where the move $1A$ indicates that one match is removed from heap A, the move $2A$ indicates that two matches are removed from heap A, and the moves $1B$ and $2B$ have similar interpretation with respect to heap B. Since games are infinite by definition, the game can never end, hence we need a dummy move to be used once there are no matches left on the table. The move 0 will be this dummy move.

Figure 2 describes the game. In the figure, each position is a vertex, and arrows indicate the moves. The name of the move is indicated above the edge.

The game starts at the root, which is the node to the left marked by $\langle\,\rangle$. In the first stage, Player I has four possible moves – to pick one or two matches from one of the two heaps.

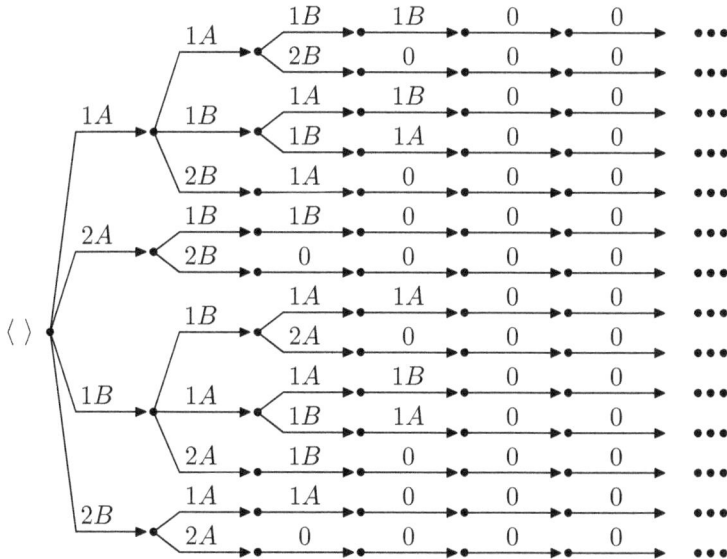

Figure 2: The game tree of Nim with two heaps containing two matches each.

The winning set is given by

$$W = [T_{\langle 1A,1A,1B,1B\rangle}] \cup [T_{\langle 1A,1B,1A,1B\rangle}] \cup [T_{\langle 1A,1B,1B,1A\rangle}] \cup [T_{\langle 1B,1A,1B,1A\rangle}]$$
$$\cup\, [T_{\langle 1B,1A,1A,1B\rangle}] \cup [T_{\langle 1B,1B,1A,1A\rangle}] \cup [T_{\langle 2A,2B\rangle}] \cup [T_{\langle 2B,2A\rangle}].$$

Remark 3.18 (Win/lose game). *The games defined in Definition 3.15 have two outcomes: a win for Player I (which is a loss for Player II) and a win for Player II (which is a loss for Player I). In Chapters 4, 5, and 6 we will study games that have more outcomes. We will then call the games in Definition 3.15* win/lose *games.*

Definition 3.19 (Controlling a position). *We will say that Player I controls all positions of even length, and that Player II controls all positions of odd length.*

A strategy of a player instructs her how to play at each position she controls. This leads us to the following definition.

Definition 3.20 (Strategy). *A strategy s_I of Player I is a function that assigns to every position $p \in T$ of even length (including the empty sequence, which has length 0) a move $a = s_I(p) \in \mathcal{A}$ such that $\langle p, a \rangle \in T$. A strategy s_{II} of Player II is a function that assigns to every position $p \in T$ of odd length a move $a = s_{II}(p) \in \mathcal{A}$ such that $\langle p, a \rangle \in T$.*

Remark 3.21 (The domain of a strategy). *A strategy of a player indicates the moves selected by her in* all *her decision nodes, even those that cannot be reached under the strategy. Namely, even if the strategy of Player I indicates to play a certain move at the root, it should indicate a move at stage 2 also in the hypothetical situation in which Player I selected a different move at the root. This definition is chosen because of its convenience, see Remark 4.26.*

The set of all strategies of a player depends on the tree and not on the winning set.

Notation 3.22 ($S_I(T)$). *We denote the set of all strategies of Player I in the game $G = (T, W)$ by $S_I(T)$.*

Every pair of strategies (s_I, s_{II}) defines a unique play, denoted $x(s_I, s_{II}) \in [T]$, which we define now.

Definition 3.23 (Play induced by a pair of strategies, $x(s_I, s_{II})$). *Let $G = (T, W)$ be a game, let $s_I \in S_I(T)$, and let $s_{II} \in S_{II}(T)$. The play induced by (s_I, s_{II}), denoted $x(s_I, s_{II})$, is the unique play $\langle a_0, a_1, \dots \rangle \in [T]$ that satisfies for every $n \in \mathbb{N}$*

$$
a_n = \begin{cases} s_I(\langle a_0, a_1, \dots, a_{n-1} \rangle), & \text{if } n \text{ is even,} \\ s_{II}(\langle a_0, a_1, \dots, a_{n-1} \rangle), & \text{if } n \text{ is odd.} \end{cases}
$$

Note that the play induced by a pair of strategies is independent of the winning set.

Definition 3.24 (Consistent play, consistent position). *A play $x \in [T]$ is* consistent *with a strategy s_i of player i, if there is a strategy s_j of the other player, player j, such that $x(s_i, s_j) = x$. A position $p \in T$ is* consistent *with a strategy s_i of player i, if it is the prefix of a play x that is consistent with s_i.*

The following result states that if the play x is consistent with the strategy s_I, and if Player II's strategy s_{II} is such that for every prefix of x of odd length, the strategy s_{II} selects the next move along x, then the play generated by s_I and s_{II} is x. The proof, which follows from the definition of consistency, is left to the reader (Exercise 11).

Lemma 3.25. *Let $G = (T, W)$ be a game, let $s_I \in S_I(T)$ be a strategy of Player I, and let $x = \langle a_0, a_1, \dots \rangle$ be a play consistent with s_I. Let $s_{II} \in S_{II}(T)$ be a strategy of Player II that satisfies the following conditions: $s_{II}(\langle a_0, a_1, \dots, a_n \rangle) = a_{n+1}$, for every even $n \in \mathbb{N}$. Then $x(s_I, s_{II}) = x$.*

We next present the concept of a winning strategy of a player, which is a strategy that guarantees that the player wins, whatever strategy the other player adopts.

Definition 3.26 (Winning strategy). *Let $G = (T, W)$ be a game. A strategy $s_{\mathrm{I}} \in S_{\mathrm{I}}(T)$ is winning for Player I in G if $x(s_{\mathrm{I}}, s_{\mathrm{II}}) \in W$ for every strategy $s_{\mathrm{II}} \in S_{\mathrm{II}}(T)$ of Player II. A strategy $s_{\mathrm{II}} \in S_{\mathrm{II}}(T)$ is winning for Player II in G if $x(s_{\mathrm{I}}, s_{\mathrm{II}}) \notin W$ for every strategy $s_{\mathrm{I}} \in S_{\mathrm{I}}(T)$ of Player I.*

An equivalent definition of a winning strategy uses the concept of consistent plays.

Definition 3.27 (Winning strategy). *Let $G = (T, W)$ be a game. A strategy $s_{\mathrm{I}} \in S_{\mathrm{I}}(T)$ is winning for Player I in G if all plays $x \in [T]$ consistent with s_{I} are in W. A strategy $s_{\mathrm{II}} \in S_{\mathrm{II}}(T)$ is winning for Player II in G if all plays $x \in [T]$ consistent with s_{II} are in W^c.*

Plainly, it cannot be that both players have winning strategies.

Definition 3.28 (Determined game). *The game (T, W) is determined if one of the players has a winning strategy.*

Remark 3.29 (Should we play determined games?). *Determined games are arguably dull–the player who has a winning strategy can guarantee to win the game, regardless of the identity of the opponent. Indeed, the game Nim that was presented in Example 3.17 is determined, and the identification of the player who has a winning strategy as well as the construction of a winning strategy is not difficult, see Maschler, Solan, and Zamir (2020, Exercise 3.14).*

There are complex games that are known to be determined, but the identity of the player who has a winning strategy is not known. In other games, the identity of the player who has a winning strategy is known, but no winning strategy was found so far, see Maschler, Solan, and Zamir (2020, Exercise 3.19).

3.3 DETERMINACY OF GAMES WITH FINITELY MANY STAGES

A game with finitely many stages is a game where the winner is determined after a bounded number of stages.

Definition 3.30 (Games with finitely many stages). *The game (T, W) is a game with finitely many stages if there exists $n \in \mathbb{N}$ such that for every two plays $x, x' \in [T]$ that satisfy $x_n = x'_n$, we have $x \in W$ if and only if $x' \in W$.*

One of the first results in Game Theory implies that games with finitely many stages are determined.

Theorem 3.31 (Zermelo, 1913). *All games with finitely many stages are determined.*

To prove Theorem 3.31 we need two definitions.

Definition 3.32 (Subgame that starts at p, G_p). *Let $G = (T, W)$ be a game, and $p \in T$ be a position. The game $G_p := (T_p, [T_p] \cap W)$ is called the subgame that starts at position p.*

Technically, the subgame G_p starts at the position $\langle\ \rangle$, yet in the first $\mathrm{len}(p)$ stages the players are required to select the moves along p, and only once the game reaches the position p, they are free to select moves.

If the game reaches a position p, and if player i has a winning strategy in the subgame G_p that starts at p, then we call p a *winning position* for player i.

Definition 3.33 (Winning position, determined position). *Let $G = (T, W)$ be a game. A position $p \in T$ is a* winning position *of player i in G if player i has a winning strategy in the subgame G_p. The position p is* determined *if it is a winning position of one of the players.*

Proof of Theorem 3.31. Let $G = (T, W)$ be a game with finitely many stages, and let $n \in \mathbb{N}$ satisfy that for every two plays $x, x' \in [T]$ such that $x_n = x'_n$, we have $x \in W$ if and only if $x' \in W$. We will prove by induction that positions in T are determined, and, in particular, the position $\langle\ \rangle$ is determined so that the game G itself is determined.

The base case of the induction is positions with a length of at least $n + 1$. Let then $p = \langle a_0, a_1, \ldots, a_m \rangle \in T$ be a position, and suppose that $m \geq n$. The condition in the theorem implies that there is a player $i \in \{\mathrm{I}, \mathrm{II}\}$ such that either all plays x that extend $\langle a_0, a_1, \ldots, a_n \rangle$ are in W, or all of them are in W^c. In particular, the position p is determined: if all plays x that extend $\langle a_0, a_1, \ldots, a_n \rangle$ are in W, then p is a winning position for Player I, while if all plays x that extend $\langle a_0, a_1, \ldots, a_n \rangle$ are in W^c, then p is a winning position for Player II.

Now suppose by induction that all positions p of length larger than m are determined, and let $p \in T$ be a position with $\mathrm{len}(p) = m$. Assume without loss of generality that m is odd, so that Player I controls the position p. By the induction hypothesis, all positions $p' \succ p$ whose length is $m + 1$ are determined. If one of these positions is a winning position of Player I, then p is a winning position for Player I: a winning strategy of Player I in G_p selects at p the move that leads to p', and then follows a winning strategy in $G_{p'}$. Otherwise, all moves of Player I at p leads to positions in which Player II has a winning strategy, so Player II has a winning strategy in G_p: this is the strategy that follows a winning strategy in $G_{p'}$, where p' is the position at stage $m + 1$. $\qquad\square$

3.4 CLOSED GAMES AND OPEN GAMES

In this section we extend Theorem 3.31 and prove that when the winning set is closed or open in the tree topology (see Definition 3.11), the game is determined.

Theorem 3.34 (Gale and Stewart, 1953). *Let $G = (T, W)$ be a game, where the set W is closed in $[T]$. Then G is determined.*

Proof. Suppose that Player II has no winning strategy. Then there is a move $a_0 \in \mathcal{A}$ such that in the subgame that starts at a_0, Player II has no winning strategy. Indeed, otherwise, for every move $a_0 \in \mathcal{A}$, Player II has a winning strategy in the subgame that starts at a_0, hence Player II has a winning strategy in the game G.

Similarly, for every $n \in \mathbb{N}$ and every position $\langle a_0, a_1, \ldots, a_{2n} \rangle$, if Player II has no winning strategy in the position $\langle a_0, a_1, \ldots, a_{2n} \rangle$, then for every move a_{2n+1} of Player II there is a move a_{2n+2} of Player I such that Player II has no winning strategy in the subgame that starts at position $\langle a_0, a_1, \ldots, a_{2n+2} \rangle$.

Define a strategy $s_I \in S_I(T)$ as follows:

- If $\langle a_0, a_1, \ldots, a_{2n-1} \rangle$ is not a winning position for Player II in G, let $s_I(\langle a_0, a_1, \ldots, a_{2n-1} \rangle)$ be a move $a_{2n} \in \mathcal{A}$ such that $\langle a_0, a_1, \ldots, a_{2n-1}, a_{2n} \rangle$ is not a winning position of Player II in G.

- If $\langle a_0, a_1, \ldots, a_{2n-1} \rangle$ is a winning position for Player II in G, let $s_I(\langle a_0, a_1, \ldots, a_{2n-1} \rangle)$ be any move $a_{2n} \in \mathcal{A}$ such that $\langle a_0, a_1, \ldots, a_{2n-1}, a_{2n} \rangle \in T$.

From the construction of s_I, for every play $x \in [T]$ that is consistent with s_I and every $n \in \mathbb{N}$, the position x_n is not a winning position of Player II. We will show that s_I is a winning strategy of Player I.

Let then $s_{II} \in S_{II}(T)$ and denote $x(s_I, s_{II}) = \langle a_0, a_1, \ldots \rangle$. We will complete the proof by showing that $x(s_I, s_{II}) \in W$. Since the set W is closed, its complement W^c is open, hence it is a union of basic open sets. Suppose, by contradiction, that $x(s_I, s_{II}) = \langle a_0, a_1, \ldots \rangle \notin W$. Then, $\langle a_0, a_1, \ldots \rangle \in W^c$. Since W^c is open, there is a basic open set $[T_p] \subseteq W^c$ such that $\langle a_0, a_1, \ldots \rangle \in [T_p]$. But then the position $x_{\mathrm{len}(p)}(s_I, s_{II})$ is winning for Player II, since all plays that extend p are not in W. The last conclusion cannot hold, because $x_n(s_I, s_{II})$ is not a winning position of Player II for every $n \in \mathbb{N}$. This implies that $x(s_I, s_{II}) \in W$. Since s_{II} is arbitrary, s_I is a winning strategy. □

Corollary 3.35. *Let $G = (T, W)$ be a game, where the set W is open in $[T]$. Then G is determined.*

Proof. The natural way to prove Corollary 3.35 is to consider the game played on the tree T where the winning set is the closed set W^c, the goal of Player II is to have the play in W^c, and the goal of Player I is to have the play in W, and to apply Theorem 3.34 to this game. There is one caveat: Theorem 3.34 applies to the case where the winning set of the *first* mover is closed, while in the game described above, the first mover is Player I, whose winning set is open and not closed. To change the identity of the first mover, we add one dummy stage at the beginning of the game, where Player II moves.

Let $a_{-1} \in \mathcal{A}$ be arbitrary, and let $\widetilde{T}$ be the following tree:

$$\widetilde{T} := \{\langle\,\rangle\} \cup \{\langle a_{-1}, p \rangle : p \in T\}.$$

Set

$$\widetilde{W} := \left\{\langle a_{-1}, a_0, a_1, \ldots \rangle \in \widetilde{T} : \langle a_0, a_1, \ldots \rangle \in W^c\right\} \subseteq \widetilde{T}.$$

Since the set W is open, the set $\widetilde{W}$ is closed. Consider the game $\widetilde{G} = (\widetilde{T}, \widetilde{W})$, where Player II is the first mover, and Player I is the second mover. By Theorem 3.34, the game $\widetilde{G}$ is determined. Since in stage 0 of $\widetilde{G}$ Player II has a single legal move – a_{-1}, a winning strategy in the game $\widetilde{G}$ translates into a winning strategy in the game G in a natural way. The lemma is thus proved. □

The following lemma asserts that when the winning set W is open, if Player I has a winning strategy and follows it, then the play essentially ends in finite time.

Lemma 3.36. *Let $G = (T, W)$ be a game where the set W is open, and suppose that Player I has a winning strategy s_I in G. Then for every strategy $s_{II} \in S_{II}(T)$ there is $n = n(s_{II}) \in \mathbb{N}$ such that*

$$[T_{x_n(s_I, s_{II})}] \subseteq W.$$

We say that the play *essentially ends in finite time* because the position $p = x_{n(s_{II})}(s_I, s_{II})$ at stage $n(s_{II})$ is such that all its extensions are in W. Therefore, to win the game, Player I needs to follow s_I for $n(s_{II})$ stages, and afterwards she can play arbitrarily and still win the game.

Proof. Fix a strategy $s_{II} \in S_{II}(T)$. Since s_I is a winning strategy of Player I, and since W is open,

$$x(s_I, s_{II}) \in W = \bigcup \{[T_p] : p \in T, \ [T_p] \subseteq W\}.$$

Hence, there is $p \in T$ such that $x(s_I, s_{II}) \in [T_p] \subseteq W$, and the claim follows. □

3.5 DETERMINACY OF GENERAL GAMES

The main result of this chapter is the extension of Theorem 3.34 to the case that W lies in the Borel sigma-algebra $\mathcal{B}(T)$.

Theorem 3.37 (Martin, 1975). *Let $G = (T, W)$ be a game. If $W \in \mathcal{B}(T)$, then the game (T, W) is determined.*

The next example shows that when the winning set W is not Borel, Theorem 3.37 may fail.

Example 3.38 (A non-determined game). *In this example we will use the following fact from Set Theory. A set X that has the cardinality of the continuum can be well-ordered in such a way that for every element $x \in X$, the set $\{y \in X : y < x\}$ has a cardinality smaller than the continuum.*

Let $T = \{0, 1\}^{<\mathbb{N}}$ be the complete binary tree, see Example 3.9. In this tree, at every stage the player who must move next has two choices: 0 and 1, and $[T] = \{0, 1\}^{\mathbb{N}}$ is the set of all infinite sequences of zeros and ones.

The set of strategies of Player I is the set of functions that assign an element in $\{0, 1\}$ to each position of even length. Since there are countably many positions in T, the set $S_I(T)$ has the cardinality of the continuum $\mathfrak{c}$. Similarly, the set $S_{II}(T)$ of strategies of Player II has the cardinality $\mathfrak{c}$. The cardinality of the set of plays $[T]$ is $\mathfrak{c}$ as well.

The game (T, W) is not determined if the following properties hold:

- *Player I has no winning strategy: for every strategy $s_I \in S_I(T)$ there is a play in W^c that is consistent with s_I.*

- *Player II has no winning strategy: for every strategy $s_{II} \in S_{II}(T)$ there is a play in W that is consistent with s_{II}.*

We note that the cardinality of the set of plays that are consistent with each strategy is $\mathfrak{c}$.

By the fact mentioned above, and using the Axiom of Choice, the sets of all strategies of the players, $S_I(T) \cup S_{II}(T)$, can be well-ordered in such a way that the set of all strategies that precede each given strategy in that well-order has cardinality smaller than $\mathfrak{c}$.

Go over the strategies in $S_I(T) \cup S_{II}(T)$ one by one. If the strategy is $s_I \in S_I(T)$, select a play in $[T]$ that is consistent with s_I and has not been selected so far, and add it to W^c. If the strategy is $s_{II} \in S_{II}(T)$, select a play in $[T]$ that has not been selected so far and is consistent with s_{II} and add it to W. Since the cardinality of the set of strategies that precede each strategy is smaller than $\mathfrak{c}$, and since the cardinality of the set of plays that are consistent with a given strategy is $\mathfrak{c}$, such a choice is indeed possible. The resulting set W satisfies that the game (T, W) is not determined.

The next example shows that when the games (T, W_1) and (T, W_2) are determined, the game $(T, W_1 \cap W_2)$ is not necessarily determined.

Example 3.39. *As in Example 3.38, let $T = \{0, 1\}^{<\mathbb{N}}$ be the complete binary tree. Let W be the set constructed in Example 3.38. The game $(T, W \cup [T_{\langle 0 \rangle}])$ is determined, because by playing 0 at stage 0, Player I guarantees that the play is in $[T_{\langle 0 \rangle}]$. Similarly, the game $(T, W \cup [T_{\langle 1 \rangle}])$ is determined. However,*

$$(W \cup [T_{\langle 0 \rangle}]) \cap (W \cup [T_{\langle 1 \rangle}]) = W,$$

and the game (T, W) is not determined.

3.6 COVERING

Since games with closed or open winning sets are determined, one approach to proving Theorem 3.37 is to construct for every game (T, W) a game $(\widetilde{T}, \widetilde{W})$ that satisfies the following properties:

- $\widetilde{W}$ is closed or open in $[\widetilde{T}]$, so that $(\widetilde{T}, \widetilde{W})$ is determined.

- The player who has a winning strategy in $(\widetilde{T}, \widetilde{W})$ has a winning strategy in (T, W).

The difficult question is how to construct a game $(\widetilde{T}, \widetilde{W})$ that satisfies these properties. In this section we introduce and study the concept of covering of a tree T, which consists of a tree $\widetilde{T}$, together with mappings $\pi : \widetilde{T} \to T$, $\varphi_I : S_I(\widetilde{T}) \to S_I(T)$, and $\varphi_{II} : S_{II}(\widetilde{T}) \to S_{II}(T)$ that satisfy various natural consistency requirements, such that $(\widetilde{T}, \pi^{-1}(W))$ satisfies the second requirement. We will then define a specific covering $\widetilde{T}$ of T such that $\pi^{-1}(W)$ is clopen, so that $(\widetilde{T}, \pi^{-1}(W))$ satisfies the first requirement.

Definition 3.40 (Covering). *Let T be a tree. A* covering *is a quadruple $(\widetilde{T}, \pi, \varphi_{\mathrm{I}}, \varphi_{\mathrm{II}})$, where $\widetilde{T}$ is a tree, $\pi : \widetilde{T} \to T$, $\varphi_{\mathrm{I}} : S_{\mathrm{I}}(\widetilde{T}) \to S_{\mathrm{I}}(T)$, and $\varphi_{\mathrm{II}} : S_{\mathrm{II}}(\widetilde{T}) \to S_{\mathrm{II}}(T)$, such that the following properties hold:*

C.1) *The function π sends positions of length n to positions of length n:*

$$\mathrm{len}(\widetilde{p}) = \mathrm{len}(\pi(\widetilde{p})), \quad \forall \widetilde{p} \in \widetilde{T}.$$

In addition, under π, two positions that extend one another are mapped to positions that extend one another:

$$\widetilde{p}, \widetilde{p}' \in \widetilde{T} \text{ and } \widetilde{p} \preceq \widetilde{p}' \quad \Rightarrow \quad \pi(\widetilde{p}) \preceq \pi(\widetilde{p}').$$

This in particular implies that π defines a function from $[\widetilde{T}]$ to $[T]$. This map is also denoted by π.

C.2) *For each $i = \mathrm{I}, \mathrm{II}$, the move $\varphi_i(\widetilde{s}_i)(p)$ depends on the behavior of $\widetilde{s}_i$ for positions of length at most $\mathrm{len}(p)$. That is, for every $\widetilde{s}_i, \widetilde{s}_i' \in S_i(\widetilde{T})$ and every $n \in \mathbb{N}$, if $\widetilde{s}_i(\widetilde{p}) = \widetilde{s}_i'(\widetilde{p})$ for every position $\widetilde{p} \in \widetilde{T}$ of length at most n that is controlled by player i, then $\varphi_i(\widetilde{s}_i)(p) = \varphi_i(\widetilde{s}_i')(p)$ for every position $p \in T$ of length at most n that is controlled by player i.*

C.3) *For each $i = \mathrm{I}, \mathrm{II}$ and every strategy $\widetilde{s}_i \in S_i(\widetilde{T})$, if the play $x \in [T]$ is consistent with $\varphi_i(\widetilde{s}_i)$, then there is a play $\widetilde{x} \in [\widetilde{T}]$ that is consistent with $\widetilde{s}_i$ and satisfies $\pi(\widetilde{x}) = x$.*

Remark 3.41 (Is π onto?). *When $(\widetilde{T}, \pi, \varphi_{\mathrm{I}}, \varphi_{\mathrm{II}})$ is a covering of T, the function π is onto. The proof of this result uses material we will cover later and is left to the reader (Exercise 3.32).*

Example 3.42. *The alphabet is $\mathcal{A} = \{a, b, c\}$. Let T be the tree that is displayed in Figure 3. In this tree, Player I has a single move at each position of even length, and Player II has a single move at each position of odd length, except at position $\langle a, a, a, a, a \rangle$, where she has two moves, b and c.*

We will now define a covering $(\widetilde{T}, \pi, \varphi_{\mathrm{I}}, \varphi_{\mathrm{II}})$ of T. Let $\widetilde{T}$ be the tree over the alphabet $\widetilde{\mathcal{A}} = \{\widetilde{a}, \widetilde{b}, \widetilde{c}\}$, which is displayed in Figure 4. In this tree, Player I has a single move at each position of even length, and Player II has a single move at each position of odd length, except at position $\langle \widetilde{a}, \widetilde{a}, \widetilde{a} \rangle$, where she has two moves, $\widetilde{b}$ and $\widetilde{c}$.

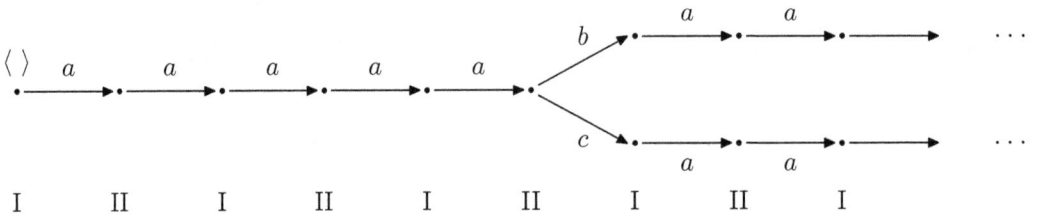

Figure 3: The tree T in Example 3.42.

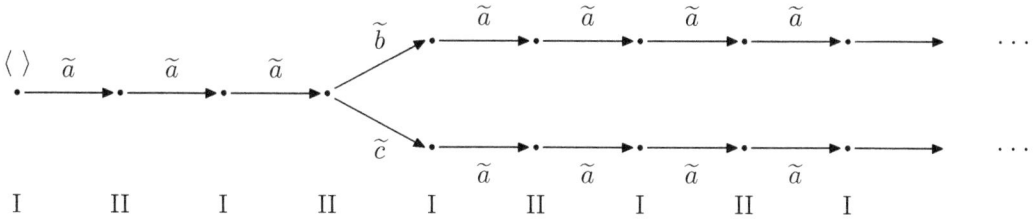

Figure 4: The tree $\widetilde{T}$ in Example 3.42.

Let $\pi : \widetilde{T} \to T$ be the following function:

$$\pi(\langle \widetilde{a} \rangle) = \langle a \rangle,$$
$$\pi(\langle \widetilde{a}, \widetilde{a} \rangle) = \langle a, a \rangle,$$
$$\pi(\langle \widetilde{a}, \widetilde{a}, \widetilde{a} \rangle) = \langle a, a, a \rangle,$$
$$\pi(\langle \widetilde{a}, \widetilde{a}, \widetilde{a}, \widetilde{b} \rangle) = \langle a, a, a, a \rangle,$$
$$\pi(\langle \widetilde{a}, \widetilde{a}, \widetilde{a}, \widetilde{c} \rangle) = \langle a, a, a, a \rangle,$$
$$\pi(\langle \widetilde{a}, \widetilde{a}, \widetilde{a}, \widetilde{b}, \widetilde{a} \rangle) = \langle a, a, a, a, a \rangle,$$
$$\pi(\langle \widetilde{a}, \widetilde{a}, \widetilde{a}, \widetilde{c}, \widetilde{a} \rangle) = \langle a, a, a, a, a \rangle,$$
$$\pi(\langle \widetilde{a}, \widetilde{a}, \widetilde{a}, \widetilde{b}, \widetilde{a}, \widetilde{a} \rangle) = \langle a, a, a, a, a, b \rangle$$
$$\pi(\langle \widetilde{a}, \widetilde{a}, \widetilde{a}, \widetilde{c}, \widetilde{a}, \widetilde{a} \rangle) = \langle a, a, a, a, a, c \rangle,$$
$$\pi(\langle \widetilde{a}, \widetilde{a}, \widetilde{a}, \widetilde{b}, \widetilde{a}, \widetilde{a}, \widetilde{a} \rangle) = \langle a, a, a, a, a, b, a \rangle,$$
$$\pi(\langle \widetilde{a}, \widetilde{a}, \widetilde{a}, \widetilde{c}, \widetilde{a}, \widetilde{a}, \widetilde{a} \rangle) = \langle a, a, a, a, a, c, a \rangle,$$
$$\pi(\langle \widetilde{a}, \widetilde{a}, \widetilde{a}, \widetilde{b}, \widetilde{a}, \widetilde{a}, \widetilde{a}, \widetilde{a} \rangle) = \langle a, a, a, a, a, b, a, a \rangle$$
$$\pi(\langle \widetilde{a}, \widetilde{a}, \widetilde{a}, \widetilde{c}, \widetilde{a}, \widetilde{a}, \widetilde{a}, \widetilde{a} \rangle) = \langle a, a, a, a, a, c, a, a \rangle.$$

Since Player I has a single move in all positions, both in T and in $\widetilde{T}$, we have $|S_I(T)| = |S_I(\widetilde{T})| = 1$, and so the definition of φ_I is trivial.

Player II has two strategies in T: one, denoted s_{II}^b, selects at $\langle a, a, a, a, a \rangle$ the move b, and the other, denoted s_{II}^c, selects at $\langle a, a, a, a, a \rangle$ the move c. She also has two strategies in $\widetilde{T}$: one, denoted $\widetilde{s}_{II}^b$, selects at $\langle \widetilde{a}, \widetilde{a}, \widetilde{a} \rangle$ the move $\widetilde{b}$, and the other, denoted $\widetilde{s}_{II}^c$, selects at $\langle \widetilde{a}, \widetilde{a}, \widetilde{a} \rangle$ the move $\widetilde{c}$. The function φ_{II} is defined by

$$\varphi_{II}(\widetilde{s}_{II}^b) := s_{II}^b, \quad \varphi_{II}(\widetilde{s}_{II}^c) := s_{II}^c.$$

The reader can verify that the quadruple $(\widetilde{T}, \pi, \varphi_I, \varphi_{II})$ is indeed a covering of T.

We next provide a more involved example of a covering.

Example 3.43 (Stopping game). *A stopping game is a game in which the possible moves are* continue *and* stop. *As long as both players select the move "continue", both moves are available; once a player selects the move "stop", the only move available to both players is "stop", so that the game effectively terminates.*

We present a stopping game as a game over the alphabet $\mathcal{A} = \{c, s\}$, where c corresponds to the move "continue" and s corresponds to the move "stop". The set of plays in this game is

$$\langle c, c, c, c, \ldots \rangle, \langle s, s, s, s, \ldots \rangle, \langle c, s, s, s, \ldots \rangle, \langle c, c, s, s, \ldots \rangle, \langle c, c, c, s, \ldots \rangle, \ldots.$$

The game tree T is displayed graphically in Figure 5.

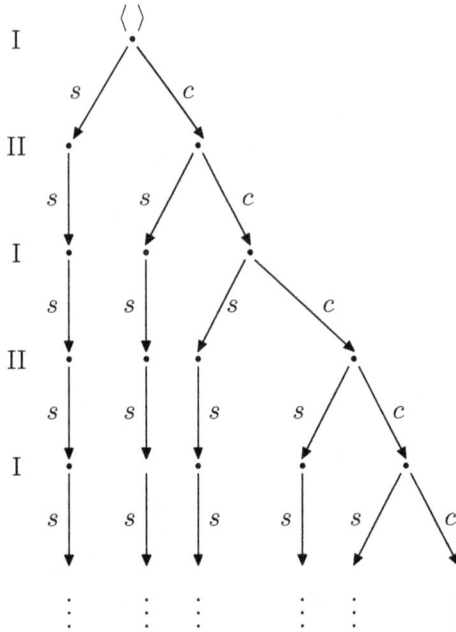

Figure 5: The tree of the stopping game in Example 3.43.

A tree $\widetilde{T}$ that is part of a covering of T is displayed in Figure 6. In this tree, Player I can make a choice only in stage 0 (where she has countably many possible moves), and Player II can make a choice only in stage 1 (where she has countably many possible moves). The interpretation of the tree $\widetilde{T}$ is that in stage 0, Player I selects her strategy for the whole stopping game, namely, the (even) stage in which she stops, or $\widetilde{c}$ if she never stops; and in stage 1 Player II selects her strategy for the whole stopping game, namely, the (odd) stage in which she stops, or $\widetilde{c}$ if she never stops. After stage 1 both players have a single possible move, denoted n. Thus, the alphabet over which the tree $\widetilde{T}$ is defined is $\widetilde{\mathcal{A}} = \{0, 1, 2, \ldots, \widetilde{c}, n\}$.

The function $\pi : \widetilde{T} \to T$ is defined as follows:

- If $\widetilde{p}$ starts with 0 (Player I stops in stage 0), then

$$\pi(\widetilde{p}) = \langle s, s, \ldots, s \rangle,$$

where the length of the image is equal to $\mathrm{len}(\widetilde{p})$.

- If $\widetilde{p}$ starts with $\langle k, 1 \rangle$ for $k \neq 0$ *(Player* I *continues in stage* 0 *and Player* II *stops in stage* 1*), then*

$$\pi(\widetilde{p}) = \langle c, s, \ldots, s \rangle,$$

where the length of the image is equal to $\mathrm{len}(\widetilde{p})$.

- If $\widetilde{p}$ starts with $\langle \widetilde{c}, \widetilde{c} \rangle$ *(both players continue throughout the game), then*

$$\pi(\widetilde{p}) = \langle c, c, c \ldots, c \rangle,$$

where the length of the image is equal to $\mathrm{len}(\widetilde{p})$.

- If $\widetilde{p}$ starts with $\langle k, l \rangle$, where either (a) $k, l \in \mathbb{N}$ and $k < l$, or (b) $k \in \mathbb{N}$ and $l = \widetilde{c}$ *(Player* I *stops before Player* II*), then*

$$\pi(\widetilde{p}) = \langle \underbrace{c, c, c \ldots, c}_{k-1 \text{ times}}, s, s, \ldots, s \rangle,$$

where the length of the image is equal to $\mathrm{len}(\widetilde{p})$, if $\mathrm{len}(\widetilde{p}) \geq k$; and

$$\pi(\widetilde{p}) = \langle c, c, c \ldots, c \rangle,$$

where the length of the image is equal to $\mathrm{len}(\widetilde{p})$, if $\mathrm{len}(\widetilde{p}) \leq k - 1$.

- If $\widetilde{p}$ starts with $\langle k, l \rangle$, where either (a) $k, l \in \mathbb{N}$ and $k > l$, or (b) $l \in \mathbb{N}$ and $k = \widetilde{c}$ *(Player* II *stops before Player* I*), then*

$$\pi(\widetilde{p}) = \langle \underbrace{c, c, c \ldots, c}_{l-1 \text{ times}}, s, s, \ldots, s \rangle,$$

where the length of the image is equal to $\mathrm{len}(\widetilde{p})$, if $\mathrm{len}(\widetilde{p}) \geq l$; and

$$\pi(\widetilde{p}) = \langle c, c, c \ldots, c \rangle,$$

where the length of the image is equal to $\mathrm{len}(\widetilde{p})$, if $\mathrm{len}(\widetilde{p}) \leq l - 1$.

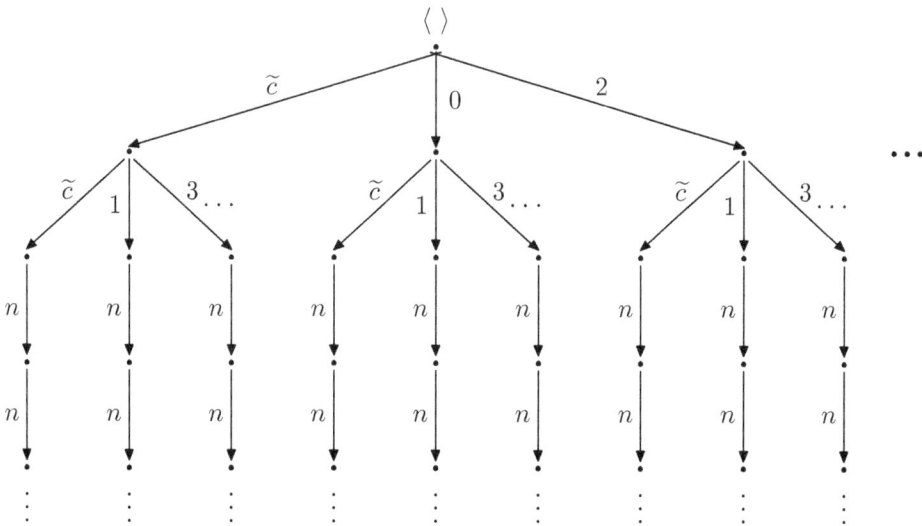

Figure 6: The covering tree $\widetilde{T}$ in Example 3.43.

What is φ_{I}? A strategy $\widetilde{s}_{\mathrm{I}} \in S_{\mathrm{I}}(\widetilde{T})$ is determined by the move of Player I in stage 0. The strategy that selects the move $k > 0$ (including $k = \infty$) at stage 0 is mapped under φ_{I} to a strategy s_{I} that plays c at all positions of even length that satisfy (a) $\mathrm{len}(p) < k$ and (b) so far no player selected the move s.

We now define the function φ_{II}. A strategy $\widetilde{s}_{\mathrm{II}} \in S_{\mathrm{II}}(\widetilde{T})$ specifies a move of Player II in stage 1 for every move of Player I in stage 0. Thus, such a strategy is given by a sequence $(l_k)_{k=0,2,\dots,\widetilde{c}}$, where each l_k is an odd number or $\widetilde{c}$, with the interpretation that if the move of Player I in stage 0 is k, then the move of Player II in stage 1 is l_k. A strategy in $S_{\mathrm{II}}(T)$ stops in a specific stage, which will be taken from $(l_k)_{k=0,2,\dots,\widetilde{c}}$. Which one will it be? For a given position $p \in T$ we will now define $\varphi_{\mathrm{II}}(\widetilde{s}_{\mathrm{II}})(p)$. The move $\varphi_{\mathrm{II}}(\widetilde{s}_{\mathrm{II}})(p)$ must depend on the behavior of s_{I} up to stage $\mathrm{len}(p)$: as soon as s_{I} did not stop before stage $\mathrm{len}(p)$, $\varphi_{\mathrm{II}}(\widetilde{s}_{\mathrm{II}})(p)$ cannot base its move at stage $\mathrm{len}(p)$ on the stage in which s_{I} will stop in the future. We therefore define $\varphi_{\mathrm{II}}(\widetilde{s}_{\mathrm{II}})(p) := \min\{l_k : k > \mathrm{len}(p)\}$. In other words, the strategy $\widetilde{s}_{\mathrm{II}}$ given by the sequence $(l_k)_{k=0,2,\dots,\widetilde{c}}$ is mapped to the strategy in $S_{\mathrm{II}}(T)$ that selects the move s in stage m (and not before), where m is the smallest integer that satisfies $m = \min\{l_k : k > l_k\}$.

We note that since in the tree $\widetilde{T}$ the players have no choices after stage 1, Condition (C.2) has no bite. The reader can verify that $(\widetilde{T}, \pi, \varphi_{\mathrm{I}}, \varphi_{\mathrm{II}})$ is indeed a covering of T (Exercise 3.15).

3.7 SKELETON OF THE PROOF OF MARTIN'S DETERMINACY THEOREM

The proof of Theorem 3.37 consists of two steps.

S.1) For every tree T and every Borel subset $W \subseteq [T]$ there is a covering $(\widetilde{T}, \pi, \varphi_{\mathrm{I}}, \varphi_{\mathrm{II}})$ of T such that the set $\pi^{-1}(W) \subseteq \widetilde{T}$ is clopen.

By Theorem 3.34, the game $(\widetilde{T}, \pi^{-1}(W))$ is determined.

S.2) Let $G = (T, W)$ be a game and let $(\widetilde{T}, \pi, \varphi_{\mathrm{I}}, \varphi_{\mathrm{II}})$ be a covering of T. If the game $\widetilde{G} = (\widetilde{T}, \pi^{-1}(W))$ is determined, then the game G is determined.

Step (S.2) follows from the definition of a covering and will be proved in Section 3.8. Step (S.1) is the challenging part of the proof, and it will be proved by transfinite induction: given a game $G = (T, W)$ where the set W is of rank α in the Borel hierarchy, there is a covering $(\widetilde{T}, \pi, \varphi_{\mathrm{I}}, \varphi_{\mathrm{II}})$ of T such that the set $\pi^{-1}(\widetilde{W})$ is clopen. As it turns out, the most difficult part will be the base of the induction, which will be proved in Section 3.10.

3.8 MORE ON COVERINGS

In this section we will prove useful properties of coverings.

Lemma 3.44. *If* $(\widetilde{T}, \pi, \varphi_{\mathrm{I}}, \varphi_{\mathrm{II}})$ *is a covering of the tree* T, *then* π *is continuous.*

Proof. Let $W \subseteq [T]$ be an open set. Then W is a union of basic open sets, and hence

$W = \bigcup\{[T_p]\colon p \in T,\ [T_p] \subseteq W\}$. Consequently,

$$\pi^{-1}(W) = \bigcup\{\pi^{-1}[T_p]\colon p \in T,\ [T_p] \subseteq W\}.$$

Condition (C.1) implies that for each position $p \in T$ there is a set of positions $\tilde{P} \subseteq \tilde{T}$, all of length $\mathrm{len}(p)$, such that

$$\pi^{-1}([T_p]) = \bigcup\left\{[\tilde{T}_{\tilde{p}}]\colon \tilde{p} \in \tilde{P}\right\}.$$

It follows that $\pi^{-1}(W)$ is a union of basic open sets, and hence open. □

The following lemma states that a composition of two coverings is a covering.

Lemma 3.45. *Let* $(T_1, \pi_1, \varphi_{1,\mathrm{I}}, \varphi_{1,\mathrm{II}})$ *be a covering of the tree* T_0, *and let* $(T_2, \pi_2, \varphi_{2,\mathrm{I}}, \varphi_{2,\mathrm{II}})$ *be a covering of the tree* T_1. *Then* $(T_2, \pi_1 \circ \pi_2, \varphi_{1,\mathrm{I}} \circ \varphi_{2,\mathrm{I}}, \varphi_{1,\mathrm{II}} \circ \varphi_{2,\mathrm{II}})$ *is a covering of* T_0.

Proof. By Definition, $(T_2, \pi_1 \circ \pi_2, \varphi_{1,\mathrm{I}} \circ \varphi_{2,\mathrm{I}}, \varphi_{1,\mathrm{II}} \circ \varphi_{2,\mathrm{II}})$ satisfies Conditions (C.1) and (C.2). Condition (C.3) also follows from the definition. Indeed, let $i \in \{\mathrm{I}, \mathrm{II}\}$, let $\tilde{s}_i \in S_i(T_2)$, and let $x_0 \in [T_0]$ be a play consistent with $\varphi_{1,i} \circ \varphi_{2,i}(\tilde{s}_i) = \varphi_{1,i}(\varphi_{2,i}(\tilde{s}_i))$. Since $(T_1, \pi_1, \varphi_{1,\mathrm{I}}, \varphi_{1,\mathrm{II}})$ is a covering of T_0, there is a play $x_1 \in [T_1]$ consistent with $\varphi_{2,i}(\tilde{s}_i)$ such that $\pi_1(x_1) = x_0$. Since $(T_2, \pi_2, \varphi_{2,\mathrm{I}}, \varphi_{2,\mathrm{II}})$ is a covering of T_1, there is a play $x_2 \in [T_2]$ consistent with $\tilde{s}_i$ such that $\pi_2(x_2) = x_1$. It follows that Condition (C.3) holds as well. □

An important property of coverings is that if the covering is determined, so is the original game. This is Step (S.2) of the proof of Theorem 3.37.

Lemma 3.46. *Let* $G = (T, W)$ *be a game, let* $(\tilde{T}, \pi, \varphi_{\mathrm{I}}, \varphi_{\mathrm{II}})$ *be a covering of* T, *and let* $i \in \{\mathrm{I}, \mathrm{II}\}$. *If player* i *has a winning strategy* $\tilde{s}_i$ *in the game* $\tilde{G} := (\tilde{T}, \pi^{-1}(W))$, *then* $\varphi_i(\tilde{s}_i)$ *is a winning strategy of player* i *in* G. *In particular, if* $\tilde{G}$ *is determined,* \ *so is* G.

Proof. Assume without loss of generality that $i = \mathrm{I}$. Fix a strategy $s_{\mathrm{II}} \in S_{\mathrm{II}}(T)$, and let $x = x(\varphi_{\mathrm{I}}(\tilde{s}_{\mathrm{I}}), s_{\mathrm{II}})$ be the play under the strategy pair $(\varphi_{\mathrm{I}}(\tilde{s}_{\mathrm{I}}), s_{\mathrm{II}})$. In particular, the play x is consistent with the strategy $\varphi_{\mathrm{I}}(\tilde{s}_{\mathrm{I}})$. By Condition (C.3), there is a play $\tilde{x} \in [\tilde{T}]$ that is consistent with the strategy $\tilde{s}_{\mathrm{I}}$ and satisfies $\pi(\tilde{x}) = x$. Since the strategy $\tilde{s}_{\mathrm{I}}$ is a winning strategy of Player I in $\tilde{G}$, we have $\tilde{x} \in \pi^{-1}(W)$. It follows that

$$x = \pi(\tilde{x}) \in W.$$

Since the strategy $s_{\mathrm{II}} \in S_{\mathrm{II}}(T)$ is arbitrary, the strategy $\varphi_{\mathrm{I}}(\tilde{s}_{\mathrm{I}})$ is indeed a winning strategy of Player I in G. □

3.9 IMPOSED SUBTREES

In this section we introduce the concept of an imposed subtree, which will be used in the construction of a covering that will satisfy (S.1). A subtree T' of a tree T is I-*imposed* if only Player I has fewer choices in T' than in T. Similarly, a subtree T' of a tree T is II-*imposed* if only Player II has fewer choices in T' than in T. Recall that all positions of even length are controlled by Player I, and all positions of odd length are controlled by Player II.

Definition 3.47 (I-imposed subtree). *Let T be a tree. A subtree T' of T is* I-imposed *if for every position $p \in T'$ of odd length and every position $p' \succ p$ that satisfies* $\text{len}(p') = \text{len}(p) + 1$ *we have $p' \in T'$.*

Example 3.48. *For every tree T, the tree T itself is a* I-*imposed subtree of T.*

Example 3.49. *Let T be a tree, and let s_I be a strategy. Let T_I be the set of all positions in T that are consistent with s_I. Then T_I is a* I-*imposed subtree of T.*

Similarly, for every set of strategies $D \subseteq S_\text{I}(T)$, the set of all positions in T that are consistent with at least one of the strategies in D is a I-*imposed subtree of T.*

The analogous property for Player II is the following.

Definition 3.50 (II-imposed subtree). *Let T be a tree. A subtree T' of T is* II-imposed *if for every position $p \in T'$ of even length and every position $p' \succ p$ that satisfies* $\text{len}(p') = \text{len}(p) + 1$ *we have $p' \in T'$.*

Denote by $\mathcal{I}_\text{I}(T)$ and $\mathcal{I}_\text{II}(T)$ the sets of all I-imposed and II-imposed subtrees of T, respectively. The following result provides an example of an imposed subtree.

Lemma 3.51. *Let $G = (T, W)$ be a game, let $i \in \{\text{I}, \text{II}\}$, and suppose that player i has a winning strategy in G. Let T_i^* be the collection of all winning positions p of player i in T that satisfy the following property: all prefixes of p are also winning positions of player i. Then T_i^* is a nonempty i-imposed subtree of T.*

Proof. We prove the claim for $i = \text{I}$. Since Player I has a winning strategy s_I in G, the initial position $\langle \, \rangle$ is in T_I^*, and hence the set T_I^* is nonempty.

By definition, (T.1) holds. We turn to prove that (T.2) holds as well.

If $p \in T_\text{I}^*$ is a winning position of even length, then there is a strategy $s_\text{I}' \in S_\text{I}(T)$ that ensures that Player I wins the subgame $G_p = (T_p, [T_p] \cap W)$. It follows that $\langle p, s_\text{I}'(p) \rangle$ is a winning position of Player I in G, and hence in T_I^*. In particular, p is not a terminal node of T_I^*.

We finally show that T_I^* is I-imposed. Indeed, if $p \in T_\text{I}^*$ is a position of odd length, then since p is a winning position of Player I, all positions in T of the form $\langle p, a \rangle$ are winning positions for Player I, and hence in T_I^*. $\qquad \square$

We now provide two properties of imposed subtrees.

Lemma 3.52. *Let T be a tree, let $W \subseteq [T]$ be an open set, and let $T_{\mathrm{I}} \subseteq T$ be a* I*-imposed subtree. At least one of the following conditions holds:*

A1.1) There is a II*-imposed subtree T_{II} of T_{I} such that $[T_{\mathrm{II}}] \subseteq W^{\mathrm{c}}$.*

A1.2) There is a position $p \in T_{\mathrm{I}}$ such that $[T_p] \subseteq W$.

Proof. If $[T_{\mathrm{I}}] \cap W = \emptyset$, then any II-imposed subtree T_{II} of T_{I} satisfies Condition (A1.1). If $[T_{\mathrm{I}}] \cap W \neq \emptyset$, then let $x \in [T_{\mathrm{I}}] \cap W$. Since W is open, $x \in [T_p]$ for some position p that satisfies $[T_p] \subseteq W$. The position p is a prefix of x, which lies in $[T_{\mathrm{I}}]$, hence it is in T_{I}. It follows that Condition (A1.2) holds with this p. □

Substituting W^{c} for W in Lemma 3.52 we obtain the following.

Corollary 3.53. *Let T be a tree, let $W \subseteq [T]$ be a closed set, and let $T_{\mathrm{I}} \subseteq T$ be a* I*-imposed subtree. At least one of the following conditions holds:*

A.1) There is a II*-imposed subtree T_{II} of T_{I} such that $[T_{\mathrm{II}}] \subseteq W$.*

A.2) There is a position $p \in T_{\mathrm{I}}$ such that $[T_p] \subseteq W^{\mathrm{c}}$.

Lemma 3.54. *Let T be a tree, let $T_{\mathrm{I}} \subseteq T$ be a* I*-imposed subtree, and let $T_{\mathrm{II}} \subseteq T_{\mathrm{I}}$ be a* II*-imposed subtree of T_{I}. Suppose that $p = \langle a_0, a_1, \ldots, a_{2n} \rangle \notin T_{\mathrm{II}}$ and $\langle a_0, a_1, \ldots, a_{2n-1} \rangle \in T_{\mathrm{II}}$. Then $p \notin T_{\mathrm{I}}$.*

Proof. Since $T_{\mathrm{II}} \subseteq T_{\mathrm{I}}$, we have

$$\langle a_0, a_1, \ldots, a_{2n-1} \rangle \in T_{\mathrm{II}} \subseteq T_{\mathrm{I}}.$$

Since T_{II} is a II-imposed subtree of T_{I}, and since Player I selects the move a_{2n} at stage $2n$, if by contradiction $\langle a_0, a_1, \ldots, a_{2n} \rangle \in T_{\mathrm{I}}$, then necessarily $\langle a_0, a_1, \ldots, a_{2n} \rangle \in T_{\mathrm{II}}$. The claim follows. □

3.10 UNRAVELING

Recall that a set is *clopen* if it is both closed and open. In this section we study coverings $(\widetilde{T}, \pi, \varphi_{\mathrm{I}}, \varphi_{\mathrm{II}})$ of trees that satisfy a special property: for a given set W, the set $\pi^{-1}(W)$ is clopen.

Definition 3.55 (Unraveling). *A covering $(\widetilde{T}, \pi, \varphi_{\mathrm{I}}, \varphi_{\mathrm{II}})$ of a tree T unravels a set $W \subseteq [T]$ if the set $\pi^{-1}(W) \subseteq [\widetilde{T}]$ is clopen.*

Since $\pi^{-1}(W^{\mathrm{c}}) = (\pi^{-1}(W))^{\mathrm{c}}$ and since the complement of a clopen set is clopen, we deduce the following.

Lemma 3.56. *Let $(\widetilde{T}, \pi, \varphi_{\mathrm{I}}, \varphi_{\mathrm{II}})$ be a covering of a tree T that unravels a set $W \subseteq [T]$. Then this covering also unravels W^{c}.*

Since a clopen set is in particular closed, Theorem 3.34 together with Lemma 3.46 imply the following.

Lemma 3.57. *Let $G = (T, W)$ be a game, and let $(\widetilde{T}, \pi, \varphi_{\mathrm{I}}, \varphi_{\mathrm{II}})$ be a covering of T that unravels W. Then the game G is determined.*

Remark 3.58 (Weakening the condition in Lemma 3.57). *In fact, we can weaken the condition in Lemma 3.57: to conclude that G is determined, it is enough to assume the existence of a covering $(\widetilde{T}, \pi, \varphi_{\mathrm{I}}, \varphi_{\mathrm{II}})$ of T where $\pi^{-1}(W)$ is closed or open. However, in the general case, the only construction we have for a covering $(\widetilde{T}, \pi, \varphi_{\mathrm{I}}, \varphi_{\mathrm{II}})$ that satisfies that $\pi^{-1}(W)$ is closed or open, satisfies that $\pi^{-1}(W)$ is in fact clopen.*

In view of Lemma 3.57, to complete the proof of Theorem 3.37 it is sufficient to prove that for every Borel set $W \subseteq [T]$ there is a covering that unravels W. This is Step (S.1) in the proof of Theorem 3.37. We will prove this claim by transfinite induction on the Borel rank of W. The next result, which is the most intricate step in the proof of Theorem 3.37, will serve as the basis of the induction.

Lemma 3.59. *Let $G = (T, W)$ be a game where the set W is closed. There exists a covering $(\widetilde{T}, \pi, \varphi_{\mathrm{I}}, \varphi_{\mathrm{II}})$ of T that unravels W.*

By Lemma 3.44, for every covering $(\widetilde{T}, \pi, \varphi_{\mathrm{I}}, \varphi_{\mathrm{II}})$ of T, the function π is continuous, hence the set $\pi^{-1}(W)$ is closed whenever W is closed. Lemma 3.59 states that there exists a covering such that $\pi^{-1}(W)$ is also open.

Proof. **Step 0:** Notations.

Recall that when $G = (T, W)$ is a game and $p \in T$ is a position, we denote by $G_p = (T_p, [T_p] \cap W)$ the subgame that starts at p.

Let $\mathcal{T}$ be the set of all subtrees of T. For every subtree T' of T, and every open or closed set $W \subseteq [T]$, the game $(T', [T'] \cap W)$ is determined (see Theorem 3.34 and Corollary 3.35). For each $i \in \{\mathrm{I}, \mathrm{II}\}$, let $\mathcal{Z}_i$ be the set of all pairs (T', W) such that $T' \in \mathcal{T}$, $W \subset [T]$ is open or closed, and player i has a winning strategy in $(T', [T'] \cap W)$. Let $s_i^* : \mathcal{Z}_i \to \bigcup_{T' \in \mathcal{T}} S_i(T')$ be a function that assigns, for every pair $(T', W) \in \mathcal{Z}_i$, a winning strategy of player i in the game $(T', [T'] \cap W)$. To ensure that such functions exist, we need to employ the Axiom of Choice.

Step 1: Definition of a tree $\widetilde{T}$.

Here we define a tree $\widetilde{T}$ that will be part of a covering of T. A typical play in $\widetilde{T}$ is a sequence

$$\langle (a_0, T_{\mathrm{I}}), (a_1, T_{\mathrm{II}}), a_2, a_3, \ldots \rangle,$$

where $\langle a_0, a_1, a_2, \ldots \rangle$ is a play in T_{II}, T_{I} is a I-imposed subtree of T, and T_{II} is a subtree of T that satisfies certain properties. In particular, $\widetilde{T}$ is defined over the alphabet $\mathcal{A} \cup (\mathcal{A} \times \mathcal{T})$.

Formally, let $\widetilde{T}$ be the tree that consists of all positions that can be generated by the following procedure. In stage 0, Player I selects a move a_0 such that $\langle a_0 \rangle \in T$, as well as a I-imposed subtree T_{I} of $T_{\langle a_0 \rangle}$. In other words, Player I selects a move and restricts her continuation strategy.

In stage 1, Player II selects a pair (a_1, T_{II}) that satisfies the following properties:

- $\langle a_0, a_1 \rangle \in T$: the move a_1 is legal at the position $\langle a_0 \rangle$.

- T_{II} is a subtree of T that satisfies $\langle a_0, a_1 \rangle \in T_{\mathrm{II}}$ and in addition (at least) one of the following conditions:

ST.1) T_{II} is a II-imposed subtree of T_{I} that satisfies $[T_{\mathrm{II}}] \subseteq W$.

or

ST.2) $T_{\mathrm{II}} = T_p$ for some $p \in T_{\mathrm{I}}$ such that $\mathrm{len}(p)$ is even, $\langle a_0, a_1 \rangle \preceq p$, and $[T_p] \subseteq W^{\mathrm{c}}$.

By Corollary 3.53, since the set W is closed, Player II has a valid move at stage 1. From stage 2 and on, both players select moves that are available to them in T_{II}.

In other words, Player I starts by choosing a move in T and a I-imposed subtree that includes this move. If in this subgame Player II can ensure that the play ends in W, she can select a II-imposed subtree that ensures that this happens (Condition (ST.1)); afterwards, each player is restricted to play moves that keep the play in T_{II}. If there is a position $p \in T_{\mathrm{I}}$ that extends the move of Player I and such that all plays that extend p are not in W (Condition (ST.2)), Player II can select such a position; afterwards both players have to follow the moves indicated by p, and once the position p is reached, they are free to play in any way (as long as their moves are valid in T_{II}).

Few comments are in order.

1. T_{II} is a subtree of T, hence if the generated play is $\langle (a_0, T_{\mathrm{I}}), (a_1, T_{\mathrm{II}}), a_2, a_3, \ldots \rangle$, then $\langle a_0, a_1, \ldots \rangle \in T_{\mathrm{II}} \subseteq T$.

2. If T_{II} satisfies Condition (ST.2), then T_{II} need not be a subtree of T_{I}; that is, once Player II selects T_{II} to satisfy Condition (ST.2), Player I is no longer restricted by her choice in stage 0.

3. If $T_{\mathrm{II}} = T_p$ satisfies Condition (ST.2), and if the generated play is $\langle (a_0, T_{\mathrm{I}}), (a_1, T_{\mathrm{II}}), a_2, a_3, \ldots \rangle$, then p is a prefix of $\langle a_0, a_1, \ldots \rangle$.

4. If Player II has several possible moves in stage 1, some satisfying Condition (ST.1) and others Condition (ST.2), then she is free to select any of these moves.

5. The tree T_{II} that Player II selects in stage 1 cannot satisfy both (ST.1) and (ST.2). In particular, in the tree $\widetilde{T}$, after observing the move of Player II in stage 1, Player I knows whether (ST.1) or (ST.2) is satisfied by T_{II}. Note that in both cases, the moves of the players from stage 2 and on must be in T_{II}.

6. Once the players select their moves $\langle (a_0, T_{\mathrm{I}}), (a_1, T_{\mathrm{II}}) \rangle$ for stages 0 and 1, we can determine whether the play $\langle (a_0, T_{\mathrm{I}}), (a_1, T_{\mathrm{II}}), a_2, a_3, \ldots \rangle$ will satisfy that $\langle a_0, a_1, \ldots \rangle \in W$ or $\langle a_0, a_1, \ldots \rangle \notin W$. The former happens if T_{II} satisfies (ST.1) and the latter if T_{II} satisfies (ST.2).

To show that $\widetilde{T}$ is part of a covering of T, we will define a function $\pi : \widetilde{T} \to T$ (Step 2) such that $\pi^{-1}(W)$ is clopen (Step 3), a function $\varphi_{\mathrm{I}} : S_{\mathrm{I}}(\widetilde{T}) \to S_{\mathrm{I}}(T)$ (Step 4), and a function $\varphi_{\mathrm{II}} : S_{\mathrm{II}}(\widetilde{T}) \to S_{\mathrm{II}}(T)$ (Step 5), and in each step prove that the relevant function satisfies the requirements imposed by the definition of a covering.

Step 2: Definition of $\pi : \widetilde{T} \to T$.

Let $\pi : \widetilde{T} \to T$ be the natural projection:

$$\pi(\langle (a_0, T_{\mathrm{I}}) \rangle) := \langle a_0 \rangle, \quad \forall (a_0, T_{\mathrm{I}}) \in \mathcal{A} \times \mathcal{T} \text{ such that } \langle (a_0, T_{\mathrm{I}}) \rangle \in \widetilde{T},$$
$$\pi(\langle (a_0, T_{\mathrm{I}}), (a_1, T_{\mathrm{II}}) \rangle) := \langle a_0, a_1 \rangle,$$
$$\forall (a_0, T_{\mathrm{I}}), (a_1, T_{\mathrm{II}}) \in \mathcal{A} \times \mathcal{T} \text{ such that } \langle (a_0, T_{\mathrm{I}}), (a_1, T_{\mathrm{II}}) \rangle \in \widetilde{T},$$
$$\pi(\langle (a_0, T_{\mathrm{I}}), (a_1, T_{\mathrm{II}}), a_2, a_3, \ldots, a_n \rangle) := \langle a_0, a_1, a_2, \ldots, a_n \rangle,$$
$$\forall (a_0, T_{\mathrm{I}}), (a_1, T_{\mathrm{II}}) \in \mathcal{A} \times \mathcal{T}, \forall a_2, \ldots, a_n \in \mathcal{A}$$
$$\text{such that } \langle (a_0, T_{\mathrm{I}}), (a_1, T_{\mathrm{II}}), a_2, \ldots, a_n \rangle \in \widetilde{T}.$$

We note that Condition (C.1) is satisfied. Since π is monotone, i.e., $\pi(p) \preceq \pi(p')$ for every two positions $p, p' \in T$ such that $p \preceq p'$, the function π induces a function $\pi : [\widetilde{T}] \to [T]$, which is continuous. Set

$$\widetilde{W} := \pi^{-1}(W) \subseteq [\widetilde{T}].$$

Step 3: The set $\widetilde{W}$ is clopen in $[\widetilde{T}]$.

Since the set W is closed and the function π is continuous, the set $\widetilde{W} = \pi^{-1}(W)$ is closed in $[\widetilde{T}]$. Since $[T_{\mathrm{II}}]$ is either a subset of W (if it satisfies (ST.1)) or is disjoint from W (if it satisfies (ST.2)), we have

$$\widetilde{W} = \pi^{-1}(W)$$
$$= \{ \langle (a_0, T_{\mathrm{I}}), (a_1, T_{\mathrm{II}}), a_2, a_3, \ldots \rangle \in [\widetilde{T}] : T_{\mathrm{I}} \in \mathcal{I}_{\mathrm{I}}(T), T_{\mathrm{II}} \text{ satisfies (ST.1)} \}$$
$$= \bigcup \left\{ [\widetilde{T}_{\langle (a_0, T_{\mathrm{I}}), (a_1, T_{\mathrm{II}}) \rangle}] : \langle (a_0, T_{\mathrm{I}}), (a_1, T_{\mathrm{II}}) \rangle \in \widetilde{T}, T_{\mathrm{II}} \text{ satisfies (ST.1)} \right\},$$

which is a union of basic open sets in $[\widetilde{T}]$, hence open in $[\widetilde{T}]$.

Interlude:

It remains to define φ_{I} and φ_{II} in such a way that Conditions (C.2) and (C.3) hold. Why is this definition not straightforward? Fix a strategy $\widetilde{s}_{\mathrm{I}} \in S_{\mathrm{I}}(\widetilde{T})$. The strategy $\varphi(\widetilde{s}_{\mathrm{I}})$ is a strategy for the tree T, and hence it should be defined for every position $\langle a_0, a_1, \ldots, a_n \rangle \in T$ of even length. Yet $\widetilde{s}_{\mathrm{I}}$ is defined for positions $\langle (a_0, T_{\mathrm{I}}), (a_1, T_{\mathrm{II}}), a_2, \ldots, a_n \rangle \in \widetilde{T}$. The definition of $\varphi(\widetilde{s}_{\mathrm{I}})(\langle \, \rangle)$ is simple: it can be the move a_0 that satisfies that $\widetilde{s}_{\mathrm{I}}(\langle \, \rangle) = (a_0, T_{\mathrm{I}})$. But how should we define $\varphi(\widetilde{s}_{\mathrm{I}})(\langle a_0, a_1 \rangle)$? Which position $\langle (a_0, T_{\mathrm{I}}), (a_1, T_{\mathrm{II}}) \rangle \in \widetilde{T}$ should we use to define it?

Condition (C.2) is fairly easy to satisfy: all that it requires is that $\varphi_{\mathrm{I}}(\widetilde{s}_{\mathrm{I}})(p)$ depends on choices of $\widetilde{s}_{\mathrm{I}}$ for positions of length at most $\mathrm{len}(p)$, for every $p \in T$. Condition (C.3) poses a greater challenge.

Suppose then that a strategy $\tilde{s}_I \in S_I(\tilde{T})$ is given. Denote by $B(\tilde{s}_I) \subseteq [T]$ the set of all plays in $[T]$ that can arise as the image under π of plays that are consistent with $\tilde{s}_I$:

$$B(\tilde{s}_I) := \{\pi(x(\tilde{s}_I, \tilde{s}_{II})): \tilde{s}_{II} \in S_{II}(\tilde{T})\} \subseteq [T].$$

For each strategy $s_I \in S_I(T)$, denote the set of plays in $[T]$ consistent with s_I by

$$B(s_I) := \{x(s_I, s_{II}): s_{II} \in S_{II}(T)\} \subseteq [T].$$

So that Condition (C.3) holds, every play that is consistent with $\varphi_I(\tilde{s}_I)$ should be the image of a play that is consistent with $\tilde{s}_I$; that is, every play in $B(\varphi_I(\tilde{s}_I))$ must be in $B(\tilde{s}_I)$:

$$B(\tilde{s}_I) \supseteq B(\varphi_I(\tilde{s}_I)). \tag{3.1}$$

This inclusion says that all plays that are consistent with $\varphi_I(\tilde{s}_I)$ are in $B(\tilde{s}_I)$, so the strategy $\varphi_I(\tilde{s}_I)$ is a winning strategy in the game $(T, B(\tilde{s}_I))$. Unfortunately, we do not know whether the set $B(\tilde{s}_I)$ is closed or open, hence we do not know whether the game $(T, B(\tilde{s}_I))$ is determined, and if so, whether Player I has a winning strategy in this game. We therefore use different games, one for the definition of φ_I and another for the definition of φ_{II}, which have open winning sets.

Note that in the definition of φ_I and φ_{II}, it is not required that $\varphi_I(\tilde{s}_I)$ is winning in (T, W); all we need is to ensure that the inclusion (3.1) holds.

Step 4: Definition of $\varphi_I : S_I(\tilde{T}) \to S_I(T)$.

Fix a strategy $\tilde{s}_I \in S_I(\tilde{T})$. We will define the strategy $\varphi_I(\tilde{s}_I) \in S_I(T)$. Strategies of Player I are defined over the set of positions $p = \langle a_0, a_1, \ldots, a_n \rangle$ where n is odd. We will therefore be particularly interested in those positions.

Denote by (a_0, T_I) the move selected by $\tilde{s}_I$ in stage 0. The subtree $T_{I,\langle a_0 \rangle}$ consists of the empty position and all positions in T_I that start with $\langle a_0 \rangle$. Consider the game

$$G' := (T_{I,\langle a_0 \rangle}, [T_{I,\langle a_0 \rangle}] \cap W^c).$$

In this game, the goal of Player I is to have the play *outside* W.

The set W is closed, hence the set $[T_{I,\langle a_0 \rangle}] \cap W^c$ is open in the relative topology on $[T_{I,\langle a_0 \rangle}]$. It follows from Corollary 3.35 that the game G' is determined.

Case 1: Definition of $\varphi_I(\tilde{s}_I)$ when Player I has a winning strategy in G'.

For every position $p = \langle a_0, a_1, \ldots, a_n \rangle$ of odd length denote by $n(p)$ the smallest odd number k such that $[T_{p_k}] \subseteq W^c$:

$$n(p) := \begin{cases} \min\{k \le n: k \text{ odd}, [T_{\langle a_0, a_1, \ldots, a_k \rangle}] \subseteq W^c\}, & \text{if } [T_p] \subseteq W^c, \\ n+1, & \text{if } [T_p] \nsubseteq W^c. \end{cases}$$

When $n(p) < \text{len}(p)$, at stage $n(p)$, whatever moves are selected in the future, the play that will be generated will be in W^c.

The strategy $\varphi_I(\tilde{s}_I)$ will be defined as follows. As long as $n(p) = \text{len}(p)$, the strategy $\varphi_I(\tilde{s}_I)$ follows a winning strategy of Player I in G'. Once a position p with $n(p) < \text{len}(p)$ is reached, the strategy $\varphi_I(\tilde{s}_I)$ coincides with $\tilde{s}_I$, assuming that Player II selected $(a_1, T_{p_{n(p)}})$ at stage 1.

Formally, let $\varphi_I(\tilde{s}_I) \in S_I(T)$ be the following strategy (see Figure 7).

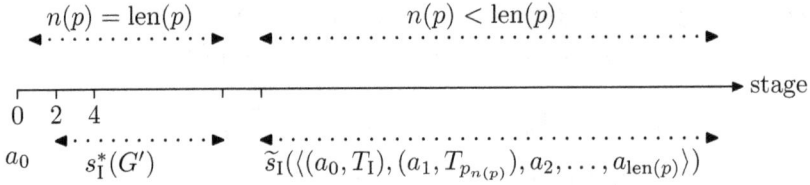

Figure 7: The strategy $\varphi_I(\tilde{s}_I)$ in Case 1.

E1.1)
$$\varphi_I(\tilde{s}_I)(\langle\,\rangle) := a_0,$$

where (a_0, T_I) is the move selected by $\tilde{s}_I$ in stage 0.

E1.2) For every position p of even length such that $n(p) = \text{len}(p)$, that is, $[T_p] \not\subseteq W^c$, we set
$$\varphi_I(\tilde{s}_I)(p) := s_I^*(G')(p).$$

E1.3) For every position p for which $n(p) < \text{len}(p)$, we set
$$\varphi_I(\tilde{s}_I)(p) := \tilde{s}_I(\langle(a_0, T_I), (a_1, T_{p_{n(p)}}), a_2, \ldots, a_{\text{len}(p)-1}\rangle).$$

This definition satisfies Conditions (C.2), as the strategy $s_I^*(G')$ depends only on the subtree T_I that is chosen in stage 0. To see that this definition also satisfies Condition (C.3), fix a play $x = \langle a_0, a_1, \ldots \rangle$ that is consistent with $\varphi_I(\tilde{s}_I)$. In particular, as long as Player II selects the moves along x, under $\varphi(\tilde{s}_I)$, Player I also selects the moves along x. Denote by $n(x)$ the smallest odd number n such that $[T_{x_{n(x)}}] \subseteq W^c$:

$$n(x) := \min\{n \in \mathbb{N} : n \text{ odd}, [T_{x_n}] \subseteq W^c\},$$

where the minimum of an empty set is ∞.

Since s_I^* is a winning strategy of Player I in G', since W^c is open, and since x is consistent with s_I^*, by Lemma 3.36 we have $n(x) < \infty$. Let $\tilde{s}_{II}$ be a strategy that selects $(a_1, T_{x_{n(x)}})$ in stage 1, and afterwards follows x. In positions that are not prefixes of x, the strategy $\tilde{s}_{II}$ selects an arbitrary move. Then

$$x(\tilde{s}_I, \tilde{s}_{II}) = \langle(a_0, T_I), (a_1, T_{x_{n(x)}}), a_2, a_3, \ldots\rangle.$$

In particular, Condition (C.3) is satisfied.

Case 2: Definition of $\varphi_I(\tilde{s}_I)$ when Player II has a winning strategy in G'.

Let T_{II}^* be the set of all positions p in T_I that satisfy the following property: all prefixes of p (including p itself) are winning positions of Player II in G'. By Lemma 3.51, T_{II}^* is a II-imposed subtree of T_I. In stage 0, the strategy $\varphi_I(\tilde{s}_I)$ will select the move selected by $\tilde{s}_I$ in stage 0, and in subsequent stages it will select the move selected by $\tilde{s}_I$, assuming Player II selects the subtree T_{II}^* in stage 1. It will do so until a position that is not in T_{II}^* is reached. This position is necessarily a winning position of Player I in G'. From that point on, $\varphi_I(\tilde{s}_I)$ will follow the construction in Case 1.

Formally, for every position $p = \langle a_0, a_1, \ldots, a_n \rangle$ define

$$n_1(p) := \begin{cases} \min\{k \leq n \colon k \text{ odd}, p_k \notin T_{\text{II}}^*\}, & \text{if there is an odd } k \leq n \\ & \text{such that } p_k \notin T_{\text{II}}^*, \\ n+1, & \text{otherwise.} \end{cases} \quad (3.2)$$

This is the first odd stage in which the play is not a winning position of Player II in the game G'. Define

$$n_2(p) := \begin{cases} \min\{k \geq n_1(p) \colon k \leq n, \, k \text{ odd}, \, [T_{p_k}] \subseteq W^c\}, & \text{if } T_p \subseteq W^c, \\ n+1, & \text{otherwise.} \end{cases} \quad (3.3)$$

This is the length of the shortest prefix p' of p, such that $\text{len}(p') \geq n_1(p)$ and $T_{p'} \subseteq W^c$.

The strategy $\varphi_{\text{I}}(\tilde{s}_{\text{I}})$ is defined as follows (see Figure 8). For every position p:

E2.1) If $p = \langle \, \rangle$,

$$\varphi_{\text{I}}(\tilde{s}_{\text{I}})(\langle \, \rangle) := a_0,$$

where (a_0, T_{I}) is the move selected by the strategy $\tilde{s}_{\text{I}}$ in stage 0.

E2.2) If $\text{len}(p) > 0$ and $n_1(p) = \text{len}(p)$,

$$\varphi_{\text{I}}(\tilde{s}_{\text{I}})(p) := \tilde{s}_{\text{I}}(\langle (a_0, T_{\text{I}}), (a_1, T_{\text{II}}^*), a_2, \ldots, a_{\text{len}(p)-1} \rangle).$$

E2.3) If $\text{len}(p) > 0$, $n_1(p) < \text{len}(p)$ and $n_2(p) = \text{len}(p)$,

$$\varphi_{\text{I}}(\tilde{s}_{\text{I}})(p) := s_{\text{I}}^*(G'_{p_{n_1(p)}})(p).$$

E2.4) If $\text{len}(p) > 0$ and $n_2(p) < \text{len}(p)$,

$$\varphi_{\text{I}}(\tilde{s}_{\text{I}})(p) := \tilde{s}_{\text{I}}(\langle (a_0, T_{\text{I}}), (a_1, T_{p_{n_2(p)}}), a_2, \ldots, a_{\text{len}(p)-1} \rangle).$$

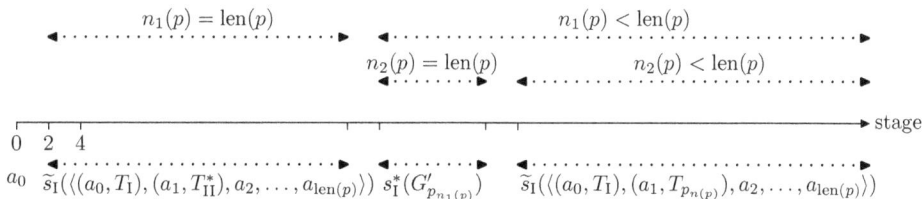

Figure 8: The strategy $\varphi_{\text{I}}(\tilde{s}_{\text{I}})$ in Case 2.

We verify that the definition in Case 2 satisfies Condition (C.2). In (E2.2), and (E2.4) the move selected at position p depends only on p and on a fictitious subtree (T_{II}^* in (E2.2), $T_{p_{n_2(p)}}$ in (E2.4)), which is determined by p. In (E2.3), the strategy $s_{\text{I}}^*(G'_{p_{n_1(p)}})$ is determined by p as well. It follows that Condition (C.2) is satisfied in this case.

We next verify that Condition (C.3) is also satisfied. Let $\tilde{s}_{\text{I}} \in S_{\text{I}}(\tilde{T})$, and let $x = \langle a_0, a_1, \ldots \rangle$ be a play that is consistent with $\varphi_{\text{I}}(\tilde{s}_{\text{I}})$. In particular, as long as Player II selects the moves along x, under $\varphi(\tilde{s}_{\text{I}})$, Player I also selects the moves along x. Denote by (a_0, T_I) the move of $\widetilde{s}_I$ in stage 0. Denote by $n_1(x)$ and $n_2(x)$ the analogous quantities to $n_1(p)$ and $n_2(p)$ that were defined in Equations (3.2)–(3.3):

$$n_1(x) := \min\{n \in \mathbb{N}: n \text{ is odd}, \langle a_0, a_1, \ldots, a_n \rangle \notin T_{II}^*\},$$
$$n_2(x) := \min\{n \geq n_1(x): n \text{ is odd}, [T_{\langle a_0, a_1, \ldots, a_n \rangle}] \subseteq W^c\},$$

where the minimum of an empty set is ∞.

Assume first that $x \in [T_{II}^*]$. Let $\widetilde{s}_{II} \in S_{II}(\widetilde{T})$ be a strategy that selects the move (a_1, T_{II}^*) in stage 1, and afterwards follows x. At positions that are not prefixes of x, the strategy $\widetilde{s}_{II}$ selects an arbitrary move. Then

$$\widetilde{x}(\widetilde{s}_I, \widetilde{s}_{II}) = \langle (a_0, T_I), (a_1, T_{II}^*), a_2, \ldots \rangle,$$

and therefore $\pi(\widetilde{x}(\widetilde{s}_I, \widetilde{s}_{II})) = x$, as required by Condition (C.3).

Assume now that $x \notin [T_{II}^*]$, so that $n_1(x) < \infty$. Since the game G' is determined, it follows that $x_{n_1(x)}$ is a winning position of Player I in the game G': at the position $x_{n_1(x)}$, Player I can ensure that the play lies in $[T_I] \cap W^c$. By the definition of $\varphi_I(\widetilde{s}_I)$, from stage $n_1(x)$ and on, Player I follows a winning strategy in the game $G'_{x_{n_1(x)}}$. Since x is consistent with $\varphi_I(\widetilde{s}_I)$, and since the set $[T_I] \cap W^c$ is open, by Lemma 3.36 we have $n_2(x) < \infty$. Let then $\widetilde{s}_{II} \in S_{II}(\widetilde{T})$ be a strategy that selects the move $(a_1, T_{x_{n_2(x)}})$ in stage 1, and follows x afterwards. In positions that are not prefixes of x, $\varphi_{II}(\widetilde{s}_{II})$ selects an arbitrary move. Then

$$\widetilde{x}(\widetilde{s}_I, \widetilde{s}_{II}) = \langle (a_0, T_I), (a_1, T_{x_{n_2(x)}}), a_2, \ldots \rangle,$$

and therefore $\pi(\widetilde{x}(\widetilde{s}_I, \widetilde{s}_{II})) = x$, as required by Condition (C.3).

Step 5: Definition of $\varphi_{II}: S_{II}(\widetilde{T}) \to S_{II}(T)$.

The definition of φ_{II} is more involved than that of φ_I. Fix a strategy $\widetilde{s}_{II} \in S_{II}(\widetilde{T})$. We will define the strategy $\varphi_{II}(\widetilde{s}_{II}) \in S_{II}(T)$. Strategies of Player II are defined over the set of positions $p = \langle a_0, a_1, \ldots, a_n \rangle$ with n even. We will therefore be particularly interested in those positions.

For every possible initial move a_0 of Player I in T, let $U(a_0)$ be the set of all positions p such that, for some I-imposed subtree T_I, under $\widetilde{s}_{II}$ Player II selects T_p in stage 1:

$$U(a_0) = U_{\widetilde{s}_{II}}(a_0) := \left\{ p \in T: \begin{array}{l} [T_p] \subseteq W^c \text{ and } \widetilde{s}_{II}(\langle (a_0, T_I) \rangle) = (a_1, T_p) \\ \text{for some } T_I \in \mathcal{I}_I(T) \text{ and } a_1 \in \mathcal{A} \end{array} \right\}. \quad (3.4)$$

Note that the set $U_{\widetilde{s}_{II}}(a_0)$ depends on $\widetilde{s}_{II}$ only through the moves $\{\widetilde{s}_{II}(\langle (a_0, T_I) \rangle), T_I \in \mathcal{I}_I(T)\}$ that $\widetilde{s}_{II}$ selects in stage 1. Note also that, in the notations of the right-hand side of Equation (3.4), we have $\langle a_0, a_1 \rangle \preceq p$. Define

$$B(a_0) = B_{\widetilde{s}_{II}}(a_0) := \bigcup\{[T_p]: p \in U_{\widetilde{s}_{II}}(a_0)\},$$

which, as a union of basic open sets, is open. We do not rule out the case $B(a_0) = \emptyset$. Consider the game $G(a_0) = G_{\widetilde{s}_{II}}(a_0) = (T_{\langle a_0 \rangle}, (B(a_0))^c)$. Thus, in stage 0 of $G(a_0)$

the only available move to Player I is a_0. Since the set $B(a_0)$ is open, Theorem 3.34 ensures that the game $G(a_0)$ is determined.

Case 1: Definition of $\varphi_{\mathrm{II}}(\widetilde{s}_{\mathrm{II}})$ when Player II has a winning strategy in $G(a_0)$.

For every position $p \in T_{\langle a_0 \rangle}$ define

$$n(p) := \min\{k \leq \operatorname{len}(p) \colon k \text{ even}, p_k \in U(a_0)\},$$

and

$$n(p) := \operatorname{len}(p)$$

otherwise. This is the first even stage along p in which the position is in $U(a_0)$. The strategy $\varphi_{\mathrm{II}}(\widetilde{s}_{\mathrm{II}})$ will follow a winning strategy in the game $G(a_0)$ until stage $n(p)$, and then follow $\widetilde{s}_{\mathrm{II}}$, assuming Player I selected in stage 0 a subtree T_{I} that leads Player II to select $T_{p_{n(p)}}$ according to $\widetilde{s}_{\mathrm{II}}$.

Formally, define the strategy $\varphi_{\mathrm{II}}(\widetilde{s}_{\mathrm{II}}) \in S_{\mathrm{II}}(T)$ as follows (see Figure 9). For every position $p \in T$ of even length:

E3.1) If $n(p) = \operatorname{len}(p)$, then

$$\varphi_{\mathrm{II}}(\widetilde{s}_{\mathrm{II}})(p) := s_{\mathrm{II}}^*(G(a_0))(p).$$

E3.2) If $n(p) < \operatorname{len}(p)$, then by the definition of the set $U(a_0)$, there are a I-imposed subtree $T_{\mathrm{I}} \in \mathcal{I}_{\mathrm{I}}(T)$ and $a_1 \in \mathcal{A}$ such that

$$\widetilde{s}_{\mathrm{II}}(\langle (a_0, T_{\mathrm{I}}) \rangle) = (a_1, T_{p_{n(p)}}). \tag{3.5}$$

Set

$$\varphi_{\mathrm{II}}(\widetilde{s}_{\mathrm{II}})(p) := \widetilde{s}_{\mathrm{II}}(\langle (a_0, T_{\mathrm{I}}), (a_1, T_{p_{n(p)}}), a_2, \ldots, a_{\operatorname{len}(p)-1} \rangle).$$

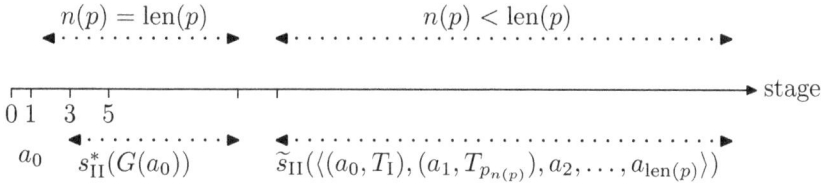

Figure 9: The strategy $\varphi_{\mathrm{II}}(\widetilde{s}_{\mathrm{II}})$ in Case 1.

This definition satisfies Conditions (C.2), as the strategy $s_{\mathrm{II}}^*(G(a_0))$ depends only on the move of Player I at stage 0. Indeed, suppose that $\widetilde{s}_{\mathrm{II}}$ and $\widetilde{s}_{\mathrm{II}}'$ are two strategies that satisfy $\widetilde{s}_{\mathrm{II}}(\widetilde{p}) = \widetilde{s}_{\mathrm{II}}'(\widetilde{p})$ for every position $\widetilde{p} \in T_{\langle a_0 \rangle}$ of length at most n. Then, in particular, $\widetilde{s}_{\mathrm{II}}(\widetilde{p}) = \widetilde{s}_{\mathrm{II}}'(\widetilde{p})$ for every position $\widetilde{p} \in T_{\langle a_0 \rangle}$ of length 1, and hence, $U_{\widetilde{s}_{\mathrm{II}}}(a_0) = U_{\widetilde{s}_{\mathrm{II}}'}(a_0)$. This in turn implies that $B_{\widetilde{s}_{\mathrm{II}}}(a_0) = B_{\widetilde{s}_{\mathrm{II}}'}(a_0)$, and therefore $G_{\widetilde{s}_{\mathrm{II}}}(a_0) = G_{\widetilde{s}_{\mathrm{II}}'}(a_0)$.

To see that this definition also satisfies Condition (C.3), let $x = \langle a_0, a_1, \ldots \rangle \in [T]$ be a play that is consistent with $\varphi_{\mathrm{II}}(\widetilde{s}_{\mathrm{II}})$, and define

$$n(x) := \min\{k \in \mathbb{N} \colon k \text{ even}, x_k \in U(a_0)\},$$

where the minimum of the empty set is ∞. Since $s_{\mathrm{II}}^*(G(a_0))$ is a winning strategy in $G(a_0)$, we have $x \in B(a_0)$. Since $B(a_0)$ is open, $n(x) < \infty$.

Let $\widetilde{s}_{\mathrm{I}} \in S_{\mathrm{I}}(\widetilde{T})$ be the strategy that selects in stage 0 a I-imposed subtree T_{I} that satisfies $\widetilde{s}_{\mathrm{II}}(\langle(a_0, T_{\mathrm{I}})\rangle) = (a_1, T_{x_{n(x)}})$, and afterwards follows x. In positions that are not prefixes of x, the strategy $\widetilde{s}_{\mathrm{I}}$ is defined arbitrarily. By Lemma 3.25,

$$x(\widetilde{s}_{\mathrm{I}}, \widetilde{s}_{\mathrm{II}}) = \langle(a_0, T_{\mathrm{I}}), (a_1, T_{x_{n(x)}}), a_2, a_3, \ldots\rangle.$$

Hence, $\pi(x(\widetilde{s}_{\mathrm{I}}, \widetilde{s}_{\mathrm{II}})) = x$, and hence Condition (C.3) is satisfied in this case.

Case 2: Definition of $\varphi_{\mathrm{II}}(\widetilde{s}_{\mathrm{II}})$ when Player I has a winning strategy in $G(a_0)$.

Let T_{I}^* be the set of all positions p in $T_{\langle a_0 \rangle}$ that satisfy the following property: all prefixes of p (including p itself) are winning positions of Player I in $G(a_0)$. By Lemma 3.51, T_{I}^* is a I-imposed subtree. The strategy $\varphi_{\mathrm{II}}(\widetilde{s}_{\mathrm{II}})$ follows $\widetilde{s}_{\mathrm{II}}$, assuming that Player I selected the move (a_0, T_{I}^*) in stage 0, until (if ever) a position not in T_{I}^* is reached, in which case the play reaches a winning position of Player II, and we apply Case 1.

We turn to the formal definition. Let $(a_1, T_{\mathrm{II}}) = \widetilde{s}_{\mathrm{II}}(\langle(a_0, T_{\mathrm{I}}^*)\rangle)$ be the move selected by $\widetilde{s}_{\mathrm{II}}$ at stage 1, assuming Player I selected the move (a_0, T_{I}^*) in stage 0. We note that the II-imposed subtree $T_{\mathrm{II}} \subseteq T_{\mathrm{I}}^*$ must satisfy Condition (ST.1). Indeed, since T_{I}^* is the set of winning positions of Player I in $G(a_0)$, whose winning set is $(B(a_0))^c$, there is no position $p \in T_{\mathrm{I}}^* \cap U(a_0)$.

For every position p, let $n_1(p)$ be the first even stage along p in which the play is not in T_{II}:

$$n_1(p) := \min\{k \le \operatorname{len}(p) \colon k \text{ even}, p_k \notin T_{\mathrm{II}}\}, \tag{3.6}$$

if there is an even number $k \le \operatorname{len}(p)$ such that $p_k \notin T_{\mathrm{II}}$, and

$$n_1(p) := \operatorname{len}(p)$$

otherwise. Lemma 3.54 implies that $p_{n_1(p)} \notin T_{\mathrm{I}}^*$. Since the game $G(a_0)$ is determined, it follows that $p_{n_1(p)}$ is a winning position of Player II in $G(a_0)$, whenever $n_1(p) < \operatorname{len}(p)$.

Let $n_2(p)$ be the first even stage along p after stage $n_1(p)$ at which a position in $U(a_0)$ is reached:

$$n_2(p) := \min\{k \ge n_1(p) \colon k \le \operatorname{len}(p), k \text{ even}, p_k \in U(a_0)\}, \tag{3.7}$$

whenever there is an even number $k \ge n_1(p)$ such that $k \le \operatorname{len}(p)$ and $p_k \in U(a_0)$, and

$$n_2(p) := \operatorname{len}(p)$$

otherwise.

For every position p of odd length define (see Figure 10):

E4.1) If $\operatorname{len}(p) = 1$,

$$(\varphi_{\mathrm{II}}(\widetilde{s}_{\mathrm{II}}))(\langle a_0 \rangle) := a_1,$$

where, as mentioned above, a_1 satisfies that $(a_1, T_{\mathrm{II}}) = \widetilde{s}_{\mathrm{II}}(\langle(a_0, T_{\mathrm{I}}^*)\rangle)$.

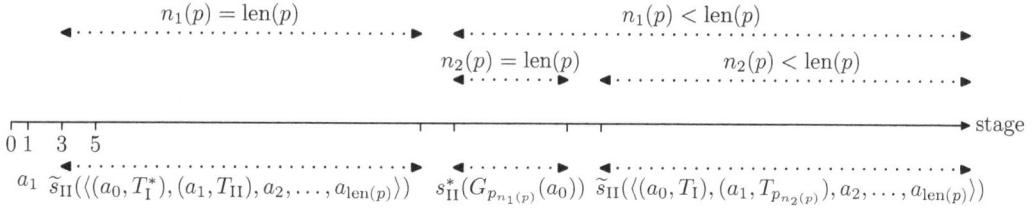

Figure 10: The strategy $\varphi_{\mathrm{II}}(\widetilde{s}_{\mathrm{II}})$ in Case 2.

E4.2) If $n_1(p) = \mathrm{len}(p)$,

$$\left(\varphi_{\mathrm{II}}(\widetilde{s}_{\mathrm{II}})\right)(p) := \widetilde{s}_{\mathrm{II}}(\langle (a_0, T_{\mathrm{I}}^*), (a_1, T_{\mathrm{II}}), a_2, \ldots, a_{\mathrm{len}(p)-1}\rangle).$$

E4.3) If $n_1(p) < \mathrm{len}(p)$ and $n_2(p) = \mathrm{len}(p)$,

$$\left(\varphi_{\mathrm{II}}(\widetilde{s}_{\mathrm{II}})\right)(p) := \left(s_{\mathrm{II}}^*(G_{p_{n_1(p)}}(a_0))\right)(p).$$

E4.4) If $n_1(p) < \mathrm{len}(p)$ and $n_2(p) < \mathrm{len}(p)$,

$$\left(\varphi_{\mathrm{II}}(\widetilde{s}_{\mathrm{II}})\right)(p) := \widetilde{s}_{\mathrm{II}}(\langle (a_0, T_{\mathrm{I}}), (a_1, T_{p_{n_2(p)}}), a_2, \ldots, a_{\mathrm{len}(p)-1}\rangle),$$

where T_{I} is the I-imposed subtree that satisfies

$$\widetilde{s}_{\mathrm{II}}(\langle (a_0, T_{\mathrm{I}})\rangle) = (a_1, T_{p_{n_2(p)}}).$$

Let us verify that the definition satisfies Condition (C.2). In (E4.2) and (E4.4) the move selected at position p depends only on p and on a fictitious subtree (T_{I}^* in (E4.2), T_{I} in (E4.4)), which is determined by p. In (E4.3), the strategy $s_{\mathrm{II}}^*(G_{p_{n_1(p)}}(a_0))$ is also determined by p. It follows that Condition (C.2) is satisfied in this case.

We now prove that Condition (C.3) holds as well. Let $x = \langle a_0, a_1, \ldots\rangle$ be a play that is consistent with $\varphi_{\mathrm{II}}(\widetilde{s}_{\mathrm{II}})$, and set

$$
\begin{aligned}
n_1(x) &:= \min\{k \in \mathbb{N}\colon k \text{ even}, x_k \notin T_{\mathrm{II}}\}, \\
n_2(x) &:= \min\{k \geq n_1(x)\colon k \text{ even}, x_k \in U(a_0)\},
\end{aligned}
$$

where the minimum of the empty set is ∞. These quantities are analogous to the quantities $n_1(p)$ and $n_2(p)$ that were defined in Equations (3.6)–(3.7).

Recall that T_{II} is the II-imposed subtree of T_{I}^* that satisfies

$$(a_1, T_{\mathrm{II}}) = \widetilde{s}_{\mathrm{II}}(\langle a_0, T_{\mathrm{I}}^*\rangle).$$

Since all positions in T_{I}^* are winning positions for Player I in $G(a_0)$, the II-imposed subtree T_{II} cannot satisfy Condition (ST.2), and hence it satisfies Condition (ST.1).

If $x \in [T_{\mathrm{I}}^*]$, then let $\widetilde{s}_{\mathrm{I}} \in S_{\mathrm{I}}(\widetilde{T})$ be a strategy in which Player I selects (a_0, T_{I}^*) in stage 0, and afterwards follows x. At positions that are not prefixes of x, the strategy $\widetilde{s}_{\mathrm{I}}$ is defined arbitrarily. It follows that the play under $(\widetilde{s}_{\mathrm{I}}, \widetilde{s}_{\mathrm{II}})$ is

$$x(\widetilde{s}_{\mathrm{I}}, \widetilde{s}_{\mathrm{II}}) = \langle (a_0, T_{\mathrm{I}}^*), (a_1, T_{\mathrm{II}}), a_2, a_3, \ldots\rangle,$$

for some II-imposed subtree $T_{II} \subseteq T_I^*$, and as a result $\pi(x(\widetilde{s}_I, \widetilde{s}_{II})) = x$. In particular, Condition (C.3) is satisfied in this case.

If $x \notin [T_I^*]$, then there is a minimal $n \in \mathbb{N}$ such that $x_n \notin T_I^*$. Since $T_{II} \subseteq T_I^*$, it follows that $n_1(x) < \infty$. By the definition of $\varphi(\widetilde{s}_{II})$, at position $x_{n_1(x)}$ the strategy $\varphi(\widetilde{s}_{II})$ starts following a winning strategy in the game $G(a_0)_{x_{n_1(x)}}$, whose winning set is $[T_{x_{n_1(x)}}] \cap (B(a_0))^c$. In particular, the play reaches a position $p \in U(a_0)$, and therefore $n_2(x) < \infty$. It follows that there is a I-imposed subtree T_I of T that satisfies Equation (3.5) with respect to $x_{n_2(x)}$, that is, $\widetilde{s}_{II}(\langle (a_0, T_I) \rangle) = (a_1, T_{x_{n_2(x)}})$. Let $\widetilde{s}_I \in S_I(\widetilde{T})$ be a strategy that selects (a_0, T_I) in stage 0, and afterwards follows x. In positions that are not prefixes of x, the strategy $\widetilde{s}_I$ is defined arbitrarily. The play under $(\widetilde{s}_I, \widetilde{s}_{II})$ is

$$x(\widetilde{s}_I, \widetilde{s}_{II}) = \langle (a_0, T_I), (a_1, T_{x_{n_2(x)}}), a_2, a_3, \ldots \rangle.$$

Therefore $\pi(x(\widetilde{s}_I, \widetilde{s}_{II})) = x$, and Condition (C.3) is satisfied in this case as well. □

We here provide several remarks regarding the proof of Lemma 3.59, which use the notations of the proof.

Remark 3.60. *If Player I has a winning strategy in the game $G = (T, W)$, then she also has a winning strategy in the game $\widetilde{G} = (\widetilde{T}, \pi^{-1}(W))$. This strategy consists of choosing in stage 0 the subtree T_I^* of all winning positions in G, and choosing the move a_n in stage n in such a way that $\langle a_0, a_1, \ldots, a_n \rangle$ is in T_I^*. The only responses of Player II in stage 1 can consist of subtrees T_{II} that satisfy (ST.1). This observation applies even when W is not closed.*

Remark 3.61. *When the set W is closed, if Player II has a winning strategy in $G = (T, W)$, then she also has a winning strategy in the game $\widetilde{G} = (\widetilde{T}, \pi^{-1}(W))$. We leave the proof of this conclusion for the reader (Exercise 3.22).*

Remark 3.62. *In the covering $(\widetilde{T}, \pi, \varphi_I, \varphi_{II})$ constructed in the proof of Lemma 3.59, the I-imposed subtree that Player I selects in stage 0 is not necessarily a subtree that is generated by a strategy of Player I; that is, there may be positions in T_I of even length at which Player I may have more than one move that keeps the play in T_I. A natural question is whether this choice is required, or maybe we could have restricted Player I to select in stage 0 a single strategy. The answer is that with such a restriction, the proof that we provided fails. Indeed, the definition of the function φ_{II} in Case 2, depends on whether the current position is in T_I^*, the subtree that consists of all winning position of Player I in the game $G(a_0)$. Once the play leaves T_I^*, the position must be a winning position of Player II in $G(a_0)$, and therefore there is an extension p of the current position that satisfies $[T_p] \subseteq W^c$, and then Player II can behave as if Player I selected in stage 0 a subtree T_I that leads Player II to select T_p in stage 1. If we restricted T_I to be the subtree of all plays that are consistent with a single strategy, then in Case 2 the subtree T_I^* would still include only winning positions of Player I in the game $G(a_0)$, but upon leaving this subtree, the position can still be a winning positions of Player I. Yet the definition of $\varphi_{II}(\widetilde{s}_{II})$ changes only once a winning position of Player II in $G(a_0)$ is reached.*

By Lemma 3.45, the composition of two coverings is a covering. The next lemma states that the composition unravels a set W as soon as the first of the two coverings unravels W.

Lemma 3.63. *Let* $(T_1, \pi_1, \varphi_{1,\mathrm{I}}, \varphi_{1,\mathrm{II}})$ *be a covering of the tree* T_0, *and let* $(T_2, \pi_2, \varphi_{2,\mathrm{I}}, \varphi_{2,\mathrm{II}})$ *be a covering of the tree* T_1. *Suppose that* $(T_1, \pi_1, \varphi_{1,\mathrm{I}}, \varphi_{1,\mathrm{II}})$ *unravels the set* W. *Then* $(T_2, \pi_1 \circ \pi_2, \varphi_{1,\mathrm{I}} \circ \varphi_{2,\mathrm{I}}, \varphi_{1,\mathrm{II}} \circ \varphi_{2,\mathrm{II}})$ *is a covering of* T_0 *that unravels the set* W.

Proof. By Lemma 3.45, $(T_2, \pi_1 \circ \pi_2, \varphi_{1,\mathrm{I}} \circ \varphi_{2,\mathrm{I}}, \varphi_{1,\mathrm{II}} \circ \varphi_{2,\mathrm{II}})$ is a covering of T_0. Since $(T_1, \pi_1, \varphi_{1,\mathrm{I}}, \varphi_{1,\mathrm{II}})$ unravels the set W, the set $\pi_1^{-1}(W)$ is clopen. By Lemma 3.44, π_2 is continuous, hence the set $\pi_2^{-1} \circ \pi_1^{-1}(W)$ is clopen. $\qquad\square$

3.11 K-COVERINGS

In this section we present a special kind of a covering, called k-*covering*. For $k \in \mathbb{N}$, a covering $(\widetilde{T}, \pi, \varphi_{\mathrm{I}}, \varphi_{\mathrm{II}})$ of a tree T is called a k-covering if the two trees T and $\widetilde{T}$ are identical in the first $2k$ levels, and the restrictions of π, φ_{I}, and φ_{II} to these levels are the identity functions.

This concept will allow us to consider sequences of trees, each one being (part of) a covering of the preceding one in the sequence, thereby extending Lemma 3.59 to winning sets that are not necessarily closed or open.

Definition 3.64 (k-covering). *Let* $k \in \mathbb{N}$. *A covering* $(\widetilde{T}, \pi, \varphi_{\mathrm{I}}, \varphi_{\mathrm{II}})$ *of a tree* T *is called a* k-*covering if the following conditions hold:*

- *$p \in T$ and* $\operatorname{len}(p) < 2k$ *if and only if* $p \in \widetilde{T}$ *and* $\operatorname{len}(p) < 2k$.

- *For every* $p \in \widetilde{T}$,

$$
\begin{aligned}
\pi(p) &= p, & \text{whenever } \operatorname{len}(p) < 2k, \\
\varphi_{\mathrm{I}}(\widetilde{s}_{\mathrm{I}})(p) &= \widetilde{s}_{\mathrm{I}}(p), & \text{whenever } \operatorname{len}(p) < 2k \text{ and } \operatorname{len}(p) \text{ is even}, \\
\varphi_{\mathrm{II}}(\widetilde{s}_{\mathrm{II}})(p) &= \widetilde{s}_{\mathrm{II}}(p), & \text{whenever } \operatorname{len}(p) < 2k-1 \text{ and } \operatorname{len}(p) \text{ is odd}.
\end{aligned}
$$

In the second bullet in Definition 3.64, the length of the position p in the condition on φ_{II} is smaller than $2k - 1$ (and not $2k$) because positions of length $2k$ in $\widetilde{T}$ and T may differ.

The analog of Lemma 3.63 for k-coverings follows. The proof of this result is similar to that of Lemma 3.63, hence omitted.

Lemma 3.65. *Let* $(T_1, \pi_1, \varphi_{1,\mathrm{I}}, \varphi_{1,\mathrm{II}})$ *be a* k-*covering of the tree* T_0, *and let* $(T_2, \pi_2, \varphi_{2,\mathrm{I}}, \varphi_{2,\mathrm{II}})$ *be a* k-*covering of the tree* T_1. *Suppose that* $(T_1, \pi_1, \varphi_{1,\mathrm{I}}, \varphi_{1,\mathrm{II}})$ *unravels the set* W. *Then* $(T_2, \pi_1 \circ \pi_2, \varphi_{1,\mathrm{I}} \circ \varphi_{2,\mathrm{I}}, \varphi_{1,\mathrm{II}} \circ \varphi_{2,\mathrm{II}})$ *is a* k-*covering of* T_0 *that unravels the set* W.

Definition 3.66 (The set of positions of length l, $T^{[l]}$). *For every tree* T *and every* $l \in \mathbb{N}$, *denote the set of all positions in* T *of length* l *by*

$$
T^{[l]} := \{p \in T : \operatorname{len}(p) = l\}.
$$

The definition of a k-covering implies that the restriction $\pi : \widetilde{T}^{[l]} \to T^{[l]}$ is a bijection, for every $l < 2k$.

In Lemma 3.59 we proved that if the winning set W of a game $G = (T, W)$ is closed, then the game has a covering $\widetilde{G} = (\widetilde{T}, \widetilde{W})$ where $\widetilde{W}$ is clopen. The next lemma states the analogous result for k-coverings.

Lemma 3.67. *Let* $G = (T, W)$ *be a game where the set* W *is closed. For every* $k \in \mathbb{N}$ *there exists a* k-*covering* $(\widetilde{T}, \pi, \varphi_{\mathrm{I}}, \varphi_{\mathrm{II}})$ *of* T *that unravels* W.

Proof. Take the construction in Lemma 3.59, but assume that in stages $0, 1, \ldots, 2k-1$ the players play as in G, in stage $2k$ Player I selects a move a_{2k} in T and a I-imposed subtree of $T_{\langle a_0, a_1, \ldots, a_{2k} \rangle}$, and in stage $2k+1$ Player II selects a subtree T_{II} that satisfies one of the conditions (ST.1) or (ST.2) and a move in T_{II}. Continue as in Lemma 3.59, to obtain a k-covering of G that unravels W. $\qquad\square$

The next result shows that the inverse limit of k-covering exists.

Lemma 3.68. *Let* T_0 *be a tree. For each* $l \in \mathbb{N}$, *let* $(T_{l+1}, \pi_{l+1}, \varphi_{l+1,\mathrm{I}}, \varphi_{l+1,\mathrm{II}})$ *be an* $(l+1)$-*covering of the tree* T_l. *Then there exists a tree* $\widetilde{T}$, *and for each* $l \in \mathbb{N}$ *there exist functions* $\widetilde{\pi}_l : \widetilde{T} \to T_l$ *and* $\widetilde{\varphi}_{l,i} : S_i(\widetilde{T}) \to S_i(T_l)$ *for* $i = \mathrm{I}, \mathrm{II}$, *such that for every* $l \in \mathbb{N}$,

IL.1) $(\widetilde{T}, \widetilde{\pi}_l, \widetilde{\varphi}_{l,\mathrm{I}}, \widetilde{\varphi}_{l,\mathrm{II}})$ *is an* $(l+1)$-*covering of* T_l.

IL.2) $\widetilde{\pi}_l = \pi_{l+1} \circ \widetilde{\pi}_{l+1}$.

IL.3) $\widetilde{\varphi}_{l,i} = \varphi_{l+1,i} \circ \widetilde{\varphi}_{l+1,i}$ *for* $i = \mathrm{I}, \mathrm{II}$.

Lemma 3.68 claims that there exist a tree $\widetilde{T}$ and functions $(\widetilde{\pi}_l, \widetilde{\varphi}_{l,\mathrm{I}}, \widetilde{\varphi}_{l,\mathrm{II}})_{l \in \mathbb{N}}$ such that the diagram in Figure 11 is commutative.

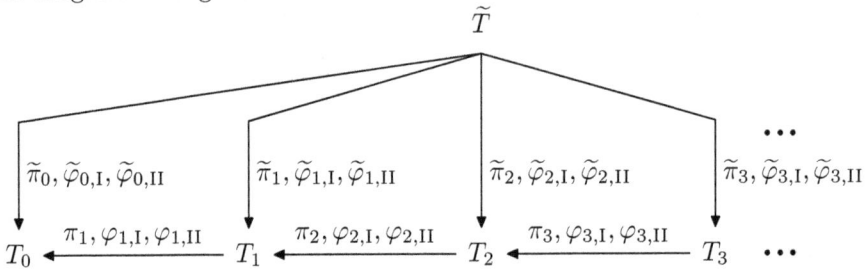

Figure 11: Graphic depiction of Lemma 3.68.

Proof of Lemma 3.68. **Step 1:** Definition of $\widetilde{T}$.

For every $l \in \mathbb{N}$ and every $k \in \mathbb{N}$, the positions of T_l of length smaller than $2l$ are the same as the positions of T_{l+k} of length smaller than $2l$, see Figure 12.

Define the tree $\widetilde{T}$, whose alphabet is the union of the alphabets of $(T_l)_{l \in \mathbb{N}}$, as follows:

$$\widetilde{T} := \bigcup_{l \in \mathbb{N}} (T_l^{[2l]} \cup T_l^{[2l+1]}).$$

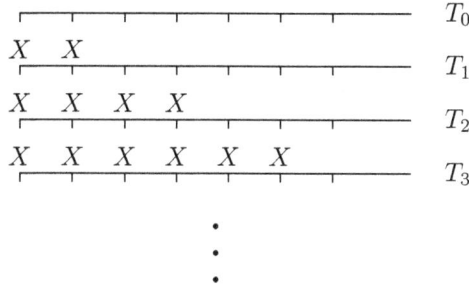

Figure 12: The levels of T_l that lie in T_{l-1}, for $l = 1, 2, 3$.

That is, the positions of $\widetilde{T}$ of length $2l$ and $2l + 1$ are the positions of T_l of length $2l$ and $2l + 1$, see Figure 13. In other words, we add to $\widetilde{T}$ the "new" positions in T_l of levels $2l$ and $2l + 1$, which do not lie in T_{l-1}.

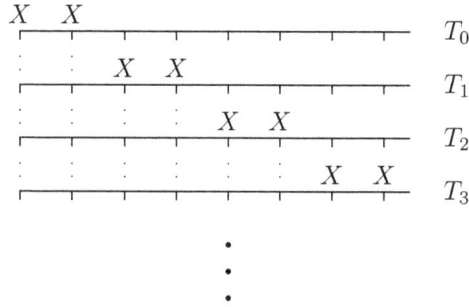

Figure 13: The construction of the tree $\widetilde{T}$.
The positions in the levels marked by X are included in $\widetilde{T}$.

Step 2: Definition of $\widetilde{\pi}_l$.

We next define the function $\widetilde{\pi}_l : \widetilde{T} \to T_l$. Let $\widetilde{p} \in \widetilde{T}$. If $\text{len}(\widetilde{p}) \leq 2l + 1$, then $\widetilde{p}$ is in T_l, and we set

$$\widetilde{\pi}_l(\widetilde{p}) := \widetilde{p}.$$

If $\text{len}(\widetilde{p}) > 2l + 1$, then $\widetilde{p}$ is in $T_{l'}$ for $l' = \lceil \text{len}(\widetilde{p})/2 \rceil - 1$, and we set

$$\widetilde{\pi}_l(\widetilde{p}) := \pi_{l+1} \circ \pi_{l+2} \circ \cdots \circ \pi_{l'}(\widetilde{p}).$$

Step 3: Definition of $\widetilde{\varphi}_{l,i}$.

Fix $i = \text{I}, \text{II}$. We turn to define the functions $\widetilde{\varphi}_{l,i} : S_i(\widetilde{T}) \to S_i(T_l)$ for $l \in \mathbb{N}$. Let then $\widetilde{s}_i \in S_i(\widetilde{T})$ and let $\widetilde{p} \in \widetilde{T}$ (where the length of $\widetilde{p}$ is even if $i = \text{I}$ and odd if $i = \text{II}$). If $\text{len}(\widetilde{p}) \leq 2l + 1$, then $\widetilde{p} \in \widetilde{T}_l$; set

$$\widetilde{\varphi}_{l,i}(\widetilde{s}_i)(\widetilde{p}) := \widetilde{s}_i(\widetilde{p}).$$

If $\text{len}(\widetilde{p}) > 2l + 1$ then, as above, $\widetilde{p}$ is in $T_{l'}$ for $l' = \lceil \text{len}(\widetilde{p})/2 \rceil - 1$; set

$$\widetilde{\varphi}_{l,i}(\widetilde{s}_i)(\widetilde{p}) := (\varphi_{l+1,i} \circ \varphi_{l+2,i} \circ \cdots \circ \varphi_{l',i}(\widetilde{s}_i))(\widetilde{p}).$$

The reader can verify that Conditions (C.1), (C.2), and (C.3) of Definition 3.40, as well as Conditions (IL.1), (IL.2), and (IL.3), are satisfied, see Exercise 3.23. □

The following lemma is a strengthening of Lemma 3.68, where G_{l+1} is a $(k+l+1)$-covering of the game G_l for some fixed $k \in \mathbb{N}$ (rather than an $(l+1)$-covering). Its proof is similar to the proof of Lemma 3.68, hence is omitted.

Lemma 3.69. *Let $k \in \mathbb{N}$ and let T_0 be a tree. For each $l \in \mathbb{N}$ consider a $(k+l+1)$-covering $(T_{l+1}, \pi_{l+1}, \varphi_{l+1,\mathrm{I}}, \varphi_{l+1,\mathrm{II}})$ of T_l. Then there exists a tree $\widetilde{T}$, and for each $l \in \mathbb{N}$ there exist functions $\widetilde{\pi}_l : \widetilde{T} \to T_l$ and $\widetilde{\varphi}_{l,i} : S_i(\widetilde{T}) \to S_i(T_l)$ for $i = \mathrm{I}, \mathrm{II}$, such that for every $l \in \mathbb{N}$,*

ILk.1) $(\widetilde{T}, \widetilde{\pi}_l, \widetilde{\varphi}_{l,\mathrm{I}}, \widetilde{\varphi}_{l,\mathrm{II}})$ *is a $(k+l+1)$-covering of T_l.*

ILk.2) $\widetilde{\pi}_l = \pi_{l+1} \circ \widetilde{\pi}_{l+1}.$

ILk.3) $\widetilde{\varphi}_{l,i} = \varphi_{l+1,i} \circ \widetilde{\varphi}_{l+1,i}$ *for $i = \mathrm{I}, \mathrm{II}$.*

3.12   PROOF OF MARTIN'S DETERMINACY THEOREM

We are now ready to prove Theorem 3.37, which will follow from the next result and Lemma 3.57.

Lemma 3.70. *For every alphabet $\mathcal{A}$, every tree T over $\mathcal{A}$, every Borel subset $W \subseteq [T]$, and every $k \in \mathbb{N}$, there is a k-covering of T that unravels W.*

Proof. We proceed by transfinite induction on the rank of W in the Borel hierarchy. Specifically, we will prove the following statement: For every countable ordinal α, every alphabet $\mathcal{A}$, every tree T over $\mathcal{A}$, every Borel set $W \subseteq [T]$ of rank α of the Borel hierarchy, and every $k \in \mathbb{N}$, there exists a k-covering of T that unravels W.

Note that the alphabet is not fixed. Rather, we simultaneously prove the result for all alphabets.

The base of the induction:

By Lemma 3.67, the claim holds for every closed set W.

By Lemma 3.56, if the claim holds for a Borel set $W \in \Pi_\alpha([T]) \cup \Sigma_\alpha([T])$, it also holds for its complement. In particular, it holds for every winning set W of rank 1 in the Borel hierarchy.

The inductive step:

Assume that α is a countable ordinal and that the claim holds for every ordinal $\beta < \alpha$. Since the sets in $\Pi_\alpha([T])$ are the complements of the sets in $\Sigma_\alpha([T])$, Lemma 3.56 implies that it is sufficient to prove the claim for sets in $\Sigma_\alpha([T])$. Figure 14 illustrates the construction that we describe in the rest of the proof.

Let T be a tree over an alphabet $\mathcal{A}$, let $W \subseteq [T]$ be a Borel subset in $\Sigma_\alpha([T])$, and let $k \in \mathbb{N}$. Then $W = \bigcup_{l \in \mathbb{N}} W_l$, where for each $l \in \mathbb{N}$ we have $W_l \in \Pi_{\beta_l}([T])$ for some $\beta_l < \alpha$.

Since $W_0 \in \Pi_{\beta_0}([T])$ and $\beta_0 < \alpha$, by the induction hypothesis, there exists a k-covering $(T_1, \pi_1, \varphi_{1,\mathrm{I}}, \varphi_{1,\mathrm{II}})$ of $T_0 := T$ (over some alphabet $\mathcal{A}_1$) that unravels W_0. In particular, $\pi_1^{-1}(W_0)$ is open.

By Lemma 3.44, the function π_1 is continuous. Therefore, and since $W_1 \in \Pi_{\beta_1}([T])$, Lemma 2.13 shows that $\pi_1^{-1}(W_1)$ is of rank β_1 in the Borel hierarchy of $[T_1]$. By the induction hypothesis, there exists a $(k+1)$-covering $(T_2, \pi_2, \varphi_{2,\mathrm{I}}, \varphi_{2,\mathrm{II}})$ of T_1 (over some alphabet $\mathcal{A}_2$) that unravels $\pi_1^{-1}(W_1)$. In particular, $\pi_2^{-1} \circ \pi_1^{-1}(W_1)$ is open.

By induction, we know that for every $l \in \mathbb{N}$ there exists a $(k+l)$-covering $(T_{l+1}, \pi_{l+1}, \varphi_{l+1,\mathrm{I}}, \varphi_{l+1,\mathrm{II}})$ of T_l (over some alphabet $\mathcal{A}_l$) that unravels $\pi_l^{-1} \circ \pi_{l-1}^{-1} \circ \cdots \circ \pi_1^{-1}(W_l)$. In particular, $\pi_{l+1}^{-1} \circ \pi_l^{-1} \circ \pi_{l-1}^{-1} \circ \cdots \circ \pi_1^{-1}(W_l)$ is open.

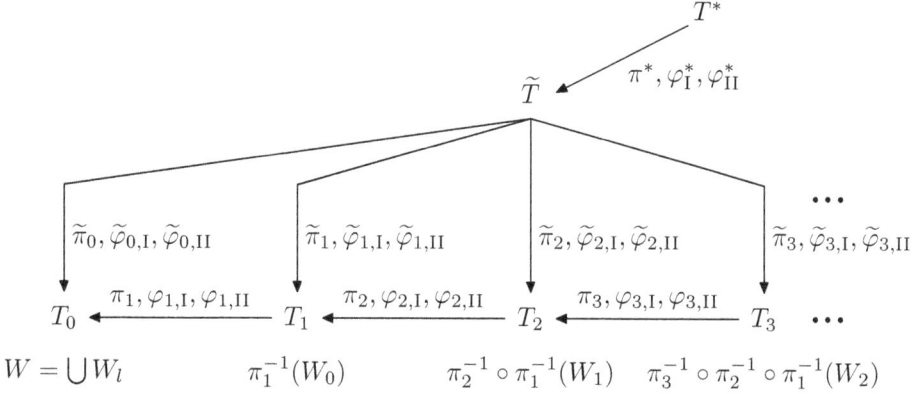

Figure 14: The construction in the proof of Lemma 3.70.

Let $\widetilde{T}$, $(\widetilde{\pi}_l)_{l \in \mathbb{N}}$, and $(\widetilde{\varphi}_{l,\mathrm{I}})_{l \in \mathbb{N}, i=1,2}$ be given by Lemma 3.69. By Lemma 3.63, the covering $(\widetilde{T}, \widetilde{\pi}_0, \widetilde{\varphi}_{0,\mathrm{I}}, \widetilde{\varphi}_{0,\mathrm{II}})$ unravels each of the sets $(W_l)_{l \in \mathbb{N}}$. For every $l \in \mathbb{N}$ the function $\widetilde{\pi}_l$ is continuous, and hence the set $\widetilde{\pi}_{l+1}^{-1} \circ \pi_{l+1}^{-1} \circ \pi_l^{-1} \circ \cdots \circ \pi_1^{-1}(W_l)$ is open. By Condition (ILk.2),

$$\widetilde{\pi}_{l+1}^{-1} \circ \pi_{l+1}^{-1} \circ \pi_l^{-1} \circ \cdots \circ \pi_1^{-1}(W_l) = \widetilde{\pi}_0^{-1}(W_l),$$

and hence $\widetilde{\pi}_0^{-1}(W_l)$ is open for every $l \in \mathbb{N}$. It follows that $\widetilde{\pi}_0^{-1}(W) = \bigcup_{l \in \mathbb{N}} \widetilde{\pi}_0^{-1}(W_l)$ is open as well.

Let $(T^*, \pi^*, \varphi_{\mathrm{I}}^*, \varphi_{\mathrm{II}}^*)$ be a k-covering of $\widetilde{T}$ that unravels $\widetilde{\pi}_0^{-1}(W)$, which exists because we already proved the claim for open sets. By Lemma 3.65, $(T^*, \widetilde{\pi}_0 \circ \pi^*, \widetilde{\varphi}_{0,\mathrm{I}} \circ \varphi_{\mathrm{I}}^*, \widetilde{\varphi}_{0,\mathrm{II}} \circ \varphi_{\mathrm{II}}^*)$ is a k-covering of T that unravels W, and the induction step is complete. $\square$

Example 3.71. *Let us explain the construction in the proof of Lemma 3.70 by an example of a set W which is the union of two closed sets (rather than countably many closed sets).*

Let $T = \{0,1\}^{<\mathbb{N}}$, and set

$$W_1 := [T_{\langle 0,0 \rangle}] = \{x = \langle a_0, a_1, \dots \rangle \in [T]: a_0 = 0, a_1 = 0\},$$
$$W_2 := [T_{\langle 0,1,1 \rangle}] = \{x = \langle a_0, a_1, \dots \rangle \in [T]: a_0 = 0, a_1 = a_2 = 1\},$$
$$W := W_1 \cup W_2,$$

see Figure 15. Player I has a winning strategy in (T, W): she selects 0 in stage 0, and, if Player II selects 1 in stage 1, she (Player I) selects 1 in stage 2. Player II *has a winning strategy in both (T, W_1) and (T, W_2): every strategy of Player II that selects 1 in stage 1 is winning in (T, W_1), and every strategy of Player II that selects 0 in stage 1 is winning in (T, W_2).*

The construction in the proof of Lemma 3.70 goes as follows.

- *In stage 0, Player I selects a pair (a_0, T_I), where $a_0 \in \{0, 1\}$ and T_I is a I-imposed subtree of T.*

- *In stage 1, Player II selects a pair (a_1, T_{II}), where $a_1 \in \{0, 1\}$ and T_{II} satisfies (ST.1) or (ST.2).*

- *In stage 2, Player I selects a pair (a_2, T_I'), where $a_2 \in \{0, 1\}$ is such that $\langle a_0, a_1, a_2 \rangle \in T_{II}$ and T_I' is a I-imposed subtree of T_{II}.*

- *In stage 3, Player II selects a pair (a_3, T_{II}'), where $a_3 \in \{0, 1\}$ is such that $\langle a_0, a_1, a_2, a_3 \rangle \in T_{II}$ and T_{II}' satisfies (ST.1) or (ST.2) with respect to T_I'.*

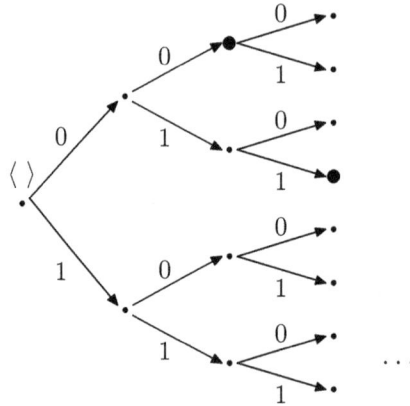

Figure 15: The situation in Example 3.71.
The positions that define W are emphasized.

To win this game, Player I selects $a_0 = 0$, and

$$T_I = T_{\langle 0,0 \rangle} \cup T_{\langle 0,1,1 \rangle}.$$

That is, Player I restricts herself by announcing in stage 0 that if Player II selects 1 in stage 1, then she (Player I) will select 1 in stage 2. As we see, this is part of the winning strategy of Player I in the game (T, W). In stage 1, Player II can select a II-imposed subtree $T_{II} \subseteq T_{\langle 0,0 \rangle}$, which satisfies (ST.1), or she can select a II-imposed subtree $T_{II} \subseteq T_{\langle 0,1,1 \rangle}$, which also satisfies (ST.1). In both cases, the generated play $\langle (a_0, T_I), (a_1, T_{II}), (a_2, T_I'), (a_3, T_{II}'), \ldots \rangle$ satisfies that $\langle a_0, a_1, \ldots \rangle \in W$.

3.13 LARGE CARDINALS AND THE DETERMINACY OF ANALYTIC GAMES

Assuming the Axiom of Choice, Theorem 3.37 states that the game is determined as soon as the winning set is Borel. Example 3.38 shows that when the winning set is not Borel, the game may not be determined. The reader may wonder whether there is a larger class of winning sets than Borel sets for which the game is determined.

In this section we will show that the game is determined as soon as the winning set is analytic. Since not all analytic sets are Borel, this result identifies a class of non-Borel winning sets for which the underlying game is determined. However, we will not prove this result within the standard axioms of Set Theory, namely, the ZFC axioms. Rather, we will impose an axiom that is part of an area in Set Theory called *large cardinals*.

The techniques used in this section differ from those introduced earlier in the chapter, and as such, this section can be read independently of the preceding material.

3.13.1 Ramsey Cardinals

The sets of natural numbers is infinite, and its cardinality is $\aleph_0$. The set of real numbers is infinite, and larger than the set of natural numbers – its cardinality is $\mathfrak{c} = 2^{\aleph_0}$. How large can infinite sets get?

In Set Theory, the area of large cardinals postulates the existence of cardinals that satisfy certain properties, which imply that they are "large." Let us illustrate this by the so-called *inaccessible cardinals*.

One property of the cardinal $\aleph_0$ is the following. Suppose we are given sets whose cardinality is smaller than $\aleph_0$; that is, they are finite. Suppose that the number of these sets that we are given is also smaller than $\aleph_0$; that is, we are given a finite collection of finite sets. Then the union of these sets has a cardinality smaller than $\aleph_0$. A second property satisfied by $\aleph_0$ is that for every cardinal ξ smaller than $\aleph_0$ (that is, a finite cardinal), we have $2^\xi < \aleph_0$.

One way to define a large cardinal uses these two observations: An uncountable cardinal κ is called *inaccessible* if (a) it is not the cardinality of a union of fewer than κ sets, each with cardinality less than κ, and (b) $\xi < \kappa$ implies $2^\xi < \kappa$. Does an inaccessible cardinal exist? Apparently, within the ZFC axioms, the existence of an inaccessible cardinal cannot be proven.

In this section we will provide a definition of large cardinals called *Ramsey cardinals*, which turn out to be larger than inaccessible cardinals: every Ramsey cardinal is inaccessible.

Ramsey's Theorem (Ramsey, 1930) states that if we color the edges of a complete graph, whose set of vertices is $\mathbb{N}$, by finitely many colors, then there is a complete infinite monochromatic subgraph. Formally, let $[\mathbb{N}]^2 := \{(i,j) \in \mathbb{N} : i < j\}$ be the set of all pairs of distinct natural numbers. Ramsey's Theorem states that for every $K \in \mathbb{N}$ and every function $f : [\mathbb{N}]^2 \to \{0, 1, \dots, K\}$ there is an infinite subset $C \subseteq \mathbb{N}$ such that $f(i,j) = f(i',j')$ whenever $(i,j), (i',j') \in [\mathbb{N}]^2$ and $i, j, i', j' \in C$.

We next present a stronger property than the one given by Ramsey's Theorem.

Definition 3.72. *(A-function, dimension) Let A be a set. An A-function is a function $f : A^n \to \mathbb{N}$, for some $n \in \mathbb{N}$. The quantity n is called the* dimension *of f.*

Definition 3.73. *(Ramsey cardinals)*[1] *An uncountable cardinal κ is called* Ramsey *if it satisfies the following property: For every countable set $\mathcal{F}$ of κ-functions, there is*

---

[1]The definition we provide here is different than the standard one, and more suitable for our application.

a subset $X \subseteq \kappa$ with cardinality κ, such that for every $n \in \mathbb{N}$, every function $f \in \mathcal{F}$ with dimension n, and every two vectors $r, r' \in X^n$, we have $f(r) = f(r')$.

3.13.2 The Main Result

We will study games in which the set of moves of both players is $\mathbb{N}$. It follows that the set of plays is $\mathbb{N}^{\mathbb{N}}$, the set of all sequences of natural numbers. The winning set will be assumed to be analytic. For simplicity, we define analytic sets only for the space $\mathbb{N}^{\mathbb{N}}$. Recall that $\mathbb{N}^{\mathbb{N}}$ is a topological space when endowed with the product topology.

Definition 3.74. *(analytic set) A subset A of $\mathbb{N}^{\mathbb{N}}$ is analytic if A is the projection of a closed subset of $\mathbb{N}^{\mathbb{N}} \times \mathbb{N}^{\mathbb{N}}$; that is, there is a closed subset $R \subseteq \mathbb{N}^{\mathbb{N}} \times \mathbb{N}^{\mathbb{N}}$ such that the projection of R on its first coordinate is A:*

$$A = \{x \in \mathbb{N}^{\mathbb{N}} \colon (x, y) \in R \text{ for some } y \in \mathbb{N}^{\mathbb{N}}\}.$$

Remark 3.75. *(analytic sets in general spaces) In general, analytic sets are defined in Polish spaces. A subset of a Polish[2] space is analytic if it is the continuous image of a Polish space.*

It is well known that every Borel set is analytic and that the complement of an analytic set is not necessarily analytic.

The main result we prove in this section is the following, which states that if a Ramsey cardinal exists, then every game $(\mathbb{N}^{<\mathbb{N}}, W)$ is determined whenever W is analytic.

Theorem 3.76 (Martin, 1970). *Let $T = \mathbb{N}^{<\mathbb{N}}$, and let $W \subseteq [T] = \mathbb{N}^{\mathbb{N}}$ be an analytic set. If there exists a Ramsey cardinal, then the game $G = (T, W)$ is determined.*

Before establishing Theorem 3.76, we introduce a total order on $\mathbb{N}^{<\mathbb{N}}$ that will be useful in its proof.

3.13.3 A Total Order on Sequences

We here define a linear order on finite sequences of natural numbers, which is close to the lexicographic order.

Definition 3.77. *(Kleene-Brouwer ordering) The* Kleene-Brouwer ordering *on $\mathbb{N}^{<\mathbb{N}}$ is defined as follows: For every two finite sequences $x = (x_0, x_1, \ldots, x_n)$ and $y = (y_0, y_1, \ldots, y_m)$, we say that $x \geq_{KB} y$ if one of the following conditions holds:*

1. y is an extension of x.

2. $y_k < x_k$, where k is the smallest integer such that $x_k \neq y_k$.

---

[2] A *Polish space* is a separable completely metrizable topological space; that is, a space homeomorphic to a complete metric space that has a countable dense subset.

Thus, the empty sequence $\emptyset$ is maximal in $\mathbb{N}^{<\mathbb{N}}$, and we have, e.g.,

$$(2) >_{\text{KB}} (1,1) >_{\text{KB}} (0) >_{\text{KB}} (0,2,1,0) >_{\text{KB}} (0,1,2) >_{\text{KB}} (0,1,0) >_{\text{KB}} (0,0,1,0).$$

Recall that in the product topology on $\mathbb{N}^{\mathbb{N}}$, a sequence $(x^l)_{l \in \mathbb{N}}$ converges to a limit if every prefix of x^* is a prefix of all elements in $(x^l)_{l \in \mathbb{N}}$ except, possibly, finitely many of them. Note that $\mathbb{N}^{\mathbb{N}}$ is not compact in this topology. Indeed, for every $l \in \mathbb{N}$ let $A^l \subseteq \mathbb{N}^{\mathbb{N}}$ be given by

$$A^l := \{x = (x_0, x_1, \dots) \in \mathbb{N}^{\mathbb{N}} \colon x_0 = l\}.$$

Then $(A^l)_{l \in \mathbb{N}}$ is an open cover of $\mathbb{N}^{\mathbb{N}}$ that does not have a finite subcover.

The next result states that a decreasing sequence of elements in $\mathbb{N}^{\mathbb{N}}$ (with respect to the Kleene-Brouwer ordering) has a convergent subsequence.

Theorem 3.78. *Let $(x^l)_{l \in \mathbb{N}}$ be a sequence of elements in $\mathbb{N}^{<\mathbb{N}}$ that is decreasing with respect to the Kleene-Brouwer ordering. Then there exist $x^* \in \mathbb{N}^{\mathbb{N}}$ and a subsequence $(x^{l_n})_{n \in \mathbb{N}}$ such that $(x^l)_{l \in \mathbb{N}}$ converges to x^*.*

Proof. Set $l_0 := 0$.

Consider the sequence $(x_0^l)_{l \in \mathbb{N}}$ of natural numbers, which is composed of the first coordinate of the elements of $(x^l)_{l \in \mathbb{N}}$. Since the sequence $(x^l)_{l \in \mathbb{N}}$ is decreasing with respect to the Kleene-Brouwer ordering, the sequence $(x_0^l)_{l \in \mathbb{N}}$ is non-increasing, and, in particular, eventually constant. Denote this constant by x_0^*, and consider the subsequence of $(x^l)_{l \in \mathbb{N}}$ in which $x_0^l = x_0^*$. Set l_1 to be the index of the first element in this subsequence (in case this index is equal to l_0, set l_1 to be the index of the second element in this subsequence).

Repeat this argument recursively with the subsequence, to generate the next elements of x^* and obtain the desired result. $\square$

3.13.4   On Infinite Trees with Finite Branches

In this subsection we state and prove a result that we need in the proof of Theorem 3.76. We start by defining the notion of a trimmed tree, which is a tree that may contain leaves.

Definition 3.79. *(trimmed tree) A* trimmed tree *(over the alphabet $\mathcal{A}$) is a set $T \subseteq \mathcal{A}^{<\mathbb{N}}$ of finite sequences in $\mathcal{A}$ with the following property: If $p \in T$ and p' is a prefix of p, then $p' \in T$.*

Definition 3.80. *(leaf) Let T be a trimmed tree. A* leaf *is a position p that has no successor: there is no $p' \in T$ such that $p \prec p'$.*

Comparing Definition 3.79 to Definition 3.2, we observe that every tree is a trimmed tree, yet while in a tree every position has a successor, in a trimmed tree some positions are allowed to be leaves.

Definition 3.81. *(a trimmed tree with finite plays) A trimmed tree T has* finite plays *if there is no $x \in \mathcal{A}^{\mathbb{N}}$ such that all prefixes of x are in T.*

The next result states that given a tree with finite plays, we can assign a countable ordinal to each position, such that (a) all leaves are assigned the ordinal 0, and (b) along each path, the ordinals are decreasing.

Theorem 3.82. *Let $T \subseteq \mathbb{N}^{<\mathbb{N}}$ be a trimmed tree with finite plays over the alphabet $\mathbb{N}$. There is a function ζ that assigns to each position in T a countable ordinal, such that*

Z.1) *$\zeta(p) = 0$ for every leaf $p \in T$.*

Z.2) *$\zeta(p) > \zeta(p')$ for every $p, p' \in T$ such that $p <_{\mathrm{KB}} p'$.*

Proof. **Step 1:** Defining a function ζ_α for every ordinal α.

For every ordinal α we define a function ζ_α that assigns an ordinal to each position in T as follows.

- $\zeta_0(p) := 0$, for every position $p \in T$.

- For every ordinal $\alpha > 0$ and every position $p \in T$ define

$$\zeta_\alpha(p) := \begin{cases} 0, & \text{if } p \text{ is a leaf,} \\ \sup\{\zeta_\beta(p') + 1 : \beta < \alpha, p' <_{\mathrm{KB}} p\}, & \text{if } p \text{ is not a leaf.} \end{cases}$$

Using transfinite induction we can prove that $\zeta_\alpha \geq \zeta_\beta$ for every two ordinals α and β such that $\alpha > \beta$; that is, $\zeta_\alpha(p) \geq \zeta_\beta(p)$ for every $p \in T$.

Step 2: For every ordinal α, the image of ζ_α contains only countable ordinals.

To prove this, assume by contradiction, that there exist an ordinal α and a position $p \in T$ such that $\zeta_\alpha(p)$ is uncountable; this can happen only if α is an uncountable ordinal. Then $\zeta_{\alpha'}(p)$ is uncountable for every $\alpha' > \alpha$. This implies that there is a position $p' \in T$ such that $p' <_{\mathrm{KB}} p$ and $\zeta_\alpha(p')$ is uncountable. Indeed, if $\zeta_\alpha(p')$ had been countable for every position $p' <_{\mathrm{KB}} p$, then $\zeta_{\alpha+1}(p)$ would have been countable, which is a contradiction. Continuing inductively, we conclude that there exists an infinite sequence $(p^n)_{n \in \mathbb{N}}$ of positions in T that is decreasing with respect to the Kleene-Brouwer ordering and satisfies that $\zeta_\alpha(p^n)$ is uncountable.

By Theorem 3.78, there is an element $x^* \in [T]$ such that every prefix of x^* is a prefix of infinitely many elements of $(p^n)_{n \in \mathbb{N}}$. Since a prefix of a position in T is a position in T, this contradicts the fact that the tree does not contain an infinite path.

Step 3: Defining ζ.

Since (a) $\zeta_\alpha \geq \zeta_\beta$ for every two ordinals α and β such that $\alpha > \beta$, and (b) the image of ζ_α is countable for every ordinal α, there is an ordinal α^* such that $\zeta_{\alpha^*+1} = \zeta_{\alpha^*}$. Since the number of positions is countable, α^* is at most the first uncountable ordinal. The definitions imply that $\zeta := \zeta_{\alpha^*}$ satisfies Conditions (Z.1) and (Z.2). $\qquad\square$

3.13.5 Proof of the Determinacy of Analytic Games

In this subsection we prove Theorem 3.76. It will be convenient to denote moves of Player I by a different symbol than moves of Player II. We therefore denote the n'th move of Player I (which is selected in stage $2n$) by a_n, and the n'th move of Player II (which is selected in stage $2n + 1$) by b_n.

If the winning set W is empty, then the result is trivial, and any strategy of Player II is winning. We will therefore assume that $W \neq \emptyset$.

Step 1: Preparations.

Since the set $W \subseteq \mathbb{N}^{\mathbb{N}}$ is analytic, there is a closed set $R \subseteq \mathbb{N}^{\mathbb{N}} \times \mathbb{N}^{\mathbb{N}}$ whose projection on the first coordinate is W. Thus, an element of R is a pair of infinite sequences, denoted $((a_0, b_0, a_1, b_1, \dots), (c_0, c_1, \dots))$.

Definition 3.83 (Secured sequence). *A finite sequence*

$$(a_0, b_0, c_0, a_1, b_1, c_1, \dots, a_{2n-1}, b_{2n-1}, c_{2n-1}) \in \mathbb{N}^{<\mathbb{N}}$$

is secured *for Player I, if it is the prefix of some element of R: there is an element $((a_0', b_0', a_1', b_1', \dots), (c_0', c_1', \dots)) \in R$ such that $(a_0', b_0', a_1', b_1', \dots)$ extends $(a_0, b_0, \dots, a_{2n-1}, b_{2n-1})$ and $(c_0', c_1', \dots)$ extends $(c_0, c_1, \dots, c_{2n-1})$.*

We note that if the sequence z is secured, then so is every prefix of z (of length which is a multiple of 3). Therefore, if z is unsecured, then so is every sequence that extends z (of length which is a multiple of 3).

Denote by $\mathbb{N}_{\text{sec}}^{<\mathbb{N}}$ the set of sequences secured for Player I. Since $W \neq \emptyset$, there is at least one element $(a_0, b_0, c_0, a_1, b_1, c_1, \dots)$ in R, and all its prefixes are secured for Player I. In particular, the set $\mathbb{N}_{\text{sec}}^{<\mathbb{N}}$ is at least countably infinite and contains at least one sequence of length $3n$, for every $n \in \mathbb{N}$. Since the set $\mathbb{N}^{<\mathbb{N}}$ of finite sequences is countably infinite, the set $\mathbb{N}_{\text{sec}}^{<\mathbb{N}}$ is at most countable. It follows that the set $\mathbb{N}_{\text{sec}}^{<\mathbb{N}}$ is countably infinite.

Let $l : \mathbb{N} \to \mathbb{N}_{\text{sec}}^{<\mathbb{N}}$ be a bijection with the property that the length of $l(n)$ is at most $3n$, for every $n \in \mathbb{N}$. Since $\mathbb{N}_{\text{sec}}^{<\mathbb{N}}$ contains at least one sequence of length $3n$, for every $n \in \mathbb{N}$, such a function l exists. The function l is an enumeration of the set $\mathbb{N}_{\text{sec}}^{<\mathbb{N}}$.

Let κ be a Ramsey cardinal. We will identify κ with the set of all ordinals of cardinality smaller than κ.

For every $n \in \mathbb{N}$, the first n sequences secured for Player I (according to the enumeration l), are $l(0), l(1), \dots, l(n-1)$. For every $n \in \mathbb{N}$ and every vector $(\xi_0, \xi_1, \dots, \xi_{n-1}) \in \kappa^n$, one can assign the ordinals $\xi_0, \xi_1, \dots, \xi_{n-1}$ to the secured sequences $l(0), l(1), \dots, l(n-1)$ in such a way that the Kleene-Brouwer monotonicity is preserved: there is a permutation $\pi = \pi(\xi_0, \xi_1, \dots, \xi_{n-1}) : \{0, 1, \dots, n-1\} \to \{0, 1, \dots, n-1\}$ such that for every $0 \leq k < m \leq n-1$,

$$\xi_{\pi(k)} > \xi_{\pi(m)} \quad \Longrightarrow \quad l(k) >_{\text{KB}} l(m).$$

The permutation π is unique when the ordinals $\xi_0, \xi_1, \dots, \xi_{n-1}$ are distinct, yet even if the ordinals are not distinct, the values $(\xi_{\pi(k)})_{k=0}^{n-1}$ are independent of the permutation.

Step 2: Definition of an auxiliary game $\widehat{G} = (\widehat{T}, \widehat{W})$.

Consider the following auxiliary game $\widehat{G}$ between two players, called 1 and 2:

- At each even stage $2n \in \mathbb{N}$, Player 1 selects a move $a_n \in \mathbb{N}$.

- At each odd stage $2n + 1 \in \mathbb{N}$, Player 2 selects (i) a move $b_n \in \mathbb{N}$, and (ii) an ordinal $\xi_n \in \kappa$, which will be assigned to the secured sequence $l(n)$.

  Thus, the tree of the game $\widehat{G}$ is

$$\widehat{T} = (\mathbb{N} \times (\mathbb{N} \times \kappa))^{<\mathbb{N}}.$$

Using the function l, the moves of Player 2 define a function $\xi^* : \mathbb{N}_{\text{sec}}^{<\mathbb{N}} \to \kappa$:

$$\xi^*(z) := \xi_{l^{-1}(z)}, \quad \forall z \in \mathbb{N}_{\text{sec}}^{<\mathbb{N}}.$$

- Instead of defining the winning set $\widehat{W}$ of Player 1, we define its complement. Consider a play $\langle a_0, (b_0, \xi_0), a_1, (b_1, \xi_1), \ldots \rangle$ in $\widehat{G}$. This play is in $\widehat{W}^c$ if the following two conditions hold:

  1. For every sequence $z = (a_0, b_0, c_0, \ldots, a_n, b_n, c_n)$ that is secured for Player I, we have $\xi^*(z) = \xi_{l^{-1}(z)} > 0$.

  2. For every two sequences $z = (a_0, b_0, c_0, \ldots, a_n, b_n, c_n)$ and $z' = (a_0, b_0, c_0', \ldots, a_k, b_k, c_k')$, if $z, z' \in \mathbb{N}_{\text{sec}}^{<\mathbb{N}}$ and $z <_{\text{KB}} z'$, then $\xi^*(z) < \xi^*(z')$ (that is, $\xi_{l^{-1}(z)} < \xi_{l^{-1}(z')}$).

Thus, in the game $\widehat{G}$, Player 1 selects moves for Player I in G, while Player 2 selects moves for Player II in G, and, in addition, she selects ordinals for all secured sequences. So that Player 2 wins in the game $\widehat{G}$, the choice of ordinals, that is, the function ξ^*, must be KB-monotone when restricted to secured sequences for Player I whose a-b coordinates agree with the play selected by the players, which implies in particular that the ordinals that are assigned to secured sequences along the actual play must be decreasing. Since any decreasing sequence of ordinals is finite, if there is a play such that all its prefixes are secured for Player I, then there must be $n \in \mathbb{N}$ such that $\xi_n = 0$. Since Player 2 loses if there is a secured sequence z with $\xi^*(z) = 0$, it follows that if Player 2 wins in the game $\widehat{G}$, then necessarily the play reaches an unsecured sequence in finite time (that is not necessarily bounded).

Step 3: The game $\widehat{G}$ is determined.

We will show that the set $\widehat{W}$ is open, and hence, by Corollary 3.35, the game $\widehat{G}$ is determined.

Indeed, the winning condition for Player 1 in $\widehat{G}$ is determined in finite time: If there are $n \in \mathbb{N}$, $k < n$, and $c_0, c_1, \ldots, c_{k-1} \in \mathbb{N}$, such that $z := l(n) = (a_0, b_0, c_0, \ldots, a_{k-1}, b_{k-1}, c_{k-1})$ is secured for Player I and $\xi_n = 0$, then at stage n we know that Player 1 will win the game $\widehat{G}$. If there are $n \in \mathbb{N}$, $n' < n$, $m \le n$, $m' \le n$, $c_0, c_1, \ldots, c_{m-1} \in \mathbb{N}$, and $c_0', c_1', \ldots, c_{m'-1}' \in \mathbb{N}$, such that $z := l(n) = (a_0, b_0, c_0, \ldots, a_{m-1}, b_{m-1}, c_{m-1})$ and $z' := l(n') = (a_0, b_0, c_0', \ldots, a_{m'-1}, b_{m'-1}, c_{m'-1}')$

are secured for Player I, $z >_{\text{KB}} z'$ yet $\xi_m \leq \xi_{m'}$, then once again at stage n we know that Player 1 will win the game $\widehat{G}$.

As a result, Player 1's winning set $\widehat{W}$ is the union of all basic open sets $\left[T_{\langle a_0, (b_0, \xi_0), a_1, (b_1, \xi_1), \ldots, a_{n-1}, (b_{n-1}, \xi_{n-1}) \rangle}\right]$ that satisfy one of the following conditions:

- $\xi_{n-1} = 0$ and there are $k \leq n - 1$ and $(c_0, c_1, \ldots, c_k)$ such that

$$l(n-1) = (a_0, b_0, c_0, a_1, b_1, c_1, \ldots, a_k, b_k, c_k).$$

- KB-monotonicity fails: There is $k < n - 1$ such that $\xi_k < \xi_{n-1}$ and $l(k) \geq_{\text{KB}} l(n-1)$.

Step 4: If Player 2 has a winning strategy in $\widehat{G}$, then Player II has a winning strategy in G.

Suppose that Player 2 has a winning strategy $\widehat{s}_2$ in $\widehat{G}$. This strategy selects at every odd stage $2n+1$ a pair $(b_n, \xi_n) \in \mathbb{N} \times \kappa$, as a function of the moves $(a_0, a_1, \ldots, a_n)$ selected by Player 1 in earlier stages. Let s_{II} be the strategy of Player II in G that selects in stage $2n + 1$ the move b_n that $\widehat{s}_2$ selects in $\widehat{G}$ in stage $2n + 1$. Formally,

$$s_{\text{II}}(\langle a_0, b_0, \ldots, a_{n-1}, b_{n-1}, a_n \rangle) = b_n,$$

where

$$(b_n, \xi_n) := \widehat{s}_2(\langle a_0, (b_0, \xi_0), \ldots, a_{n-1}, (b_{n-1}, \xi_{n-1}), a_n \rangle).$$

Thus, b_0 and ξ_0 are determined by a_0; b_1 and ξ_1 are determined by (a_0, b_0, ξ_0, a_1), and therefore by a_0, a_1; etc. In particular, s_{II} is well-defined.

We will show that s_{II} is a winning strategy of Player II in G. Assume to the contrary that s_{II} is not a winning strategy of Player II in G. Then there is a strategy s_{I} of Player I in G such that $x(s_{\text{I}}, s_{\text{II}}) = \langle a_0, b_0, a_1, b_1, \ldots \rangle \in W$. This implies that there is a vector $(c_0, c_1, \ldots) \in \mathbb{N}^{\mathbb{N}}$ such that

$$((a_0, b_0, a_1, b_1, \ldots), (c_0, c_1, \ldots)) \in R.$$

By definition, all the prefixes of $((a_0, b_0, a_1, b_1, \ldots), (c_0, c_1, \ldots))$ are secured for Player I: for every $n \in \mathbb{N}$, the sequence $(a_0, b_0, c_0, a_1, b_1, c_1, \ldots, a_n, b_n, c_n)$ is secured for Player I.

Let $\widehat{s}_1$ be the strategy of Player 1 in $\widehat{G}$ that selects in each stage $2n$ the move a_n that s_{I} selects in G (which means that it ignores the ordinals $(\xi_0, \ldots, \xi_{n-1})$ that Player 2 selected in $\widehat{G}$). Consider the sequence of ordinals that the strategy $\widehat{s}_2$ assigns to the prefixes of $((a_0, b_0, a_1, b_1, \ldots), (c_0, c_1, \ldots))$. Since $\widehat{s}_2$ is winning in $\widehat{G}$, this sequence is decreasing. Since any decreasing sequence of ordinals is finite, there is a prefix for which $\widehat{s}_2$ assigns the ordinal 0. This contradicts the assumption that $\widehat{s}_2$ is winning.

Step 5: If Player 1 has a winning strategy in $\widehat{G}$, then Player I has a winning strategy in G.

Let $\hat{s}_1$ be a winning strategy of Player 1 in $\hat{G}$. This strategy cannot be easily adapted to the original game G. Indeed, when Player I selects her move in G at stage $2n$, she knows the current position $\langle a_0, b_0, a_1, b_1, \ldots, a_{n-1}, b_{n-1} \rangle$. However, to know what $\hat{s}_1$ selects in stage $2n$, Player I also needs to know $(\xi_0, \xi_1, \ldots, \xi_{n-1})$, information that does not exist in G. To construct s_I we will use the assumption that κ is a Ramsey cardinal.

Step 5.1: A definition of a collection $\mathcal{F}$ of κ-functions.

We here define a countable collection $\mathcal{F}$ of κ-functions: for every position $p = \langle a_0, b_0, a_1, b_1, \ldots, a_{n-1}, b_{n-1} \rangle \in T$ this set will contain a κ-function $f_p : \kappa^n \to \mathbb{N}$, which we define now.

Let $l(0), l(1), \ldots, l(n-1)$ be the first n sequences in $\mathbb{N}_{\text{sec}}^{<\mathbb{N}}$. As mentioned before, for every vector $(\xi_0, \xi_1, \ldots, \xi_{n-1}) \in \kappa^n$, the ordinals $\xi_0, \xi_1, \ldots, \xi_{n-1}$ can be assigned to the sequences $l(0), l(1), \ldots, l(n-1)$ in an (essentially) unique way, so as to satisfy monotonicity.

Define a function $f_p : \kappa^n \to \mathbb{N}$ as follows: For each vector $(\xi_0, \xi_1, \ldots, \xi_{n-1}) \in \kappa^n$,

$$f_p(\xi_0, \xi_1, \ldots, \xi_{n-1}) := \hat{s}_1(\langle a_0, (b_0, \xi_{\pi(0)}), a_1, (b_1, \xi_{\pi(1)}), \ldots, a_{n-1}, (b_{n-1}, \xi_{\pi(n-1)}) \rangle),$$

where $\pi = \pi(\xi_0, \xi_1, \ldots, \xi_{n-1})$ is a permutation that ensures KB-monotonicity. That is, $f_p(\xi_0, \ldots, \xi_{n-1})$ is the move selected by $\hat{s}_1$ when Player 2 selects in the first n odd rounds of $\hat{G}$ the ordinals $\xi_0, \ldots, \xi_{n-1}$ properly ordered.

Let

$$\mathcal{F} := \{ f_p : p \in T \}.$$

Step 5.2: Defining a strategy s_I in G.

The set $\mathcal{F}$ is countably infinite, and all its members are κ-functions. Since κ is a Ramsey cardinal, there is a subset $X \subseteq \kappa$ with cardinality κ, such that for every $n \in \mathbb{N}$, every $p \in T$ of length n, and every two elements $z, z' \in X^n$, we have $f_p(z) = f_p(z')$.

We are now ready to define the strategy s_I in G. For each position $p \in T$, set $s_I(p)$ to be the common value of f_p; namely, $s_I(p) = f_p(z)$, for every $z \in X^n$. In words, s_1 selects the move that $\hat{s}_1$ selects against a large set of possible moves of Player 2 in $\hat{G}$.

Step 5.3: s_I is a winning strategy of Player I in G.

Suppose, by contradiction, that s_I is not a winning strategy of Player I in G. Then there is a strategy s_{II} of Player II in G such that

$$x(s_I, s_{II}) = \langle a_0^*, b_0^*, a_1^*, b_1^*, \ldots \rangle \notin W.$$

This implies that for every sequence $(c_0, c_1, \ldots) \in \mathbb{N}^{\mathbb{N}}$, we have

$$((a_0^*, b_0^*, a_1^*, b_1^*, \ldots), (c_0, c_1, \ldots)) \notin R. \tag{3.8}$$

Since R^c is open, for every sequence $(c_0, c_1, \ldots) \in \mathbb{N}^{\mathbb{N}}$ there is a prefix $(a_0^*, b_0^*, c_0, a_1^*, b_1^*, c_1, \ldots, a_n^*, b_n^*, c_n)$ of the infinite sequence on the left-hand side of Equation (3.8) that is unsecured for Player I.

Consider all finite sequences $(a_0, b_0, c_0, a_1, b_1, c_1, \ldots, a_n, b_n, c_n) \in \mathbb{N}^{<\mathbb{N}}$ as a tree. We trim from this tree the following sequences:

- For every sequence $(a_0, b_0, c_0, a_1, b_1, c_1, \ldots, a_n, b_n, c_n)$ such that $(a_n, b_n) \neq (a_n^*, b_n^*)$, we trim all its *descendants*.

- For every $(c_0, c_1, \ldots, c_n)$ such that the finite sequence $(a_0^*, b_0^*, c_0, a_1^*, b_1^*, c_1, \ldots, a_n^*, b_n^*, c_n)$ is unsecured for Player I, we trim all *descendants* of $(a_0^*, b_0^*, c_0, a_1^*, b_1^*, c_1, \ldots, a_n^*, b_n^*, c_n)$.

In other words, we trim from the tree all positions with a strict prefix that is unsecured for Player I or with a strict prefix whose a-b coordinates differ from $(a_0^*, b_0^*, a_1^*, b_1^*, \ldots)$.

By the previous paragraph, there are no infinite paths in the trimmed tree. By Theorem 3.82, there exists a function ζ that assigns a countable (finite or infinite) ordinal to each position in the trimmed tree, such that $\zeta(y) = 0$ for all leaves y (and, in particular, for all unsecured sequences that were not trimmed), and $\zeta(y) > \zeta(y')$ for every y, y' such that $y >_{\text{KB}} y'$. Let ψ be a monotone function from the image of ζ to X such that $\psi(0) = 0$, and define $\xi(y) = \psi(\zeta(y))$ for every secured sequence y.

To complete the proof, we will define a strategy $\widehat{s}_2$ of Player 2 in $\widehat{G}$ that is winning against $\widehat{s}_1$, contradicting the assumption that $\widehat{s}_1$ is a winning strategy of Player 1 in $\widehat{G}$.

Step 5.4: Definition of a strategy $\widehat{s}_2$ that wins against $\widehat{s}_1$.

Fix a position $\langle (a_0, (b_0, \xi_0)), (a_1, (b_1, \xi_1)), \ldots, (a_{n-1}, (b_{n-1}, \xi_{n-1})) \rangle$ in $\widehat{G}$. The strategy $\widehat{s}_2$ should select $b_n \in \mathbb{N}$ and $\xi_n \in \kappa$. The choice of b_n is determined by s_{II}:

$$b_n := s_{\text{II}}(\langle a_0, b_0, \ldots, a_{n-1}, b_{n-1} \rangle).$$

To determine the ordinal ξ_n, recall that it will be assigns to some sequence secured for Player I whose length does not exceed $3n$, namely, to a sequence $(a_0', b_0', c_0', \ldots, a_k', b_k', c_k')$, where $k < n$. If $(a_0, b_0, a_1, b_1, \ldots, a_k, b_k) \neq (a_0', b_0', a_1', b_1', \ldots, a_k', b_k')$, then the choice of ξ_n is irrelevant. If $(a_0, b_0, a_1, b_1, \ldots, a_k, b_k) = (a_0', b_0', a_1', b_1', \ldots, a_k', b_k')$, then we set

$$\xi_n := \xi(a_0, b_0, c_0', a_1, b_1, \ldots, a_k, b_k, c_k').$$

We claim that the strategy $\widehat{s}_2$ wins in the game $\widehat{G}$ against $\widehat{s}_1$. Indeed, by definition, for every sequence $(c_0, c_1, \ldots, c_n)$, if $z = (a_0, b_0, c_0, \ldots, a_n, b_n, c_n)$ is secure for Player I, then $\xi(z) > 0$. In addition, for every two sequences $(c_0, c_1, \ldots, c_n)$ and $(c_0', c_1', \ldots, c_k')$, if $z = (a_0, b_0, c_0, \ldots, a_n, b_n, c_n)$ and $z' = (a_0, b_0, c_0', \ldots, a_k, b_k, c_k')$ are secured for Player I, and $z <_{\text{KB}} z'$, then once again by definition we have $\xi^*(z) < \xi^*(z')$.

3.14 DISCUSSION

Gale and Stewart (1953) were the first who presented and studied infinite alternating-move games with a winning set, and proved Theorem 3.34. A rich literature extended their result to games with winning sets in higher levels in the Borel hierarchy, until Martin (1975) proved Theorem 3.37: as soon as the winning set is Borel, the game is determined. This result is valid regardless of what the sets of moves of the players are: they may be arbitrary sets and they may depend on the position. The only requirements are that (a) the players move alternately, and (b) the winning set is Borel, namely, it belongs to the tree sigma-algebra. To prove the result, we defined the notion of a covering, and showed that for every Borel winning set there is a covering of the tree such that, under this covering, the inverse image of the winning set is clopen. The proof of this result was achieved by induction on the rank of the winning set in the Borel hierarchy. The claim for sets of rank 1 turned out to be the most difficult. The proof that we provided is based on Martin (1975, 1985) and on the expositions of Martin's proof by Bryant (2002) and Gowers (2013). Other expositions of the proof can be found in Moschovakis (2009), Buffard, Levrel, and Mayo (2024), and the unpublished manuscript Martin (2020).

As Example 3.38 shows, when the winning set is not Borel, the game may not be determined. In Section 3.13 we showed that under a large cardinal axiom, when the winning set is analytic the game is determined. Within the Zermelo-Fraenkel framework, the determinacy of games with analytic winning sets is not provable (Mycielski, 1964).

The reader may wonder whether there are other classes of winning sets, which include all Borel sets, for which the corresponding games are determined. Such extensions were found, e.g., by Martin (1990) within ZFC, and by Neeman (2010) and Montalbán and Shore (2012) within other axiomatic systems.

There are various natural extensions to Theorem 3.37:

- Can the result be formulated (and proved) for games where the outcome is not win/lose, or when the players do not necessarily have opposite goals?

- How about simultaneous-move games?

- And about games where the outcome is not necessarily win/lose, the players move simultaneously, and each player has a different goal?

- And finally, about games where the players do not observe each other's moves immediately, but rather after some delay.

Those extensions will be studied in Chapters 4, 5, 6, and 7, respectively.

The determinacy of games was used to prove various results in mathematics, see, e.g., the surveys by Debs and Saint Raymond (1996) and Marks (2016), and Exercise 3.30, which is based on Blackwell (1967). This result will also help us in the study of games where the outcome is not necessarily win-lose; in the study of nonzero-sum games, where the gain of one player is not necessarily the loss of the other; and in the study of simultaneous-move games.

3.15   EXERCISES

The solution to Exercise 3.13 requires Exercise 3.12. The solution to Exercise 3.30 requires Exercise 3.14. The solution to Exercises 3.33, 3.34, and 3.35 requires Exercise 3.32.

I thank Auriel Rosenzweig for suggesting Exercise 3.27. Exercise 3.29 is adapted from Arieli and Levy (2011), and Exercise 3.30 from Blackwell (1967). I thank Sean Landsberg for finding the solution to Exercise 3.32.

Exercises marked with a star are more challenging than unmarked ones, while those marked with two stars are even more difficult.

1. Prove that the subtree T_p as defined in Definition 3.6 is a tree according to Definition 3.2.

2. Prove that the tree topology introduced in Definition 3.11 is indeed a topology.

3. Let $T = \{0,1\}^{<\mathbb{N}}$. Provide an example of a set $A \subseteq [T]$ that is closed and not open.

4. Let $T = \{0,1\}^{<\mathbb{N}}$ and let

$$A = \{\langle a_0, a_1, \ldots \rangle \in [T] : a_n = 1 \text{ infinitely often}\}.$$

Is A a Borel set? If so, what is the rank of A in the Borel hierarchy of T? Justify your answer.

5. Repeat Exercise 3.4 with the set

$$A = \left\{ \langle a_0, a_1, \ldots \rangle \in [T] : \limsup_{n \to \infty} \frac{1}{n} \sum_{k=0}^{n-1} a_k \geq \frac{1}{3} \right\}.$$

6. Let L be a given finite collection of finite sequences of zeros and ones. For example, L can be $\{(1,1,0,0),(0,1,0),(1,1,1,0,0,1,0,1,1,0)\}$. Alternately Anne and Bertha select numbers in $\{0,1\}$, to create an infinite sequence x of zeros and ones. Anne wins if one of the sequences in L can be found along x; otherwise Bertha wins. With the example of L provided above, if Anne and Bertha chose the sequence $x = (1,1,1,1,1,1,0,1,0,0,0,0,0,0,\ldots)$ then Anne wins because the sequence $(0,1,0) \in L$ appears along x. Do the following.

   (a) Present the situation as a game; that is, write down the data that define the game.

   (b) Is the winning set of Anne open, closed, or neither? Justify your answer.

   (c) There is a second finite collection of finite sequences of zeros and ones, denoted M. Suppose that Anne wins if the infinite sequence x generated by Anne and Bertha contains one of the elements of L, and contains no element of M. Find α such that Anne's winning set is a Borel set of rank α.

7. In this exercise we will study simple relations between determined (and non-determined) games.

   (a) Suppose that the game (T,W) is not determined. Is the game (T,W^c) necessarily not determined?

   (b) Suppose that Player I has a winning strategy in the game (T,W). Is it necessary that in the game (T,W^c) Player II has a winning strategy?

(c) Suppose that Player I has a winning strategy in the game (T, W_1), and Player II has a winning strategy in the game (T, W_2). For each of the games $(T, W_1 \cup W_2)$, $(T, W_1 \cap W_2)$, $(T, W_1 \setminus W_2)$, and $(T, W_2 \setminus W_1)$, determine whether Player I necessarily has a winning strategy and whether Player II necessarily has a winning strategy.

8. Let $T = \{0, 1\}^{<\mathbb{N}}$ be the complete binary tree. The Lebesgue measure on T is the measure λ that satisfies $\lambda([T_p]) = \frac{1}{2^{1+\operatorname{len}(p)}}$, for every position $p \in T$. Does there exist a Borel set W with Lebesgue measure 0 such that Player I has a winning strategy in the game (T, W)? Justify your answer.

9. Let $T = \{0, 1\}^{\mathbb{N}}$, and let $W \subseteq [T]$ be a countable set.

   (a) Is W a Borel set? If so, what is the rank of W in the Borel hierarchy of T?
   (b) Which player has a winning strategy in the game (T, W)?

10. The set $S_{\mathrm{I}}(T)$ of strategies of Player I is a topological space with the following topology: for each position $p \in T$ of even length and each move $a \in \mathcal{A}$, we have the basic open set $\{s_{\mathrm{I}} \in S_{\mathrm{I}}(T) : s_{\mathrm{I}}(p) = a\}$. The set $S_{\mathrm{II}}(T)$ of Player II is a topological space with an analogous topology. It follows that the product set $S_{\mathrm{I}}(T) \times S_{\mathrm{II}}(T)$ is a topological space with the product topology.

    Prove that the function $x : S_{\mathrm{I}}(T) \times S_{\mathrm{II}}(T) \to [T]$ that assigns to each pair of strategies $(s_{\mathrm{I}}, s_{\mathrm{II}}) \in S_{\mathrm{I}}(T) \times S_{\mathrm{II}}(T)$ the play that they induce is continuous.

11. Prove Lemma 3.25: Let $G = (T, W)$ be a game, let $s_{\mathrm{I}} \in S_{\mathrm{I}}(T)$ be a strategy of Player I, and let $x = \langle a_0, a_1, \dots \rangle$ be a play consistent with s_{I}. Let $s_{\mathrm{II}} \in S_{\mathrm{II}}(T)$ be a strategy of Player II such that $s_{\mathrm{II}}(\langle a_0, a_1, \dots, a_n \rangle) = a_{n+1}$, for every even $n \in \mathbb{N}$. Prove that $x(s_{\mathrm{I}}, s_{\mathrm{II}}) = x$.

12. Prove that a play x is consistent with a strategy s_{I} if and only if all its prefixes are consistent with s_{I}.

13. Suppose that the alphabet $\mathcal{A}$ is finite. Prove that for every strategy $s_{\mathrm{I}} \in S_{\mathrm{I}}(T)$, the set of plays that are consistent with s_{I} is compact.

14. The outcome of a game as defined in Definition 3.15 is a win to one of the players. In this exercise we consider games that can also end with a draw. A *3-outcome game* is a triplet $G = (T, W_{\mathrm{I}}, W_{\mathrm{II}})$ where T is a tree and W_{I} and W_{II} are disjoint subsets of $[T]$. The game is played as an alternating-move game (see Definition 3.15), except for the outcome: if the play $x = \langle a_0, a_1, \dots \rangle$ that the players generate is in W_{I} (resp., in W_{II}) Player I (resp., Player II) wins, and otherwise the outcome is a *draw*. A strategy $s_{\mathrm{I}} \in S_{\mathrm{I}}(T)$ is a winning strategy if $x(s_{\mathrm{I}}, s_{\mathrm{II}}) \in W_{\mathrm{I}}$ for every $s_{\mathrm{II}} \in S_{\mathrm{II}}(T)$, and the strategy *guarantees at least a draw for Player* I if $x(s_{\mathrm{I}}, s_{\mathrm{II}}) \notin W_{\mathrm{II}}$ for every $s_{\mathrm{II}} \in S_{\mathrm{II}}(T)$. Winning strategies of Player II and strategies that guarantee at least a draw for Player II are defined analogously.

    Prove that if the sets W_{I} and W_{II} are Borel, then exactly one of the following three alternatives holds:

(i) Player I has a winning strategy.

(ii) Player II has a winning strategy.

(iii) Both players have a strategy that guarantees at least a draw.

15. Prove that the covering $(\widetilde{T}, \pi, \varphi_{\mathrm{I}}, \varphi_{\mathrm{II}})$ that was defined in Example 3.43 is indeed a covering.

16. In Example 3.42 we showed that $\widetilde{T}$ is part of a covering of T. Find a second covering of T that involves the tree $\widetilde{T}$.

17. (*) Let T be a tree over the alphabet $\mathcal{A}$. Let $\widetilde{T}$ be the tree over the alphabet $\mathcal{A} \cup S_{\mathrm{I}}(T) \cup S_{\mathrm{II}}(T)$, which consists of all prefixes of the following plays:

$$\langle s_{\mathrm{I}}, s_{\mathrm{II}}, a_0, a_1, \ldots \rangle,$$

where $s_{\mathrm{I}} \in S_{\mathrm{I}}(T)$, $s_{\mathrm{II}} \in S_{\mathrm{II}}(T)$, and $\langle a_0, a_1, \ldots \rangle = x(s_{\mathrm{I}}, s_{\mathrm{II}})$. That is, in stage 0 Player I selects a strategy for the tree T, in stage 1 Player II selects a strategy for the tree T, and afterwards the players have a single possible move in each stage, and these moves generate the play $x(s_{\mathrm{I}}, s_{\mathrm{II}})$. Let $W \subseteq [T]$, and define

$$\widetilde{W} := \{\langle s_{\mathrm{I}}, s_{\mathrm{II}}, a_0, a_1, \ldots \rangle \colon \langle a_0, a_1, \ldots \rangle \in W\} \subseteq [\widetilde{T}].$$

(a) Suppose that Player I has a winning strategy in (T, W). Does she necessarily have a winning strategy in $(\widetilde{T}, \widetilde{W})$?

(b) Suppose that Player II has a winning strategy in (T, W). Does she necessarily have a winning strategy in $(\widetilde{T}, \widetilde{W})$?

(c) Suppose that Player II does not have a winning strategy in (T, W). Can she have a winning strategy in $(\widetilde{T}, \widetilde{W})$?

(d) Is it necessarily true that $\widetilde{T}$ is part of a covering of T?

18. In Example 3.43, identify all sets $W \subseteq [T]$ such that the covering described in the example unravels (T, W).

19. In this exercise we provide an alternative proof to Lemma 3.44. A *metric* on a set X is a function $d : X \times X \to [0, \infty)$ with the following properties:

- $d(x, y) = 0$ if and only if $x = y$.
- Symmetry: $d(x, y) = d(y, x)$.
- The triangle inequality: $d(x, y) + d(y, z) \geq d(x, z)$ for every $x, y, z \in X$.

Let T be a tree. Define a metric on $[T]$ by:

$$d_T(x, y) := 2^{-\min\{n \in \mathbb{N} \colon x_n \neq y_n\}}.$$

Let T_1, T_2 be two trees. A function $\pi : T_1 \to T_2$ is *monotone* if $\mathrm{len}(\pi(p)) = \mathrm{len}(p)$ for every $p \in T_1$, and $\pi(p) \prec \pi(p')$ whenever $p \prec p'$. For every two positions $p = \langle a_0, a_1, \ldots, a_k \rangle, p' = \langle a'_0, a'_1, \ldots, a'_m \rangle \in T$ define

$$d_T(p, p') := 2^{-\min\{n \in \mathbb{N} \colon a_n \neq a'_n\}},$$

where by convention $a_n \neq a'_n$ whenever $n > \min\{k, m\}$.

(a) Define basic open sets in terms of the metric d_T.

(b) Let T_1, T_2 be two trees. Prove that every monotone function $\pi : T_1 \to T_2$ is 1-Lipschitz.

(c) Prove Lemma 3.44 using items (a) and (b).

20. In Lemma 3.36 we proved that when the winning set W is open and Player I has a winning strategy s_I, for every strategy $s_{II} \in S_{II}(T)$ there is $n = n(s_{II}) \in \mathbb{N}$ such that $[T_{x_n(s_I, s_{II})}] \subseteq W$. Is it true that there is $M > 0$ such that $n(s_{II}) \leq M$ for every $s_{II} \in S_{II}(T)$? Prove or provide a counterexample.

21. Let $G = (T, W)$ be a game where the set W is Borel. Let x be a play such that x_n is a winning position for Player I, for every $n \in \mathbb{N}$. Is it necessarily true that $x \in W$? Prove or provide a counterexample.

22. Let $G = (T, W)$ be a tree, where the set W is closed, and suppose that Player II has a winning strategy in G. Prove that in the game $\widetilde{G} = (\widetilde{T}, \pi^{-1}(W))$ that was defined in the proof of Lemma 3.59, Player II has a winning strategy.

23. Complete the proof of Lemma 3.68: Namely, prove that the functions $(\widetilde{\pi}_l, \widetilde{\varphi}_{l,I}, \widetilde{\varphi}_{l,II})_{l \in \mathbb{N}}$ satisfy Conditions (C.1), (C.2), and (C.3) of Definition 3.40, as well as Conditions (IL.1), (IL.2), and (IL.3).

24. (*) Tyrone argues that one can present the proof of Lemma 3.70 without transfinite induction. He claims that what we actually proved is that the collection of all subsets W of $[T]$ that satisfy the following property contains the closed sets and is closed under complements and countable unions: For every $k \in \mathbb{N}$ there is a k-covering of T that unravels W.

It then follows that this collection is a sigma-algebra that contains all closed sets, and in particular contains all Borel sets.

Is Tyrone correct? Explain your answer.

25. (*) Let $G = (T, W)$ be a game and assume that Player I has a winning strategy s_I in G. Prove that there exists a closed set $W' \subseteq W$ such that s_I is a winning strategy of Player I in (T, W').

26. (*) Let T be a tree over an alphabet $\mathcal{A}$, and let W and $(W_n)_{n \in \mathbb{N}}$ be subsets of $[T]$ such that $W = \bigcap_{n \in \mathbb{N}} W_n$. For each of the following statements, prove whether it is true or provide a counterexample.

(a) If Player I has a winning strategy in (T, W), then she has a winning strategy in (T, W_n), for every $n \in \mathbb{N}$.

(b) If Player II has a winning strategy in (T, W), then she has a winning strategy in (T, W_n), for every $n \in \mathbb{N}$ sufficiently large.

27. (*) Let T be a tree over a finite alphabet $\mathcal{A}$. Let $(W_n)_{n \in \mathbb{N}}$ be a decreasing sequence of closed subsets of $[T]$, and define $W = \bigcap_{n \in \mathbb{N}} W_n$. Prove that if

Player I has a winning strategy in the game (T, W_n) for every $n \in \mathbb{N}$, then she has a winning strategy in the game (T, W).

28. (*) Let (T, W) be a game, where W is not necessarily Borel. Prove that Player I has a winning strategy in the game (T, W) if and only if she has a winning strategy in the game (T, W'), for some Borel subset $W' \subseteq W$.

29. (*) In this exercise we study Borel Games in which there are two possible winning sets, and Nature selects at random one of them; Player I knows which winning set was chosen, while Player II does not.

    A *game with incomplete information on one side* is given by a tree T, two Borel sets $W_0, W_1 \subseteq [T]$, and a number $q \in [0, 1]$. The game is played as follows. Nature selects one of the sets W_0 (with probability q) or W_1 (with probability $1 - q$), and informs Player I (but not Player II) of the choice. The players then alternately select moves along T, as in Definition 3.15, thereby generating a play $\langle a_0, a_1, a_2, \ldots \rangle$. Player I wins if $\langle a_0, a_1, a_2, \ldots \rangle$ is in the set chosen by Nature, and Player II wins otherwise. We denote this game by (T, W_0, W_1, q)

    A strategy of Player II in (T, W_0, W_1, q) is a function s_{II} that assigns to each position p of odd length a move a such that $\langle p, a \rangle \in T$. The set of strategies of Player II is then $S_{\mathrm{II}}(T)$.

    A strategy of Player I in (T, W_0, W_1, q) is a pair $s_{\mathrm{I}} = (s_{\mathrm{I},0}, s_{\mathrm{I},1}) \in S_{\mathrm{I}}(T) \times S_{\mathrm{I}}(T)$, with the interpretation that Player I adopts $s_{\mathrm{I},0}$ when the selected set is W_0, and she adopts $s_{\mathrm{I},1}$ when the selected set is W_1. The set of strategies of Player I in (T, W_0, W_1, q) is then $\widetilde{S}_{\mathrm{I}}(T) := S_{\mathrm{I}}(T) \times S_{\mathrm{I}}(T)$.

    For every pair of strategies $(s_{\mathrm{I}}, s_{\mathrm{II}}) \in \widetilde{S}_{\mathrm{I}}(T) \times S_{\mathrm{II}}(T)$, let $\mathbf{P}(q, s_{\mathrm{I}}, s_{\mathrm{II}})$ denote the probability that Player I wins:

    $$\mathbf{P}(q, s_{\mathrm{I}}, s_{\mathrm{II}}) := q \cdot \mathbf{1}_{\{x(s_{\mathrm{I},0}, s_{\mathrm{II}}) \in W_0\}} + (1 - q) \cdot \mathbf{1}_{\{x(s_{\mathrm{I},0}, s_{\mathrm{II}}) \in W_1\}}.$$

    The game (T, W_0, W_1, q) is *determined* if

    $$\min_{s_{\mathrm{II}} \in S_{\mathrm{II}}(T)} \max_{s_{\mathrm{I}} \in \widetilde{S}_{\mathrm{I}}(T)} \mathbf{P}(q, s_{\mathrm{I}}, s_{\mathrm{II}}) = \max_{s_{\mathrm{I}} \in \widetilde{S}_{\mathrm{I}}(T)} \min_{s_{\mathrm{II}} \in S_{\mathrm{II}}(T)} \mathbf{P}(q, s_{\mathrm{I}}, s_{\mathrm{II}}). \tag{3.9}$$

    If the game is determined, then the common value in Equation (3.9) is called the *value* of the game.

    (a) Prove that if Player I has a winning strategy in both (T, W_0) and (T, W_1), then the game (T, W_0, W_1, q) is determined and its value is 1.

    (b) Prove that if Player I has a winning strategy in (T, W_0), and Player II has a winning strategy in (T, W_1), then (T, W_0, W_1, q) is determined and its value is q.

    (c) Prove that if Player II has a winning strategy in (T, W_0), and Player I has a winning strategy in (T, W_1), then (T, W_0, W_1, q) is determined and its value is $1 - q$.

(d) Prove that if Player II has a winning strategy in $(T, W_0 \cup W_1)$, then (T, W_0, W_1, q) is determined and its value is 0.

(e) Prove that if Player II has a winning strategy in both (T, W_0) and (T, W_1), but she does not have a winning strategy in $(T, W_0 \cup W_1)$, then (T, W_0, W_1, q) is determined and its value is $\min\{q, 1 - q\}$.

(f) Are all parts of the exercise correct without the assumption that W_0 and W_1 are Borel? Justify your answer.

30. (**) Denote by $T = \mathbb{N}^{<\mathbb{N}}$ the complete tree over the alphabet $\mathbb{N}$. Let Y be a Polish space. A subset $A \subseteq Y$ is *analytic* if there is a continuous function $f : [T] \to Y$ whose image is A. It is well known that every Borel set is analytic.

In this exercise we will prove the following result, due to Kuratowski (1936): For every two analytic sets $A, B \subseteq Y$ there are two analytic sets $A', B' \subseteq Y$ such that (a) $A \subseteq A'$ and $B \subseteq B'$, (b) $A' \cup B' = Y$, and (c) $A' \cap B' = A \cap B$. That is, one can extend A and B to two analytic sets that cover Y without changing their intersection.

Let $A, B \subseteq Y$ be analytic sets and let $f_A, f_B : [T] \to Y$ be two continuous functions whose images are A and B, respectively. For every $p \in T$ denote by $C_A(p)$ the closure of the set $f_A([T_p])$, and by $C_B(p)$ the closure of the set $f_B([T_p])$

For every $y \in Y$ define a game G^y over the tree T with three possible outcomes: a win for Player I, a win for Player II, and a draw (see Exercise 3.14).

- A play x is a win for Player I if there is $n \in \mathbb{N}$ such that $y \in C_A(\langle a_0, a_2, \ldots, a_{2n}\rangle)$ and $y \notin C_B(\langle a_1, a_3, \ldots, a_{2n+1}\rangle)$.
  Note: $\langle a_0, a_2, \ldots, a_{2n}\rangle$ is a position that is made only of moves of Player I, yet we interpret it as a position in the tree of length $n + 1$.
- A play x is a win for Player II if there is $n \in \mathbb{N}$ such that $y \notin C_A(\langle a_0, a_2, \ldots, a_{2n}\rangle)$ and $y \in C_B(\langle a_1, a_3, \ldots, a_{2n-1}\rangle)$.
  Note: by convention, if $y \notin C_A(\langle a_0\rangle)$, Player II wins G^y.
- A play x is a draw if $y \in C_A(\langle a_0, a_2, \ldots, a_{2n}\rangle)$ and $y \in C_B(\langle a_1, a_3, \ldots, a_{2n+1}\rangle)$ for every $n \in \mathbb{N}$.

Denote by $A' \subseteq Y$ the set of all $y \in Y$ such that Player I can guarantee at least a draw in G^y. Denote by $B' \subseteq Y$ the set of all $y \in Y$ such that Player II can guarantee at least a draw in G^y.

(a) Prove that the game G^y is well-defined for every $y \in Y$.

(b) Prove that the set

$$\{(y, s_\mathrm{I}) \in Y \times S_\mathrm{I}(T): s_\mathrm{I} \text{ guarantees at least a draw in } G^y\}$$

is Borel. You will need to define a suitable topology on $S_\mathrm{I}(T)$.

(c) Prove that the sets A' and B' are analytic.

(d) Prove that $A \subseteq A'$ and $B \subseteq B'$.

(e) Prove that the winning set of Player I in G^y is open. Deduce that $A' \cup B' = Y$.

(f) Prove that $A' \cap B' = A \cap B$.

31. (*) Let (T, W) be a game, and suppose that $(s_{\mathrm{I}}^n)_{n \in \mathbb{N}}$ are all (possibly different) winning strategies of Player I. Let s_{I} be the strategy that selects at stage $2n$ the move selected by s_{I}^n at that stage, that is,

$$s_{\mathrm{I}}(p) := s_{\mathrm{I}}^{\mathrm{len}(p)/2}(p), \quad \forall p \text{ such that } \mathrm{len}(p) \text{ is even.}$$

Is s_{I} necessarily a winning strategy of Player I? Justify your answer.

32. (*) Let $(\widetilde{T}, \pi, \varphi_{\mathrm{I}}, \varphi_{\mathrm{II}})$ be a covering of T. In this exercise we prove that the function π is onto.

Assume, by contradiction, that there is a position p that is not in the image of π, and let x be a play that extends p. For every play y, say that Player I (resp., Player II) *deviates from* x if $y_0 = x_0, y_1 = x_1, \ldots, y_k = x_k, y_{k+1} \neq x_{k+1}$ and k is odd (resp., k is even). Let A be the set of all plays where Player I deviates from x, and let B be the set of all plays where Player II deviates from x.

(a) Prove that A and B are open.

(b) Prove that in the game $(T, B \cup \{x\})$ Player I has a winning strategy.

(c) Prove that in the game (T, B) Player II has a winning strategy.

(d) Use the previous items to derive a contradiction.

33. (*) In Example 3.42 we provided two trees T and $\widetilde{T}$, and we showed that $\widetilde{T}$ is part of a covering of T. Is T part of a covering of $\widetilde{T}$? Prove your answer.

34. (*) Repeat Exercise 3.33 for the trees T and $\widetilde{T}$ defined in Example 3.43.

35. (*) Let T and $\widetilde{T}$ be the two trees in Example 3.42. Let $x = \langle a, a, a, a, a, b, a, a, a, \ldots \rangle$, and let $\pi : \widetilde{T} \to T$ be the function that maps each position $\widetilde{p} \in \widetilde{T}$ to $x_{\mathrm{len}(\widetilde{p})}$. Can $(\widetilde{T}, \pi)$ be extended to a covering of T?

36. (*) Let $D \subseteq S_{\mathrm{I}}(T) \cup S_{\mathrm{II}}(T)$ be a set of strategies, and let $W \subseteq [T]$ be the set of all plays that are consistent with some strategy in D; formally, $x \in W$ if and only if for every $n \in \mathbb{N}$ there is a strategy $s \in D$ such that x_n is consistent with s. Is the set W necessarily open? Is it necessarily closed? Is there a set $D \subseteq S_{\mathrm{I}}(T) \cup S_{\mathrm{II}}(T)$ for which W is open? Is there a set $D \subseteq S_{\mathrm{I}}(T) \cup S_{\mathrm{II}}(T)$ for which W is closed? Justify your answers.

37. (*) This exercise explores the relation between the winner in the game and the winner in related games with finitely many stages.

Let T be a tree, and let $u : T \to \{0,1\}$ be some function such that for every $x \in [T]$, the limit $\lim_{n\to\infty} u(x_n)$ exists. Denote

$$W_n := \{x \in [T] : u(x_n) = 1\}, \quad \forall n \in \mathbb{N},$$

and

$$W := \left\{x \in [T] : \lim_{n\to\infty} u(x_n) = 1\right\}.$$

Is it true that Player I has a winning strategy in (T, W) if and only if she has a winning strategy in (T, W_n) for every $n \in \mathbb{N}$ sufficiently large? Justify your answer.

38. (*) In the proof of Theorem 3.76, did we use the assumption that the set of moves is countably infinite? If so, where? Is the theorem valid when the set of moves is finite? Justify your answer.

39. (*) In Step 5 of the proof of Theorem 3.76 where did we use the assumption that the enumeration l satisfies the property that the length of $l(n)$ is at most n, for every $n \in \mathbb{N}$?

40. (**) Gödel's Incompleteness Theorem states that for every set of axioms that is sufficiently rich, and, in particular, for ZFC, there is a theorem that can be neither proved nor disproved. Such a theorem will have the form

$$\mathcal{T} = \exists x_0 \forall x_1 \exists x_2 \forall x_3 \cdots \exists x_{2n} \forall x_{2n+1} P(x_0, x_1, \ldots, x_{2n+1}),$$

where P is some formula with free variables $x_0, x_1, \ldots, x_{2n+1}$. Consider the following game G, which lasts $2n + 2$ rounds. For each $k \in \{0, 1, \ldots, n\}$,

- In round $2k$, Player I selects x_{2k}.
- In round $2k + 1$, Player II selects x_{2k+1}.

Player I wins the game if $P(x_0, x_1, \ldots, x_{2n+1})$ is true, and loses otherwise.

(a) Show that $\mathcal{T}$ is true if and only if Player I has a winning strategy in G, and $\mathcal{T}$ is false if and only if Player II has a winning strategy in G.

(b) Does Theorem 3.37 contradict Gödel's Incompleteness Theorem? Explain your answer.

CHAPTER 4

Games with General Payoff Functions

Not all games end in a win for a player. Some games, like chess, go, and mancala, can also end with a draw. Other games have a richer set of outcomes. For example, in games that involve gambling, the outcome depends on the amounts on which the players gamble. In this chapter we provide two extensions of Theorem 3.37. In Section 4.1 we study games that have a general set of outcomes, and in Section 4.2 we study games where the goals of the players are not antagonistic.

4.1 ZERO-SUM GAMES

In this section we study two-player zero-sum games. In these games, the outcome is not a win or a loss, but rather an amount that Player II pays to Player I. In case this amount is negative, it is interpreted as an amount that Player I pays to Player II. In particular, the goal of Player I is to maximize this amount, and the goal of Player II is to minimize it.

Definition 4.1 (Zero-sum game, payoff function). *A zero-sum game is a pair $G = (T, f)$ where T is a tree and $f : [T] \to \mathbb{R}$. The function f is called the* payoff function *of the game.*

   Zero-sum games are played as the games that we studied in Chapter 3: Player I selects a move in even stages, and Player II selects a move in odd stages. In contrast to games where the outcome is a win for Player I or a win for Player II, depending on whether the generated play x is in the winning set or outside of it, in two-player zero-sum games the outcome is the real number $f(x)$.

Remark 4.2 (Payoff function). *As mentioned above, the outcome $f(x)$ can be viewed as an amount that Player II pays to Player I. Thus, the sum of the payments of the two players is zero, which explains the appellation "zero-sum" of this type of games.*

Remark 4.3 (Games and zero-sum games). *To distinguish between games with a winning set as studied in Chapter 3 and games with a payoff function, we will call the former* games *(or* win/lose games*) and the latter* zero-sum games.

DOI: 10.1201/9781003582106-4

Example 4.4. *Every game (T, W) is equivalent to the zero-sum game (T, f), where $f := \mathbf{1}_W$ is the characteristic function of the winning set W; that is,*

$$f(x) := \begin{cases} 1, & \text{if } x \in W, \\ 0, & \text{if } x \notin W. \end{cases}$$

Recall that $x(s_\mathrm{I}, s_\mathrm{II})$ is the play that is generated by the pair of strategies $(s_\mathrm{I}, s_\mathrm{II}) \in S_\mathrm{I}(T) \times S_\mathrm{II}(T)$.

Definition 4.5 (Strategy guarantees). *Let $G = (T, f)$ be a zero-sum game, and let $z \in \mathbb{R}$. A strategy $s_\mathrm{I} \in S_\mathrm{I}(T)$ guarantees z in G if*

$$f(x(s_\mathrm{I}, s_\mathrm{II})) \geq z, \quad \forall s_\mathrm{II} \in S_\mathrm{II}(T). \tag{4.1}$$

A strategy $s_\mathrm{II} \in S_\mathrm{II}(T)$ guarantees z in G if

$$f(x(s_\mathrm{I}, s_\mathrm{II})) \leq z, \quad \forall s_\mathrm{I} \in S_\mathrm{I}(T). \tag{4.2}$$

Remark 4.6 (The direction of the inequality sign in Definition 4.5). *Note the different directions of the inequality sign in Equations (4.1) and (4.2), which correspond to the different goals of the players: Player I's goal is to maximize the payoff, while Player II's goal is to minimize it.*

Remark 4.7 (Guaranteeing in games). *When $G = (T, W)$ is a game, a strategy s_I is winning in G if and only if it guarantees 1 in the zero-sum game $(T, \mathbf{1}_W)$. Similarly, a strategy s_II is winning in G if and only if it guarantees 0 in the zero-sum game $(T, \mathbf{1}_W)$.*

The value of a strategy of Player I (resp., Player II) is the maximal (resp., minimal) amount it guarantees to Player I (resp., Player II).

Definition 4.8 (Value of strategy). *Let $G = (T, f)$ be a zero-sum game. The* value *of a strategy $s_\mathrm{I} \in S_\mathrm{I}(T)$ is*

$$\mathrm{val}\,(s_\mathrm{I}) := \inf_{s_\mathrm{II} \in S_\mathrm{II}(T)} x(f(s_\mathrm{I}, s_\mathrm{II})).$$

The value *of a strategy $s_\mathrm{II} \in S_\mathrm{II}(T)$ is*

$$\mathrm{val}\,(s_\mathrm{II}) := \sup_{s_\mathrm{I} \in S_\mathrm{I}(T)} x(f(s_\mathrm{I}, s_\mathrm{II})).$$

Definition 4.9 (Maxmin value). *The highest outcome that Player I can guarantee in the zero-sum game $G = (T, f)$ is*

$$\underline{v}(G) := \sup_{s_\mathrm{I} \in S_\mathrm{I}(T)} \mathrm{val}\,(s_\mathrm{I}) = \sup_{s_\mathrm{I} \in S_\mathrm{I}(T)} \inf_{s_\mathrm{II} \in S_\mathrm{II}(T)} f(x(s_\mathrm{I}, s_\mathrm{II})). \tag{4.3}$$

The quantity $\underline{v}(G)$ is called the maxmin value *of the zero-sum game $G = (T, f)$.*

Definition 4.10 (Minmax value). *The lowest amount that Player* II *can guarantee in* (T, f) *is*

$$\overline{v}(G) := \inf_{s_{II} \in S_{II}(T)} \text{val}\,(s_{II}) = \inf_{s_{II} \in S_{II}(T)} \sup_{s_I \in S_I(T)} f(x(s_I, s_{II})). \tag{4.4}$$

The quantity $\overline{v}(G)$ *is called the* minmax value *of the zero-sum game* $G = (T, f)$.

Since we did not require that the function f is bounded, the minmax value and the maxmin value may be $+\infty$ or $-\infty$.

It is quite natural that the amount that Player I can guarantee is not larger than the amount that Player II can guarantee. This is summarized by the following result.

Theorem 4.11. *For every zero-sum game* $G = (T, f)$,

$$\underline{v}(G) \leq \overline{v}(G).$$

Proof. For every pair of strategies $(\widehat{s}_I, \widehat{s}_{II}) \in S_I(T) \times S_{II}(T)$,

$$\inf_{s_{II} \in S_{II}(T)} f(x(\widehat{s}_I, s_{II})) \leq f(x(\widehat{s}_I, \widehat{s}_{II})). \tag{4.5}$$

Taking the supremum over all $\widehat{s}_I \in S_I(T)$ in both sides of Equation (4.5), we obtain

$$\underline{v}(G) = \sup_{s_I \in S_I(T)} \inf_{s_{II} \in S_{II}(T)} f(x(s_I, s_{II})) \leq \sup_{s_I \in S_I(T)} f(x(s_I, \widehat{s}_{II})). \tag{4.6}$$

Since Equation (4.6) holds for every strategy $\widehat{s}_{II} \in S_{II}(T)$, we deduce that

$$\underline{v}(G) \leq \inf_{s_{II} \in S_{II}(T)} \sup_{s_I \in S_I(T)} f(x(s_I, s_{II})) = \overline{v}(G),$$

as claimed. □

Definition 4.12 (Value). *Let* $G = (T, f)$ *be a zero-sum game. The real number* $v = v(G)$ *is the* value *of* G *if*

$$v = \underline{v}(G) = \overline{v}(G).$$

When the payoff function f is bounded, an equivalent definition for the value involves the concept of guaranteeing a quantity (see Exercise 4.1).

Definition 4.13 (Value). *The quantity* v *is the* value *of the zero-sum game* $G = (T, f)$ *if for every* $\varepsilon > 0$ *there is a strategy* $s_I^\varepsilon \in S_I(T)$ *that guarantees* $v - \varepsilon$, *and a strategy* $s_{II}^\varepsilon \in S_{II}(T)$ *that guarantees* $v + \varepsilon$.

Thus, the existence of the value implies that the players can guarantee that the outcome will be arbitrarily close to v.

The following example shows that a player does not necessarily have a strategy that guarantees the value.

Example 4.14. *Consider the complete binary tree $T = \{a, b\}^{<\mathbb{N}}$. For each play $x = \langle a_0, a_1, \dots \rangle \in [T]$, denote the first stage in which Player I selects the move b by*

$$n(x) := \min\{n \in \mathbb{N}: n \text{ even}, a_n = b\}.$$

The payoff function f is defined by

$$f(x) := \begin{cases} -1, & \text{if } n(x) = \infty, \\ -\frac{1}{n(x)}, & \text{if } n(x) < \infty. \end{cases}$$

Thus, Player II is a dummy player: the payoff function f is independent of her choices. The goal of Player I is to play b as late as possible. The value of this zero-sum game is 0. For every $n \in \mathbb{N}$, the strategy s_I^n that selects b in stage $2n$, and a in all other stages, guarantees $-\frac{1}{2n}$ for Player I. However, there is no strategy $s_I \in S_I(T)$ that guarantees 0 for Player I.

The main result of this section is the following.

Theorem 4.15. *Every zero-sum game $G = (T, f)$, where f is bounded and Borel measurable, has a value.*

Remark 4.16 (Games with three outcomes). *In Exercise 3.14 we considered games (T, W_I, W_{II}) that have three outcomes: win for Player I, win for Player II, and a draw. We showed that in this case exactly one of the following three alternatives hold:*

D.1) *Player I has a winning strategy.*

D.2) *Player II has a winning strategy.*

D.3) *Both players have a strategy that guarantees at least a draw.*

Let us define a function $f : [T] \to \mathbb{R}$ by

$$f(x) = \begin{cases} 1, & \text{if } x \in W_I, \\ -1, & \text{if } x \in W_{II}, \\ 0, & \text{if } x \in (W_I \cup W_{II})^c. \end{cases}$$

Alternative (D.1) holds when the value of the zero-sum game (T, f) is 1, Alternative (D.2) holds when the value of the zero-sum game (T, f) is -1, and Alternative (D.3) holds when the value of the zero-sum game (T, f) is 0.

Remark 4.17 (Zero-sum games with unbounded payoffs). *When f is not bounded, Definitions 4.12 and 4.13 are not necessarily equivalent, see Exercise 4.2. Nevertheless, Theorem 4.15 extends to zero-sum games with unbounded payoffs when the value is defined as in Definition 4.12, see Exercise 4.6.*

Proof of Theorem 4.15. For every $z \in \mathbb{R}$ define a game $G_z = (T, W_z)$ over the tree T and with a winning set W_z defined by

$$W_z := \{x \in [T]: f(x) \geq z\} = f^{-1}([z, \infty)).$$

Since the function f is Borel measurable, W_z is a Borel set. By Theorem 3.37, for every $z \in \mathbb{R}$ the game G_z is determined.

If $z < z'$, then $W_z \supseteq W_{z'}$. In particular, any winning strategy of Player I in $G_{z'}$ (if such a strategy exists) is a winning strategy in G_z as well. Similarly, if $z > z'$, then any winning strategy of Player II in $G_{z'}$ (if such a strategy exists) is a winning strategy in G_z as well.

The payoff function f is bounded, say, between $-M$ and M. If $z > M$ then the set W_z is empty, and all strategies of Player II are winning strategies in G_z. If $z < -M$ then $W_z = [T]$, and all strategies of Player I are winning strategies in G_z. Denote

$$v := \sup\{z \in \mathbb{R}\colon \text{ Player I has a winning strategy in } G_z\}.$$

By the above discussion, $v \in [-M, M]$. Moreover,

- For every $z < v$, Player I has a winning strategy in G_z.

- For every $z > v$, Player II has a winning strategy in G_z.

We claim that v is the value of G. Indeed, fix $\varepsilon > 0$ and let $z \in (v - \varepsilon, v)$. In the game G_z Player I has a winning strategy s_I^z, hence

$$x(s_I^z, s_{II}) \in W_z, \quad \forall s_{II} \in S_{II}(T).$$

Since $W_z = f^{-1}([z, \infty))$, we have

$$f(x(s_I^z, s_{II})) \geq z > v - \varepsilon, \quad \forall s_{II} \in S_{II}(T).$$

In particular, the strategy s_I^z guarantees $v - \varepsilon$ for Player I in the zero-sum game G. Analogously, for every $\varepsilon > 0$ Player II has a winning strategy in the game $G_{v+\varepsilon}$, and every such strategy guarantees $v + \varepsilon$ for Player II in the zero-sum game G. It follows that v is the value of G. □

Definition 4.18 (ε-optimal strategy). *Let $G = (T, f)$ be a zero-sum game, let v be its value, and let $\varepsilon \geq 0$. A strategy $s_I \in S_I(T)$ is ε-optimal for Player I in G if it guarantees $v - \varepsilon$ for Player I. A strategy $s_{II} \in S_{II}(T)$ is ε-optimal for Player II in G if it guarantees $v + \varepsilon$ for Player II.*

The proof of Theorem 4.15 allows us to identify ε-optimal strategies for Player I in a zero-sum game (T, f) with winning strategies of Player I in a certain game.

Theorem 4.19. *Let (T, f) be a zero-sum game where f is bounded and Borel measurable, let v be its value, and let $\varepsilon \geq 0$. Define*

$$W_{v-\varepsilon} := \{x \in [T]\colon f(x) \geq v - \varepsilon\} = f^{-1}([v - \varepsilon, \infty)).$$

Every winning strategy of Player I in the game $(T, W_{v-\varepsilon})$ is ε-optimal for Player I in the zero-sum game (T, f), and vice versa.

The analogous result for Player II is the following.

Theorem 4.20. *Let (T, f) be a determined zero-sum game, let v be its value, and let $\varepsilon \geq 0$. Define*

$$W_{v+\varepsilon} := \{x \in [T] \colon f(x) \leq v + \varepsilon\} = f^{-1}([v + \varepsilon, \infty)).$$

Then every winning strategy of Player II in the game $(T, W_{v+\varepsilon})$ is ε-optimal for Player II in the zero-sum game (T, f), and vice versa.

As in Chapter 3, we will be interested below in the version of the game that starts at some given position p. Such a zero-sum game is equivalent to the zero-sum game in which in the first $\mathrm{len}(p)$ stages the player whose turn it is to select a move has a single choice – the move indicated by p. We now formally define this game, analogously to Definition 3.32.

Definition 4.21 (Subgame that starts at position p). *Let $G = (T, f)$ be a zero-sum game, and let $p \in T$. The subgame that starts at position p is the zero-sum game $G_p := (T_p, f|_{[T_p]})$, where $f|_{[T_p]}$ is the restriction of f to $[T_p]$.*

The following result states that as long as Player I follows an ε-optimal strategy, the value of the current position cannot drop by much.

Lemma 4.22. *Let $G = (T, f)$ be a zero-sum game where f is bounded and Borel measurable, let v denote the value of G, let $\varepsilon \geq 0$, and let $s_\mathrm{I} \in S_\mathrm{I}(T)$ be an ε-optimal strategy of Player I. Then for every play $x \in [T]$ that is consistent with s_I and every $n \in \mathbb{N}$ we have*

$$v(G_{x_n}) \geq v - \varepsilon.$$

Note that since f is bounded and Borel measurable, by Theorem 4.15, for every position $p \in T$ the zero-sum game G_p has a value $v(G_p)$.

Proof. Let $x = \langle a_0, a_1, \ldots \rangle \in [T]$ be a play that is consistent with s_I, and let $n \in \mathbb{N}$. Fix for the moment $\delta > 0$. Let $s_\mathrm{II}^\delta \in S_\mathrm{II}(T)$ be a δ-optimal strategy of Player II in the zero-sum game G_{x_n}. Let $\widehat{s}_\mathrm{II} \in S_\mathrm{II}(T)$ be the following strategy of Player II:

- For every odd $k < n$,

$$\widehat{s}_\mathrm{II}(x_{k-1}) := a_k.$$

- If the position x_n is reached, then from stage n and on, Player II follows the strategy s_II^δ:

$$\widehat{s}_\mathrm{II}(p) := s_\mathrm{II}^\delta(p), \quad \forall p \in T \text{ such that } \mathrm{len}(p) \text{ is odd and } p \succeq x_n.$$

- The definition of $\widehat{s}_\mathrm{II}$ at all other positions is arbitrary.

Under $(s_\mathrm{I}, \widehat{s}_\mathrm{II})$ the position x_n is reached, and afterwards, Player II follows s_II^δ. Since s_I is ε-optimal for Player I in G,

$$v - \varepsilon \leq f(x(s_\mathrm{I}, \widehat{s}_\mathrm{II})) \leq v(G_{x_n}) + \delta. \tag{4.7}$$

Since $\delta > 0$ is arbitrary, the claim follows. □

The following corollary of Lemma 4.22 states that if both players follow ε-optimal strategies, then the value of each position that is visited is close to the value of the game, and by deviating, no player can increase by much the value of the positions that are reached. The proof is left to the reader (Exercise 4.8).

Corollary 4.23. *Let* $G = (T, f)$ *be a zero-sum game where* f *is bounded and Borel measurable, let* $\varepsilon \geq 0$, *and let* $s_{\mathrm{I}}^{\varepsilon}$ *and* $s_{\mathrm{II}}^{\varepsilon}$ *be* ε-*optimal strategies of the two players. Then for every position* p *that is consistent with* $s_{\mathrm{I}}^{\varepsilon}$ *we have*

$$v(G_p) \geq v - \varepsilon,$$

and for every position p *that is consistent with* $s_{\mathrm{II}}^{\varepsilon}$ *we have*

$$v(G_p) \leq v + \varepsilon.$$

In particular, denoting by $x = x(s_{\mathrm{I}}^{\varepsilon}, s_{\mathrm{II}}^{\varepsilon})$ *the play generated by* $s_{\mathrm{I}}^{\varepsilon}$ *and* $s_{\mathrm{II}}^{\varepsilon}$, *we have*

$$v - \varepsilon \leq v(G_{x_n}) \leq v + \varepsilon, \quad \forall n \in \mathbb{N}.$$

Lemma 4.22 states that if Player I follows an ε-optimal strategy, then whatever strategy Player II adopts, the play cannot reach a position whose value is much higher than the value of the game. However, such a strategy may miss opportunities: it might happen that Player II plays in such a way that leads the play to a position p with $v(G_p) > v + \varepsilon$, yet the ε-optimal strategy of Player I will only ensure the payoff v rather than the higher payoff $v(G_p)$.

Example 4.24. *Consider the zero-sum game* G *displayed in Figure 16, where the alphabet is* $\mathcal{A} = \{a, b, c\}$ *and* $f(x)$ *is displayed to the right of each play* x. *In stages 0 and 1, the player whose turn it is to move has two possible moves; in all other stages, the player whose turn it is to move has a single possible move.*

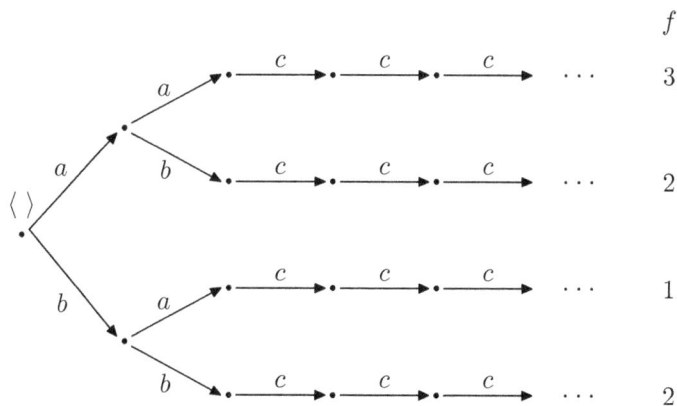

Figure 16: The zero-sum game G in Example 4.24.

The value of this zero-sum game is 2. The unique 0-optimal strategy s_{II} *for Player* II *is*

$$s_{\mathrm{II}}(\langle c \rangle) = b.$$

Player I has two 0-optimal strategies, s_I and s_I', given by:

$$s_I(\langle c, a \rangle) = b, \qquad\qquad s_I'(\langle c, a \rangle) = a,$$
$$s_I(\langle c, b \rangle) = b, \qquad\qquad s_I'(\langle c, b \rangle) = b.$$

The play generated by both (s_I, s_{II}) and (s_I', s_{II}) is the same:

$$x(s_I, s_{II}) = x(s_I', s_{II}) = \langle c, b, b, c, c, c, \ldots \rangle,$$

which leads to the payoff 2.

The strategy s_I is 0-optimal, but it might miss opportunities for obtaining a payoff higher than the value. For example, suppose that at the position $\langle c \rangle$ Player II selects the move a rather than b. The strategy s_I then selects the move b and yields the payoff 2, while if Player I selected the move a the payoff would be 3, which is better from the point of view of Player I. It is true that when Player II follows a 0-optimal strategy, the play will not reach a position where Player I can obtain an outcome higher than 2, but if Player II makes mistakes, then such opportunities may arise.

We next define the concept of a strategy that does not miss opportunities that arise due to mistakes of the other player. In fact, the concept we will define will be even stronger: it will require that the player follows an ε-optimal strategy in *all* subgames: those that arise due to mistakes of the other player, as well as those that arise due to past mistakes of the player herself.

Definition 4.25 (Play induced by a pair of strategies in a subgame, $x(s_I, s_{II}; p)$). *Let T be a tree, let $s_I \in S_I(T)$ and $s_{II} \in S_{II}(T)$ be strategies of the two players, and let $p \in T$ be a position. Denote by $x(s_I, s_{II}; p)$ the play under (s_I, s_{II}) in the subgame G_p. This is the play $\langle a_0, a_1, \ldots \rangle \in [T]$ such that*

- $\langle a_0, a_1, \ldots, a_{\mathrm{len}(p)-1} \rangle = p$,

- $a_{2n} = s_I(\langle a_0, a_1, \ldots, a_{2n-1} \rangle)$, *for every $n \in \mathbb{N}$ such that $2n \geq \mathrm{len}(p)$, and*

- $a_{2n+1} = s_{II}(\langle a_0, a_1, \ldots, a_{2n} \rangle)$, *for every $n \in \mathbb{N}$ such that $2n + 1 \geq \mathrm{len}(p)$.*

Remark 4.26 (On the domain of strategies). *A strategy of a player is a function, whose domain is the set of all positions that the player controls. In particular, it is defined for positions inconsistent with itself. This choice, which was already discussed in Remark 3.21, allows us to define $x(s_I, s_{II}; p)$ in all subgames. It also implies that for all positions p, a strategy in the whole game induces a strategy in the subgame G_p.*

Definition 4.27 (Subgame ε-optimal strategy). *Let $G = (T, f)$ be a zero-sum game where f is bounded and Borel measurable, and let $\varepsilon > 0$. A strategy $s_I \in S_I(T)$ is* subgame ε-optimal *if for every position p and every strategy $s_{II} \in S_{II}(T_p)$,*

$$f(x(s_I, s_{II}; p)) \geq v(G_p) - \varepsilon, \quad \forall n \in \mathbb{N}. \tag{4.8}$$

A strategy $s_{II} \in S_{II}(T)$ is subgame ε-optimal *if for every position p and every strategy $s_I \in S_I(T_p)$,*

$$f(x(s_I, s_{II}; p)) \leq v(G_p) + \varepsilon, \quad \forall n \in \mathbb{N}. \tag{4.9}$$

Remark 4.28 (On Definition 4.27). *Note that Equations (4.8) and (4.9) depend only on the restriction of s_I and s_{II} to T_p.*

The following result states that a subgame ε-optimal strategy exists as soon as the payoff function is bounded and Borel measurable.

Theorem 4.29. *Let $G = (T, f)$ be a zero-sum game where f is bounded and Borel measurable. For every $\varepsilon > 0$, both players have a subgame ε-optimal strategy.*

Proof. We will prove the result for Player I. The proof for Player II is analogous.

Step 0: The idea of the proof.
We start by explaining the idea of the proof. Let s_I be an ε-optimal strategy for Player I in (T, f). For every position $p = \langle a_0, a_1, \ldots, a_{n-1} \rangle \in T$, this strategy induces a strategy in G_p, denoted s_I^p, which follows the moves along p in all prefixes of p, and follows s_I once p is reached:

$$s_I^p(p') = \begin{cases} a_{2k}, & \text{if } p' = \langle a_0, a_1, \ldots, a_{2k-1} \rangle, 1 \leq k < n/2, \\ s_I(p'), & \text{if } p' \succeq p. \end{cases}$$

In which positions p might the restriction of s_I to T_p fail to be ε-optimal in G_p? There are two cases in which this can happen:

- In positions p that are inconsistent with s_I, the definition of s_I does not affect the realized play, and hence $\mathrm{val}\,(s_I^p)$ in G_p may be low.

- Since $\mathrm{val}\,(s_I) \leq v$, in position p that are consistent with s_I and such that $v(G_p) > v + \varepsilon$, the strategy s_I might not be ε-optimal.

The way to turn s_I into a subgame ε-optimal strategy is to redefine it, in such a way that whenever a position p that satisfies one of the two bullets above is reached, the strategy starts following an ε-optimal strategy in G_p. Since, when Player I follows the ε-optimal strategy in G_p, the play might reach again a position $p' \succ p$ that satisfies one of the two bullets above (namely, p' might be inconsistent with the ε-optimal strategy in G_p, or $v(G_{p'})$ might be larger than $v(G_p) + \varepsilon$), the redefinition will be done recursively.

Step 1: Persistent position.
Fix $\varepsilon > 0$. For every position $p \in [T]$, let s_I^p be an ε-optimal strategy in G_p.
Let $p \preceq p'$ be two positions. We say that p' is *persistent* relative to p if the following two conditions are satisfies:

G.1) p' is consistent with s_I^p in G_p.

G.2) $v(p') \leq v(p) + \varepsilon$.

In particular, each position is persistent relative to itself.

Step 2: Definition of a strategy $s_I^* \in S_I(T)$.

We will define recursively a sequence of sets of positions $(Z_k)_{k \in \mathbb{N}}$. Set

$$Z_0 := \{\langle \, \rangle\}$$

be the set that contains only the initial position. For every $k > 0$, let Z_k be the set of all positions $p' = \langle a_0, a_1, \ldots, a_{2n+1} \rangle \in T$ that satisfy the following property: there is a position $p \in Z_k$ such that p' is not persistent relative to p but $\langle a_0, a_1, \ldots, a_{2n-1} \rangle$ is persistent relative to p. That is, Z_{k+1} contains all even-length positions that are minimal non-persistent extensions of positions in Z_k. In particular, any position in Z_{k+1} extends some position in Z_k.

We are now ready to define the strategy s_{I}^*, which will be shown in the next step to be subgame 2ε-optimal. Fix a position $p \in T$. Since $Z_0 = \{\langle \, \rangle\}$, at least one prefix of p lies in $Z := \bigcup_{k \in \mathbb{N}} Z_k$. Let $\widehat{p}$ be the maximal prefix of p that lies in Z. In particular, p is consistent with $s_{\mathrm{I}}^{\widehat{p}}$ in $G_{\widehat{p}}$. Define:

$$s_{\mathrm{I}}^*(p) := s_{\mathrm{I}}^{\widehat{p}}(p).$$

That is, the play is defined in blocks of varying size: Along each block, Player I follows an ε-optimal strategy in the subgame that starts at the position at the beginning of the block, the first block starts at the initial position, and a new block starts at the first position that is not persistent relative to the position at the beginning of the previous block.

Step 3: The strategy s_{I}^* is subgame 2ε-optimal.

Let $s_{\mathrm{II}} \in S_{\mathrm{II}}(T)$ be arbitrary and let $p \in T$ be any position. Set $x = x(s_{\mathrm{I}}^*, s_{\mathrm{II}}; p)$. We will prove that

$$f(x) \geq v(G_p) - 2\varepsilon. \tag{4.10}$$

We first argue that the number of non-persistent positions along x is finite; that is, the number of prefixes of x that lie in Z is finite. For each $k \in \mathbb{N}$, at most one prefix of x can lie in Z_k. If such a prefix exists, denote it by x_{n_k}, so that n_k is odd. Thus, $x_{n_{k+1}}$ is non-persistent relative to x_{n_k}, but $x_{n_{k+1}-2}$ is persistent relative to x_{n_k}. In particular, $x_{n_{k+1}-2}$ is consistent with $s_{\mathrm{I}}^{x_{n_k}}$ in $G_{x_{n_k}}$. In fact, $x_{n_{k+1}}$ is consistent with $s_{\mathrm{I}}^{x_{n_k}}$ in $G_{x_{n_k}}$. Indeed, $x_{n_{k+1}-2}$ is consistent with $s_{\mathrm{I}}^{x_{n_k}}$, at stage $n_{k+1}-2$ Player I selects her move according to $s_{\mathrm{I}}^{x_{n_k}}$, and hence $x_{n_{k+1}-1}$ is consistent with $s_{\mathrm{I}}^{x_{n_k}}$, and at stage $n_{k+1}-1$ Player II selects a move, and hence $x_{n_{k+1}}$ is consistent with $s_{\mathrm{I}}^{x_{n_k}}$.

Let $k_0 \in \mathbb{N}$ be the maximal integer such that $x_{n_{k_0}} \preceq p$. In particular, starting at stage n_{k_0}, Player I follows s_{I}^*. Note that before stage n_{k_0}, Player I may have made mistakes and not follow s_{I}^*.

The definition of k_0 implies that p is persistent with $x_{n_{k_0}}$, and hence

$$v(G_p) \leq v(G_{x_{n_{k_0}}}) + \varepsilon. \tag{4.11}$$

Since for every k the position $x_{n_{k+1}}$ is consistent with s_{I}^* in $G_{x_{n_k}}$ but not persistent relative to x_{n_k}, Condition (G.2) implies that

$$v(G_{x_{n_k}}) < v(G_{x_{n_{k+1}}}) - \varepsilon, \quad \forall k \geq k_0. \tag{4.12}$$

Equation (4.12) implies that since payoffs are bounded, the number of k's for which n_{k+1} is well-defined and must be finite.

Denote by K the maximal k for which n_k is defined. Then, starting at stage n_K, along the play x the strategy s_{I}^* coincides with $s_{\mathrm{I}}^{x_{n_K}}$. Since $s_{\mathrm{I}}^{x_{n_K}}$ is ε-optimal in $G_{x_{n_K}}$, and by Equations (4.11)–(4.12),

$$f(x) \geq v(G_{x_{n_K}}) - \varepsilon \geq v(G_{x_{n_{k_0}}}) - \varepsilon \geq v(G_p) - 2\varepsilon,$$

as claimed. □

4.2   NONZERO-SUM GAMES

In Section 4.1 we studied zero-sum games, where the goals of the players were antagonistic: the goal of Player I was to maximize the outcome, and the goal of Player II was to minimize the outcome. In this section we will study nonzero-sum games, where the goals of the players are not opposing.

Definition 4.30 (Nonzero-sum game). *A nonzero-sum game is a triplet $(T, f_{\mathrm{I}}, f_{\mathrm{II}})$, where T is a tree (over an alphabet $\mathcal{A}$), and $f_{\mathrm{I}}, f_{\mathrm{II}} : [T] \to \mathbb{R}$.*

Nonzero-sum games are played as zero-sum games that we studied in Section 4.1: Player I selects a move in even stages, and Player II selects a move in odd stages. The only difference between the two models is in the goals of the players: in zero-sum games, Player I's goal is to maximize the payoff and Player II's goal is to minimize this amount. In nonzero-sum games, the goal of Player I is to maximize $f_{\mathrm{I}}(x)$, where x is the play that was generated by the players, and the goal of Player II is to maximize $f_{\mathrm{II}}(x)$. For each $i \in \{\mathrm{I}, \mathrm{II}\}$, the function f_i is called the *payoff function* of player i.

Example 4.31 (Zero-sum games). *Every zero-sum game (T, f) is in particular a nonzero-sum game, with $f_{\mathrm{I}} = f$ and $f_{\mathrm{II}} = -f$.*

Remark 4.32 (More than two players). *A nonzero-sum game involves two players. The results that we will prove in this section extend to nonzero-sum games that involve more than two players, each wishing to maximize a different payoff function, see Exercises 4.12, 4.13, and 4.15.*

The solution concepts that we studied for zero-sum games were the value and ε-optimal strategies. The analogous concepts for nonzero-sum games are the equilibrium payoff and ε-equilibrium strategies, which we define now (the analogy between the concepts is studied in Exercise 4.7).

Definition 4.33 (ε-equilibrium). *Let $(T, f_{\mathrm{I}}, f_{\mathrm{II}})$ be a nonzero-sum game and let $\varepsilon \geq 0$. A pair of strategies $(s_{\mathrm{I}}^*, s_{\mathrm{II}}^*) \in S_{\mathrm{I}}(T) \times S_{\mathrm{II}}(T)$ is an ε-equilibrium if*

$$
\begin{aligned}
f_{\mathrm{I}}(x(s_{\mathrm{I}}^*, s_{\mathrm{II}}^*)) &\geq f_{\mathrm{I}}(x(s_{\mathrm{I}}, s_{\mathrm{II}}^*)) - \varepsilon, \quad \forall s_{\mathrm{I}} \in S_{\mathrm{I}}(T), && (4.13) \\
f_{\mathrm{II}}(x(s_{\mathrm{I}}^*, s_{\mathrm{II}}^*)) &\geq f_{\mathrm{II}}(x(s_{\mathrm{I}}^*, s_{\mathrm{II}})) - \varepsilon, \quad \forall s_{\mathrm{II}} \in S_{\mathrm{II}}(T). && (4.14)
\end{aligned}
$$

The pair $(f_{\mathrm{I}}(x(s_{\mathrm{I}}^*, s_{\mathrm{II}}^*)), f_{\mathrm{II}}(x(s_{\mathrm{I}}^*, s_{\mathrm{II}}^*))) \in \mathbb{R}^2$ is called an ε-equilibrium payoff.

According to Equations (4.13) and (4.14), a pair of strategies (s_I^*, s_{II}^*) is an ε-equilibrium if no player can increase her payoff by more than ε by unilaterally deviating from (s_I^*, s_{II}^*).

Remark 4.34 (Equilibrium as stable behavior). *The concept of ε-equilibrium expresses stability: under an ε-equilibrium, each player acts to her best (up to ε) possible advantage with respect to the behavior of the other player. Another way to express the property of stability is that, if there is an "agreement" to play a particular ε-equilibrium, then, even if the agreement is not binding, it will not be breached: no player will violate the agreement, because any unilateral violation will not improve the deviator's payoff by more than ε.*

Example 4.14 shows that a 0-equilibrium need not exist. While the value of a zero-sum game is unique, a nonzero-sum game may have several equilibrium payoffs, as exhibited by the following example.

Example 4.35 (A game with two equilibria). *Consider the nonzero-sum game that is displayed in Figure 17, where the alphabet is $\mathcal{A} = \{a, b, c\}$.*

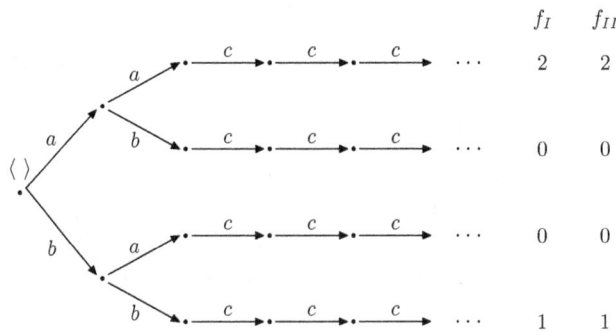

Figure 17: The nonzero-sum game in Example 4.35.

In this game, Player I has two strategies and Player II has four strategies:

- *In stage 0, Player I can select between two moves, a and b. Denote by s_I^a (resp., s_I^b) the strategy in which Player I selects the move a (resp., b) in stage 0.*

- *In stage 1, Player II has two moves, a and b. The four strategies are denoted s_{II}^{aa}, s_{II}^{ab}, s_{II}^{ba}, and s_{II}^{bb}, and the moves they select in stage 1 are displayed in the following table:*

strategy	position $\langle a \rangle$	position $\langle b \rangle$
s_{II}^{aa}	a	a
s_{II}^{ab}	a	b
s_{II}^{ba}	b	a
s_{II}^{bb}	b	b

There are two 0-equilibria in this game:

- (s_I^a, s_{II}^{aa}), *with equilibrium payoff* $(2, 2)$.

- (s_I^b, s_{II}^{bb}), *with equilibrium payoff* $(1, 1)$.

Example 4.36 (Stopping game). *In Example 3.43 we presented the class of two-player stopping games, in which the possible moves are* continue *and* stop*. As long as both players select the move "continue", both moves are available; once a player selects the move "stop", the only move available to both players is "stop".*

Formally, a stopping game is a game over the alphabet $\mathcal{A} = \{c, s\}$*, where* c *corresponds to the move "continue" and* s *corresponds to the move "stop". The set of plays in this game is*

$$\langle c, c, c, c, \ldots \rangle, \langle s, s, s, s, \ldots \rangle, \langle c, s, s, s, \ldots \rangle, \langle c, c, s, s, \ldots \rangle, \langle c, c, c, s, \ldots \rangle, \ldots.$$

Consider a nonzero-sum stopping game where the payoffs are as follows, see Figure 18:

- *If both players always continue, the payoff to each player is* 0*:*

$$f_{\mathrm{I}}(\langle c, c, c, \ldots \rangle) = 0, \quad f_{\mathrm{II}}(\langle c, c, c, \ldots \rangle) = 0.$$

- *The player who stops first gets* 1*, and the other player gets* 2*:*

$$f_{\mathrm{I}}(\langle c, c, \ldots, c, s, s, s \ldots \rangle) = 1, \quad f_{\mathrm{II}}(\langle c, c, \ldots, c, s, s, s \ldots \rangle) = 2,$$

if the first player who selects s *is Player I; and*

$$f_{\mathrm{I}}(\langle c, c, \ldots, c, s, s, s \ldots \rangle) = 2, \quad f_{\mathrm{II}}(\langle c, c, \ldots, c, s, s, s \ldots \rangle) = 1,$$

if the first player who selects s *is Player II.*

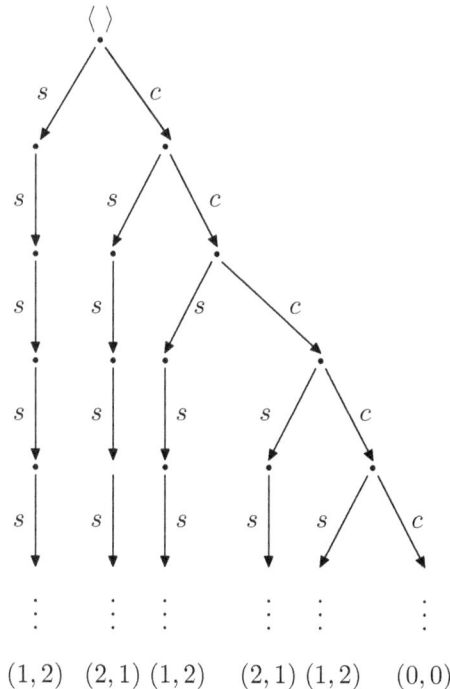

Figure 18: The nonzero-sum stopping game in Example 4.36.

Let us calculate the 0-equilibria of this game. Denote a strategy of a player by the first stage in which she selects "stop". Thus, the strategy 6 is the strategy in which Player I selects "stop" for the first time in stage 6, the strategy 11 is the strategy in which Player II selects "stop" for the first time in stage 11, and the strategy ∞ is the strategy of either Player I or Player II in which they always select "continue".

The strategy pair (∞, ∞) is not a 0-equilibrium, since both players can profit by deviating to a strategy that selects "stop" in finite time.

For every even number $n \in \mathbb{N}$, the strategy pair (n, ∞) is a 0-equilibrium. Similarly, for every odd number $m \in \mathbb{N}$, the strategy pair (∞, m) is a 0-equilibrium.

For every even number $n \in \mathbb{N}$ and odd number $m \in \mathbb{N}$, the strategy pair (n, m) is not a 0-equilibrium, because the player who stops first (namely, Player I if $n < m$, or Player II if $m < n$) profits by deviating to the strategy ∞ (because by this deviation she increases her payoffs from 1 to 2).

As the next result states, when the payoff functions are bounded and Borel measurable, for every $\varepsilon > 0$ an ε-equilibrium exists.

Theorem 4.37 (Mertens, 1987). *In every nonzero-sum game $G = (T, f_\mathrm{I}, f_\mathrm{II})$, where f_I and f_II are bounded and Borel-measurable functions, an ε-equilibrium exists for every $\varepsilon > 0$.*

Proof. The idea behind the proof is as follows. For every function $f : [T] \to \mathbb{R}$, denote by $G(f) = (T, f)$ the zero-sum game with payoff function f. Suppose that Player I follows a subgame ε-optimal strategy in the zero-sum game $G(f_\mathrm{I})$, and Player II follows a subgame ε-optimal strategy in the zero-sum game $G(-f_\mathrm{II})$. Note that in the latter game, the goal of Player II is to minimize $-f_\mathrm{II}$, which is equivalent to maximizing f_II, her payoff in the original game G. Let x^* be the play generated by this pair of strategies. Since both players follow subgame ε-optimal strategies, $f_\mathrm{I}(x^*) \geq v(G(f_\mathrm{I})) - \varepsilon$ and $f_\mathrm{II}(x^*) \geq -v(G(-f_\mathrm{II})) - \varepsilon$: if the generated play is x^*, then the payoff of each player is more than she can guarantee by herself in G. We will devise a 3ε-equilibrium where each player follows the play x^*, and, if a deviation from x^* is observed, the deviator is punished. The fact that x^* is generated by subgame ε-optimal strategies will ensure that the threat of punishment is effective, and no player has the incentive to deviate from x^*.

We now turn to the formal proof. We will use the following notation. For every position p and every bounded and Borel-measurable function $f : [T] \to \mathbb{R}$, denote by

$$G_p(f) = (T_p, f|_{[T_p]})$$

the *zero-sum* subgame that starts at position p and where the outcome is given by the function f restricted to the subtree $[T_p]$. By Theorem 4.15, this zero-sum game has a value, which is denoted by $v_p(f)$.

For every position $p \in T$ consider two auxiliary zero-sum games:

- The zero-sum game $G_p(f_\mathrm{I})$, which is the subgame that starts at p, and where Player I maximizes her payoff in G, while Player II minimizes Player I's payoff in G.

- The zero-sum game $G_p(-f_{\mathrm{II}})$, which is the subgame that starts at p, and where Player II maximizes her payoff in G, while Player I minimizes Player II's payoff in G.

Fix $\varepsilon > 0$. By Theorem 4.29, in the zero-sum game $G_{\langle\,\rangle}(f_{\mathrm{I}})$ Player I has a subgame ε-optimal strategy $\widehat{s}_{\mathrm{I}}^{\varepsilon} \in S_{\mathrm{I}}(T)$. This strategy satisfies

$$f_{\mathrm{I}}(x) \geq v(G_{x_n}(f_{\mathrm{I}})) - \varepsilon, \quad \forall x \in [T], \forall n \in \mathbb{N}. \tag{4.15}$$

Similarly, Player II has a subgame ε-optimal strategy $\widehat{s}_{\mathrm{II}}^{\varepsilon} \in S_{\mathrm{II}}(T)$, such that

$$f_{\mathrm{II}}(x) \geq -v(G_{x_n}(-f_{\mathrm{II}})) - \varepsilon, \quad \forall x \in [T], \forall n \in \mathbb{N}. \tag{4.16}$$

For every position $p \in T$, let $s_{\mathrm{II}}(f_{\mathrm{I}}, p) \in S_{\mathrm{II}}(T_p)$ be an ε-optimal strategy of Player II in the zero-sum game $G_p(f_{\mathrm{I}})$. Then we have, in particular,

$$f_{\mathrm{I}}(x(s_{\mathrm{I}}, s_{\mathrm{II}}(f_{\mathrm{I}}, p); p)) \leq v(G_p(f_{\mathrm{I}})) + \varepsilon, \quad \forall s_{\mathrm{I}} \in S_{\mathrm{I}}(T). \tag{4.17}$$

Similarly, for every $p \in T$ let $s_{\mathrm{I}}(f_{\mathrm{II}}, p) \in S_{\mathrm{I}}(T_p)$ be an ε-optimal strategy of Player I in the zero-sum game $G_p(-f_{\mathrm{II}})$. This strategy satisfies

$$f_{\mathrm{II}}(x(s_{\mathrm{I}}(f_{\mathrm{II}}, p), s_{\mathrm{II}}; p)) \leq -v(G_p(-f_{\mathrm{II}})) + \varepsilon, \quad \forall s_{\mathrm{II}} \in S_{\mathrm{II}}(T). \tag{4.18}$$

Let $x^* := x(\widehat{s}_{\mathrm{I}}^{\varepsilon}, \widehat{s}_{\mathrm{II}}^{\varepsilon}) \in [T]$ be the play that is generated when Player I (resp., Player II) adopts her subgame ε-optimal strategy in the zero-sum game $G_{\langle\,\rangle}(f_{\mathrm{I}})$ (resp., $G_{\langle\,\rangle}(-f_{\mathrm{II}})$), and denote $x^* = \langle a_0^*, a_1^*, \ldots \rangle$. By Equations (4.15) and (4.16),

$$f_{\mathrm{I}}(x^*) \geq v(G_{x_n^*}(f_{\mathrm{I}})) - \varepsilon, \quad f_{\mathrm{II}}(x^*) \geq -v(G_{x_n^*}(-f_{\mathrm{II}})) - \varepsilon, \quad \forall n \in \mathbb{N}. \tag{4.19}$$

For every position $p = \langle a_0, a_1, \ldots, a_n \rangle \in T$, the first stage in which a player deviates from x^* is given by

$$n(p) := \min\{k \leq n \colon a_k \neq a_k^*\},$$

where the minimum of an empty set is ∞. Thus, $p_{n(p)}$ is the first position along p that is not a prefix of x^*, provided $n(p) < \infty$.

Let $s_{\mathrm{I}}^* \in S_{\mathrm{I}}(T)$ be the strategy of Player I that follows $\widehat{s}_{\mathrm{I}}^{\varepsilon}$ until Player II deviates from x^*, and then it switches to a strategy that lowers Player II's payoff. Formally, for every position $p = \langle a_0, a_1, \ldots, a_n \rangle \in T$ of even length,

$$s_{\mathrm{I}}^*(p) = \begin{cases} a_{n+1}^*, & \text{if } p = x_n^*, \\ \left(s_{\mathrm{I}}(-f_{\mathrm{II}}, p_{n(p)})\right)(p), & \text{if } p \neq x_n^*. \end{cases}$$

Let $s_{\mathrm{II}}^* \in S_{\mathrm{II}}(T)$ be the analogous strategy of Player II: for every position $p = \langle a_0, a_1, \ldots, a_n \rangle \in T$ of odd length,

$$s_{\mathrm{II}}^*(p) = \begin{cases} a_{n+1}^*, & \text{if } p = x_n^*, \\ \left(s_{\mathrm{II}}(f_{\mathrm{I}}, p_{n(p)})\right)(p), & \text{if } p \neq x_n^*. \end{cases}$$

We will prove that the strategy pair (s_I^*, s_{II}^*) is a 3ε-equilibrium in G. To this end, we will show that Player I cannot profit more than 3ε by deviating from s_I^*. The proof that Player II cannot profit more than 3ε by deviating from s_{II}^* is analogous.

By the definition of s_I^* and s_{II}^*,

$$x(s_I^*, s_{II}^*) = x^* = x(\widehat{s}_I^\varepsilon, \widehat{s}_{II}^\varepsilon).$$

Let now $s_I \in S_I(T)$ be arbitrary. If $x(s_I, s_{II}^*) = x^*$, then

$$f_I(x(s_I, s_{II}^*)) = f_I(x^*) = f_I(x(s_I^*, s_{II}^*)),$$

and so Player I does not profit by deviating to s_I.

If $x(s_I, s_{II}^*) \neq x^*$, denote by $n+1$ the first stage in which these two plays differ:

$$n + 1 := \min\{k \in \mathbb{N} \colon x_k(s_I, s_{II}^*) \neq x_k^*\}.$$

Note that $n+1$ is necessarily even, because Player I has made the move that differs from a_{n+1}^*. We have

$$
\begin{aligned}
f_I(x(s_I, s_{II}^*)) &\leq v(G_{x_{n+1}(s_I, s_{II}^*)}(f_I)) + \varepsilon &\quad (4.20)\\
&\leq v(G_{x_n^*}(f_I)) + 2\varepsilon &\quad (4.21)\\
&\leq f_I(x^*) + 3\varepsilon, &\quad (4.22)
\end{aligned}
$$

where Equation (4.20) holds because from stage $n+1$ on the strategy s_{II}^* coincides with an ε-optimal strategy of Player II in $G_{x_{n+1}(s_I, s_{II}^*)}(f_I)$, Equation (4.21) holds because $x_n(s_I, s_{II}^*) = x_n^*$, and because $\widehat{s}_I$ is a subgame ε-optimal strategy in $G_{\langle \rangle}(f_I)$, and Equation (4.22) holds by Equation (4.19). It follows that Player I cannot profit more than 3ε by deviating from s_I^*, as we wanted to prove. □

4.3 DISCUSSION

The concept of 0-equilibrium was defined by Nash (1950), and is the most studied solution concept in Game Theory. It underlies modern economic theory, which assumes that agents maximize their utility, and areas in operations research, computer science, and political science, where it is assumed, e.g., that customers minimize their waiting time, servers minimize the time needed for the information they send to reach its destination, and voters maximize the probability that their favorite candidate will win.

A fundamental difference between an ε-optimal strategy and an ε-equilibrium is that while the former relates to strategies of a single player in a zero-sum game, the latter is a pair of strategies, one for each player. Thus, a strategy of a player is ε-optimal if it generates a good outcome for the player, whatever the other player does, while an ε-equilibrium is a pair of strategies that generates a good outcome for *both* players, assuming no player deviates. In particular, constructing an ε-optimal strategy is sometimes a difficult task, as this strategy has to be good against all possible behaviors of the opponent. This is in contrast to an ε-equilibrium, where all that is needed is that no player will have a profitable deviation. Surprisingly, there are models of games where identifying the value and 0-optimal strategies in the zero-sum case is easy, while finding equilibria in the nonzero-sum case is difficult.

As we have seen in the proof of Theorem 4.37, to ensure that no player has a profitable deviation, we can utilize threats of punishment: if a player detects a deviation of the other player, she switches to a punishment strategy, which lowers the deviator's payoff. Thus, often, ε-equilibrium strategies have the following structure:

- An *equilibrium path* that players should follow, and which generates a high payoff to each player.

- A *test* that verifies that no player deviates from the equilibrium path.

- *Punishment strategies*, which are employed once a deviation is detected.

In the proof of Theorem 4.37, the equilibrium play is induced by a pair of subgame ε-optimal strategies, and deviations from the equilibrium play are easily detected. This approach works since the players select their moves deterministically, and hence identifying deviations is straightforward. In Chapter 6, where we will study simultaneous-move nonzero-sum games, this approach no longer works, and we will develop new techniques to construct the equilibrium play and more elaborate ways to detect deviations.

4.4 EXERCISES

Exercise 4.14 was inspired by Das, Fishman, Simmons, and Urbański (2017, 2024). Exercise 4.16 is adapted from Solan and Vieille (2003).

Exercises marked with a star are more challenging than unmarked ones, while those marked with two stars are even more difficult.

1. Let $G = (T, f)$ be a zero-sum game. Prove that when f is bounded, Definitions 4.12 and 4.13 for the value are equivalent.

2. (*) Let $G = (T, f)$ be a zero-sum game. Show that when f is not bounded, Definitions 4.12 and 4.13 are not necessarily equivalent.

3. Let (T, f) be a zero-sum game, where f is bounded and Borel measurable. Let $\varepsilon > 0$, and let s_I be an ε-optimal strategy of Player I in (T, f). Prove that for every bounded and Borel-measurable function $g : [T] \to \mathbb{R}$, the strategy s_I is an $(\varepsilon + 2\|f - g\|_\infty)$-optimal strategy in (T, g).

4. Let $G = (T, f)$ be a zero-sum game where f is Borel measurable and its image is finite.

   (a) Prove that the value of G is in the image of f.

   (b) Prove that both players have 0-optimal strategies.

   (c) Prove that Parts (a) and (b) do not necessarily hold if the image of f is countable.

5. (*) Let $G = (T, f)$ be a zero-sum game where the function f is bounded and Borel measurable.

   (a) Prove that at least one player has a 0-optimal strategy.

   (b) Can it happen that both players have a 0-optimal strategy? Justify your answer.

6. (*) In this exercise we extend Theorem 4.15 to zero-sum games with unbounded payoff function.

   (a) Prove that every zero-sum game $G = (T, f)$, where f is unbounded and Borel measurable, has a value according to Definition 4.12.

   (b) Provide an example of a zero-sum game with an unbounded and Borel measurable payoff function, where Player I does not have an ε-optimal strategy, for any $\varepsilon > 0$.

7. In this exercise we study the relation between ε-optimal strategies and ε-equilibria.

   Let (T, f) be a zero-sum game, and let $\varepsilon > 0$.

   (a) Prove that if $(\sigma_I^*, \sigma_{II}^*)$ is an ε-equilibrium in the nonzero-sum game $(T, f, -f)$, then σ_I^* is a 2ε-optimal strategy of Player I in (T, f), and σ_{II}^* is a 2ε-optimal strategy of Player II in (T, f).

   (b) Prove that if σ_I^* is an ε-optimal strategy of Player I in (T, f) and σ_{II}^* is an ε-optimal strategy of Player II in (T, f), then $(\sigma_I^*, \sigma_{II}^*)$ is a 2ε-equilibrium in the nonzero-sum game $(T, f, -f)$.

8. Prove Corollary 4.23.

9. Let T be a tree over an alphabet $\mathcal{A}$, and let $(f_n)_{n \in \mathbb{N}}$ be a sequence of Borel-measurable functions from $[T]$ to $[0, 1]$. For each of the following statements, provide a proof or a counterexample.

   (a) If the sequence $(f_n)_{n \in \mathbb{N}}$ converges uniformly to a limit f, then the sequence of values of the zero-sum games (T, f_n) converges to the value of the zero-sum game (T, f).

   (b) If the sequence $(f_n)_{n \in \mathbb{N}}$ converges pointwise to a limit f, then the sequence of values of the zero-sum games (T, f_n) converges to the value of the zero-sum game (T, f).

10. (*)

    (a) Provide an example of a nonzero-sum game whose set of 0-equilibrium payoffs is $\{(1, 1), (2, 2)\}$.

    (b) Let $U \subseteq \mathbb{R}^2$ be a non-empty bounded Borel set. Provide a nonzero-sum game whose set of 0-equilibrium payoffs is U.

11. Let T be the complete binary tree over the alphabet $\mathcal{A} = \{0,1\}$. Let $G = (T, f_\mathrm{I}, f_\mathrm{II})$ be the nonzero-sum game where

$$f_\mathrm{I}(x) = \begin{cases} 0, & \text{if } a_1 = 0, \\ 1, & \text{if } a_0 = 0, a_1 = 1, \\ 2, & \text{if } a_0 = 1, a_1 = 1, \end{cases} \qquad f_\mathrm{II}(x) = \begin{cases} 1, & \text{if } a_0 = 0, \\ 2, & \text{if } a_0 = 1. \end{cases}$$

Identify all 0-equilibria and all 0-equilibrium payoffs in the game.

12. (*) In this exercise we generalize Theorem 4.37 to games with more than two players.

Let $L \in \{1, 2, \ldots\}$. A game played by L players is specified by a vector $(T, f_0, f_1, \ldots, f_{L-1})$, where T is a tree and $f_i : [T] \to \mathbb{R}$ for each $i \in \{0, 1, \ldots, L-1\}$. For each i, player i selects a move in every stage n such that $n = i - 1 \bmod L$. Such a game is called a *multiplayer game*. A strategy s_i of player i is a function that assigns a move for each position $p \in T$ such that $\mathrm{len}(p) = i \bmod L$. Denote by $x(s_0, s_1, \ldots, s_{L-1})$ the play that is induced by the vector of strategies $(s_0, s_1, \ldots, s_{L-1})$. Given $\varepsilon \geq 0$, a vector of strategies $(s_0^*, s_1^*, \ldots, s_{L-1}^*)$ is an *ε-equilibrium* if for every $i \in \{0, 1, \ldots, L-1\}$ and every $s_i \in S_i(T)$,

$$f_i(x(s_0^*, s_1^*, \ldots, s_{L-1}^*)) \geq f_i(x(s_0^*, s_1^*, \ldots, s_{i-1}^*, s_i, s_{i+1}^*, \ldots, s_{L-1}^*)) - \varepsilon.$$

For each player $i \in \{0, 1, \ldots, L-1\}$ define her *minmax value* in G by

$$\bar{v}_i(G) := \inf_{s_{-i} \in \prod_{j \neq i} S_j(T)} \sup_{s_i \in S_i(T)} f_i(s_i, s_{-i}),$$

and her *maxmin value* by

$$\underline{v}_i(G) := \sup_{s_i \in S_i(T)} \inf_{s_{-i} \in \prod_{j \neq i} S_j(T)} f_i(s_i, s_{-i}).$$

Suppose that the functions $(f_i)_{i=0}^{L-1}$ are bounded and Borel measurable.

(a) Prove that $\underline{v}_i(G) = \bar{v}_i(G)$, for every player $i \in N$.

(b) Prove that the multiplayer game admits an ε-equilibrium, for every $\varepsilon > 0$.

13. (*) Let T be the tree that corresponds to a stopping game. This is the tree that is defined over the alphabet $\{s, c\}$, and at each position p, there are two moves $\{c, s\}$ if $p = \langle c, c, \ldots, c \rangle$, and there is a unique move s otherwise.

Let G be the three-player stopping game $G = (T, f_\mathrm{I}, f_\mathrm{II}, f_\mathrm{III})$, where

- Player I selects moves in stages n such that $n \bmod 3 = 0$, Player II selects moves in stages n such that $n \bmod 3 = 1$, and Player III selects moves in stages n such that $n \bmod 3 = 2$.

- If Player I is the first player who selects s, then her payoff is 1, that of Player II is 3, and that of Player III is 0.

- If Player II is the first player who selects s, then the payoff of Player I is 0, that of Player II is 1, and that of Player III is 3.

- If Player III is the first player who selects s, then the payoff of Player I is 3, that of Player II is 0, and that of Player III is 1.

- If the three players always select c, the payoff of each player is 0.

(a) Draw the tree T, and indicate the payoffs of the three players.

(b) Find all 0-equilibria and all 0-equilibrium payoffs in this game.

14. (*) For each nonempty set $W \subseteq \{0,1\}^{\mathbb{N}}$, consider the following zero-sum game G_W.

- At every even stage $2n$, the move B_n of Player I is a nonempty subset of $\{0,1\}$.

- At every odd stage $2n+1$, the move a_n of Player II is one of the elements of B_n.

- The payoff function is

  - $f(\langle B_0, a_0, B_1, a_1, \ldots \rangle) := -1$, if $(a_0, a_1, \ldots) \notin W$.
  - $f(\langle B_0, a_0, B_1, a_1, \ldots \rangle) := \liminf_{N \to \infty} \frac{1}{N} \sum_{n=0}^{N-1} \log_2(|B_n|)$, if $(a_0, a_1, \ldots) \in W$.

In words, along the play Player II constructs an element of $\{0,1\}^{\mathbb{N}}$. At every stage, Player I dictates which bits Player II can select from at that stage. Player I loses (her payoff is -1) if the generated number is not in the winning set W, and her payoff is higher if she gives Player II more freedom in her (Player II's) choices.

(a) What is the alphabet over which G_W is defined?

(b) What is the value of the game G_W when W is a nonempty open set?

(c) Find a set W for which the value of the game G_W is $\frac{1}{2}$. Prove your claim.

(d) For each real number $r \in [0,1]$, find a set W for which the value of the game G_W is r.

(e) Set

$$W_1 := \{(a_0, a_1, \ldots): a_{2n} + a_{2n+1} = 1, \quad \forall n \in \mathbb{N}\}, \qquad (4.23)$$
$$W_2 := \{(a_0, a_1, \ldots): \max\{a_{2n}, a_{2n+1}\} = 1, \quad \forall n \in \mathbb{N}\}. \quad (4.24)$$

What is the value of the games G_{W_1} and G_{W_2}?

(f) Consider the variation $\widehat{G}$ of G, in which the players select two bits in every stage rather than one. That is, in each even stage $2n$, Player I selects a nonempty subset $B_n \subseteq \{(0,0), (0,1), (1,0), (1,1)\}$, in each odd stage $2n+1$, Player II selects an element $(a_{2n}, a_{2n+1}) \in B_n$, where a_{2n} is the first bit in the pair and a_{2n+1} is the second bit in the pair, and the payoff function is

$$f(\langle B_0, (a_0, a_1), B_1, (a_2, a_3), \dots \rangle)$$

$$:= \begin{cases} -1, & \text{if } (a_0, a_1, \dots) \notin W, \\ \liminf_{N \to \infty} \frac{1}{N} \sum_{n=0}^{N-1} \log_4(|B_n|), & \text{if } (a_0, a_1, \dots) \in W. \end{cases}$$

What is the value of the games $\widehat{G}_{W_1}$ and $\widehat{G}_{W_2}$, where W_1 and W_2 are the sets defined in Equations (4.23) and (4.24)?

Das, Fishman, Simmons, and Urbański (2017, 2024) related the Hausdorff dimension of the set W to a variation of the game G_W (different from $\widehat{G}_W$).

15. (*) In this exercise we generalize Theorem 4.37 (and Exercise 12) to games with countably many players.

A game with countably many players is a triplet $(T, \iota, (f_i)_{i \in \mathbb{N}})$, where T is a tree, $\iota : T \to \mathbb{N}$, and $f_i : [T] \to [0, 1]$ for each $i \in \mathbb{N}$. The interpretation of such a game is that at each position $p \in T$ player $\iota(p)$ selects a move. Thus, there is no specific order in which the players move, the order by which the players move may depend on their past choices, and some players may move only once along the game, while other may move infinitely often.

A strategy s_i of player i is a function that assigns a move for each position $p \in T$ such that $\iota(p) = i$. The set of strategies of player i is denoted $S_i(T, \iota)$. Denote by $s = (s_i)_{i \in \mathbb{N}}$ a vector of strategies, one for each player, and by $x(s)$ the play that is induced by the vector of strategies s. Given $\varepsilon \geq 0$, a vector of strategies s^* is an *ε-equilibrium* if

$$f_i(x(s^*)) \geq f_i(x(s_i^*, s_{-i}^*)) - \varepsilon, \quad \forall i \in \mathbb{N}, \forall s_i \in S_i(T, \iota).$$

Prove that every game with countably many players, where the functions $(f_i)_{i \in \mathbb{N}}$ are Borel measurable, admits an ε-equilibrium, for every $\varepsilon > 0$.

16. (*) In Theorem 4.29 we proved that in every zero-sum game with bounded and Borel-measurable payoff function, each player has a subgame ε-optimal strategy. In this exercise we show that this result does not extend to ε-equilibria in nonzero-sum games.

Consider the nonzero-sum stopping game where the payoff functions of the two players are as follows:

- The payoff if both players always continue is $(0,0)$; that is, $f_\mathrm{I}(\langle c, c, c, \dots \rangle) = f_\mathrm{II}(\langle c, c, c, \dots \rangle) = 0$.
- The payoff if Player I is the first to stop is $(-1, 2)$; that is, $f_\mathrm{I}(x) = -1$ and $f_\mathrm{II}(x) = 2$ for every play x that starts with an even number of c's followed by an s.

- The payoff if Player II is the first to stop is $(-2, 1)$; that is, $f_{\mathrm{I}}(x) = -2$ and $f_{\mathrm{II}}(x) = 1$ for every play x that starts with an odd number of c's followed by an s.

Prove that for every strategy pair $(s_{\mathrm{I}}, s_{\mathrm{II}})$ there is a position p such that $(s_{\mathrm{I}}, s_{\mathrm{II}})$, restricted to the subgame Γ_p, is not an ε-equilibrium, for every $\varepsilon \in (0, 1)$.

17. (**) Archer complains that the results we proved in this chapter lead to a contradiction: "Let $G = (T, f)$ be a zero-sum game, where f is bounded and Borel measurable and with value $v = 0$," she says, "and suppose that Player I wins the auxiliary game G_0 that was defined in the proof of Theorem 4.15. Let G^{CM} and G^{MC} be two copies of this game. Suppose that Catherine and Massego play the following zero-sum game $\widehat{G}$, which is composed of a simultaneous play of G^{CM} and G^{MC}: in G^{CM}, Catherine is Player I and Massego Player II, and in G^{MC} Massego is Player I and Catherine Player II (the appellations CM and MC indicate the identity of the first and second mover in each game).

That is, first Catherine selects her move for stage 0 in G^{CM}, then Massego selects his moves for stage 1 in G^{CM} and for stage 0 in G^{MC}, then Catherine selects her moves for stage 2 in G^{CM} and for stage 1 in G^{MC}, and so on. The payoff of $\widehat{G}$ is the payoff in G^{CM} minus the payoff in G^{MC}.

"The value of the zero-sum game $\widehat{G}$ is 0: Catherine can guarantee that the outcome in G^{CM} is non-negative, and for each $\varepsilon > 0$ Catherine can guarantee that the outcome in G^{MC} is at most ε. Similarly, for each $\varepsilon > 0$ Massego can guarantee that the outcome in G^{CM} is at most ε, and he can guarantee that the outcome in G^{MC} is non-negative.

"Now, one of the players wins the auxiliary game $\widehat{G}_0$ that was defined in the proof of Theorem 4.15. Suppose without loss of generality that it is Massego, and let $\widehat{s}_{\mathrm{M}}$ be his winning strategy in $\widehat{G}_0$. So $\widehat{s}_{\mathrm{M}}$ guarantees that the outcome in $\widehat{G}$ is non-positive. In G_0^{CM} Catherine is acting as Player I, so Catherine has a winning strategy, and therefore Massego has no strategy that guarantees that the payoff in G^{CM} is non-positive. It follows that Catherine has a response to $\widehat{s}_{\mathrm{M}}$ that ensures that the outcome in G^{CM} is positive, say δ. And if Catherine plays a $(\delta/2)$-optimal strategy in G^{MC}, Catherine guarantees that the payoff in G^{MC} is at most $\delta/2$. But this means that Catherine can guarantee that the payoff in $\widehat{G}$ is at least $\delta/2$, which contradicts the choice of $\widehat{s}_{\mathrm{M}}$."

Is Archer correct? If she is, point at the error(s) in the proofs we presented in this chapter. If she is not, explain what is the flaw in her argument.

18. (**) In this exercise we study stopping games and prove a result due to Dynkin (1967) using the theory developed so far.

Let $(\Omega, \mathcal{F}, \mathbf{P})$ be a probability space, let $\mathcal{F}_0 \subseteq \mathcal{F}_1 \subseteq \cdots \subseteq \mathcal{F}$ be a filtration, and let $(X_n)_{n \in \mathbb{N}}$ and $(R_n)_{n \in \mathbb{N}}$ be two real-valued stochastic processes adapted to $(\mathcal{F}_n)_{n \in \mathbb{N}}$. Consider the following zero-sum game:

- The set of strategies of Player I is the set of all stopping times τ_I for the filtration $(\mathcal{F}_n)_{n \in \mathbb{N}}$ such that $\mathbf{P}(R_{\tau_I} \leq 0) = 1$.
- The set of strategies of Player II is the set of all stopping times τ_{II} for the filtration $(\mathcal{F}_n)_{n \in \mathbb{N}}$ such that $\mathbf{P}(R_{\tau_{II}} > 0) = 1$.
- The payoff is $\mathbf{E}\left[\mathbf{1}_{\{\min\{\tau_I, \tau_{II}\} < \infty\}} \cdot X_{\min\{\tau_I, \tau_{II}\}}\right]$.

Verbally, in this game players decide when to stop, and the payoff is determined once the first among them decides to stop. Player I can stop whenever the process $(R_n)_{n \in \mathbb{N}}$ is non-positive, and Player II can stop whenever this process is positive.

(a) Present the stopping game as a zero-sum game (as defined in Definition 4.1).

(b) Using the theory developed in this section, identify conditions on $(\Omega, \mathcal{F}, \mathbf{P})$, $(\mathcal{F}_n)_{n \in \mathbb{N}}$, $(X_n)_{n \in \mathbb{N}}$, and $(R_n)_{n \in \mathbb{N}}$ (the more general conditions, the better) that ensure the value of the game exists. Justify your answer.

CHAPTER 5

The Value of Zero-Sum Blackwell Games

In Chapters 3 and 4 we studied alternating-move games: the players selected their moves in turns, one after the other. In this chapter, we study simultaneous-move games, where in each turn the players select their moves at the same time. Rock-Scissors-Paper is an example of such a game. We will define the value for two-player zero-sum simultaneous-move games, and prove that such games have a value, provided the set of moves of each player is finite and the payoff function is bounded and Borel measurable. Games that satisfy these conditions will be called *Blackwell games*. This result can be extended to the case in which at every position, at least one player has a finite set of available moves.

Before studying two-player zero-sum simultaneous-move games, we recall basic results on two-player zero-sum one-shot games.

5.1 TWO-PLAYER ZERO-SUM ONE-SHOT GAMES

So far we have studied games with infinite horizon – the games never end. Here we define games that are played for a single round, and in which the two players move simultaneously: each player selects a move, and the pair of moves determine the payoff that Player II pays to Player I. In the language of Game Theory, such games are called *strategic-form games* or *normal-form games*. We will call them *one-shot games*, to emphasize that they last only one round.

Definition 5.1 (Two-player zero-sum one-shot game). *A two-player zero-sum one-shot game is a triple (X_I, X_{II}, h), where X_I and X_{II} are nonempty finite or countably infinite sets and $h \colon X_I \times X_{II} \to \mathbb{R}$.*

The interpretation of a two-player zero-sum one-shot game is as follows. The sets of moves of the two players are X_I and X_{II}. The players select moves only once; Player I selects a move a_I in X_I and Player II selects a move a_{II} in X_{II}. The choices are made simultaneously. Once the moves are selected, Player II pays to Player I the amount $h(a_I, a_{II})$. We will often refer to the one-shot game by h, without mentioning

DOI: 10.1201/9781003582106-5

explicitly the sets of moves of the two players. The function h will be called the *payoff function* of the game.

Remark 5.2 (On the cardinality of the sets of moves). *In Definition 5.1 we assumed that the sets of moves X_{I} and X_{II} are finite or countably infinite. The model can be extended to incorporate general sets of moves.*

Example 5.3 (Matching Pennies). *Suppose that $X_{\mathrm{I}} = X_{\mathrm{II}} = \{a, b\}$, and the payoff function h is given by*

$$h(a, a) = h(b, b) = 1, \quad h(a, b) = h(b, a) = -1.$$

The function h can be presented in a matrix form:

		Player II	
		a	b
Player I	a	1	-1
	b	-1	1

This matrix provides all the data that defines the game: Player I's set of moves corresponds to the rows of the matrix, Player II's set of moves corresponds to the columns of the matrix, and the numbers in the matrix correspond to the payoff function h.

In this game, Player I wants to match the move of Player II, and Player II wants to mismatch the move of Player I. How should the players play in this game? Against the move $a_{\mathrm{I}} = a$, Player II could guarantee to pay -1 by selecting $a_{\mathrm{II}} = b$. Similarly, against the move $a_{\mathrm{I}} = b$, Player II could guarantee to pay -1 by selecting $a_{\mathrm{II}} = a$. To guarantee on expectation more than -1, Player I has to randomly select her move. If, for example, she selects each move a or b with probability $\frac{1}{2}$, then, since the players make their choices simultaneously, whatever move Player II selects, the moves of the two players will match with probability $\frac{1}{2}$, and hence the expected payoff will be 0.

An analogous analysis holds for Player II: for any move $a_{\mathrm{II}} \in X_{\mathrm{II}}$, Player I has a response that yields a payoff 1. To ensure that the payoff is less than 1, Player II has to randomly select her move. By selecting each move a or b with probability $\frac{1}{2}$, whatever move Player I selects, the moves of the two players will mismatch with probability $\frac{1}{2}$, and hence the expected payoff will be 0.

Thus, by using randomization, each player can guarantee the amount 0, and therefore this is the value of the game when randomization is allowed.

Notation 5.4 ($\vec{a}$). *We will denote an element of $X_{\mathrm{I}} \times X_{\mathrm{II}}$ by $(a_{\mathrm{I}}, a_{\mathrm{II}})$ or $\vec{a}$.*

Example 5.3 shows that when the players move simultaneously, they may benefit from selecting their moves randomly. Recall that the set of probability distributions over a measurable space X is denoted by $\Delta(X)$ (Notation 2.18). This leads us to the definition of mixed moves.

Definition 5.5 (Mixed move). *Let $(X_{\mathrm{I}}, X_{\mathrm{II}}, h)$ be a two-player zero-sum one-shot game, and let $i \in \{\mathrm{I}, \mathrm{II}\}$ where X_{I} and X_{II} are measurable spaces. A mixed move for player i is an element of $\Delta(X_i)$.*

The interpretation of a mixed move $x_i \in \Delta(X_i)$ is that player i selects her move randomly, according to the probability distribution x_i. The actual move that is selected is revealed to the other player only after she, the other player, already selected her own move.

Notation 5.6 (x_i, $x_i(A)$, $x_i(a_i)$, $x_I \otimes x_{II}$, $\mathbf{E}_{x_I \otimes x_{II}}$). *We will denote mixed moves of player i by x_i. In this case, the probability that player i selects a specific move $a_i \in X_i$ will be denoted by $x_i(a_i)$, and the probability under which the selected move belongs to the set $A \subseteq X_i$ will be denoted by $x_i(A)$.*

When given a specific mixed move, for example, the one that selects the move a with probability $\frac{1}{2}$, the move b with probability $\frac{1}{3}$, and the move c with probability $\frac{1}{6}$, we will use the notation $[\frac{1}{2}(a), \frac{1}{3}(b), \frac{1}{6}(c)]$.

Every pair of mixed moves (x_I, x_{II}) defines a product probability distribution on $X_I \times X_{II}$. We will denote this product probability distribution by $x_I \otimes x_{II}$, and the corresponding expectation operator by $\mathbf{E}_{x_I \otimes x_{II}}[\cdot]$.

Notation 5.7 ($h(x_I, x_{II})$). *The expected payoff when the two players select their moves at random according to x_I and x_{II}, respectively, is*

$$h(x_I, x_{II}) := \sum_{a_I \in X_I} \sum_{a_{II} \in X_{II}} x_I(a_I) x_{II}(a_{II}) h(a_I, a_{II}).$$

Definition 5.8 (Value of mixed move, maxmin value, minmax value). *Let (X_I, X_{II}, h) be a two-player zero-sum one-shot game. The* value *of the mixed move $x_I \in \Delta(X_I)$ is*

$$\text{val}_O(x_I) := \inf_{x_{II} \in \Delta(X_{II})} h(x_I, x_{II}) = \inf_{a_{II} \in X_{II}} h(x_I, a_{II}). \tag{5.1}$$

The maxmin value *of the game is*

$$\underline{v}_O(h) := \sup_{x_I \in \Delta(X_I)} \text{val}_O(x_I) = \sup_{x_I \in \Delta(X_I)} \inf_{x_{II} \in \Delta(X_{II})} h(x_I, x_{II}). \tag{5.2}$$

The value *of the mixed move $x_{II} \in \Delta(X_{II})$ is*

$$\text{val}_O(x_{II}) := \sup_{x_I \in \Delta(X_I)} h(x_I, x_{II}) = \sup_{a_I \in X_I} h(a_I, x_{II}). \tag{5.3}$$

The minmax value *of the game is*

$$\overline{v}_O(h) := \inf_{x_{II} \in \Delta(X_{II})} \text{val}_O(x_{II}) = \inf_{x_{II} \in \Delta(X_{II})} \sup_{x_I \in \Delta(X_I)} h(x_I, x_{II}). \tag{5.4}$$

Remark 5.9 (The subscript O). *The subscript O in the notations $\text{val}_O(x_I)$, $\text{val}_O(x_{II})$, $\underline{v}_O(h)$, and $\overline{v}_O(h)$ reminds us that these quantities correspond to one-shot games, and not to games that last infinitely many stages.*

Remark 5.10 (Finite set of moves). *When the set X_I is finite, the supremum in Equations (5.2), (5.3), and (5.4) is in fact a maximum. When the set X_{II} is finite, the infimum in Equations (5.1), (5.2), and (5.4) is in fact a minimum.*

Definition 5.11 (Value). *Let (X_I, X_{II}, h) be a two-player zero-sum one-shot game. The real number $v_O = v_O(h)$ is the* value *of the game if*

$$v_O = \underline{v}_O(h) = \overline{v}_O(h).$$

The following result states two simple properties of the value operator. The proof is left to the reader (Exercise 5.1).

Theorem 5.12. *Let (X_I, X_{II}, h) and $(X_I, X_{II}, \widehat{h})$ be two two-player zero-sum one-shot games that share the same sets of moves, and let $c > 0$. Suppose that $v_O(h)$ and $v_O(\widehat{h})$ exist.*

1. *If $h(\vec{a}) \geq \widehat{h}(\vec{a})$ for every $\vec{a} \in X_I \times X_{II}$, then $v_O(h) \geq v_O(\widehat{h})$.*

2. *If $h(\vec{a}) = \widehat{h}(\vec{a}) + c$ for every $\vec{a} \in X_I \times X_{II}$, then $v_O(h) = v_O(\widehat{h}) + c$.*

Definition 5.13. *(ε-optimal mixed move) Let (X_I, X_{II}, h) be a two-player zero-sum one-shot game, and let $\varepsilon \geq 0$. A mixed move $x_I^* \in \Delta(X_I)$ is ε-optimal for Player I in (X_I, X_{II}, h) if*

$$\mathrm{val}_O(x_I^*) \geq \underline{v}_O(h) - \varepsilon.$$

A mixed move $x_{II}^* \in \Delta(X_{II})$ is ε-optimal for Player II in (X_I, X_{II}, h) if

$$\mathrm{val}_O(x_{II}^*) \leq \overline{v}_O(h) + \varepsilon.$$

Example 5.3, continued. As we have seen, the value of the Matching Pennies game is $\frac{1}{2}$, and a 0-optimal mixed move of each player is $[\frac{1}{2}(a), \frac{1}{2}(b)]$. ▽

Example 5.14. *Consider the following two-player zero-sum one-shot game:*

		Player II	
		a	b
Player I	a	3	2
	b	0	1

The value of this game is 2. A 0-optimal mixed move for Player I is $[1(a), 0(b)]$, and a 0-optimal mixed move for Player II is $[0(a), 1(b)]$.

Remark 5.15 (Domination). *In the game in Example 5.14, the move a of Player I is always better than the move b. We say that a* dominates b *for Player I, or that b is* dominated *by a. If we remove a dominated move from the game, that is, we consider the smaller two-player zero-sum one-shot game that is similar to the original game except that the dominated move is removed, we obtain a game that has the same value.*

The definition of domination for Player II is analogous: the move a dominates the move b for Player II, if for any move c of Player I, $h(c, a) < h(c, b)$.

A related concept to domination is weak domination. *We say that the move a* weakly dominates *the move b for Player I, if for any move c of Player II, $h(a, c) \geq$*

$h(b, c)$, and for some move c of Player II this inequality is strict. The reader can verify that removing a weakly dominated move from the game does not alter the game's value either. Removing dominated and weakly dominated moves from a game is the first step in calculating the game's value.

Example 5.16. Consider the following two-player zero-sum one-shot game, where one of the payoffs is a parameter:

		Player II	
		a	b
Player I	a	z	1
	b	1	0

When $z > 1$, the move a dominates the move b for Player I, and the move b dominates the move a for Player II. In this case, the value of the game is 1, and the unique 0-optimal mixed moves of the two players are a and b, respectively.

When $z = 1$, the move a weakly dominates the move b for Player I, and the move b weakly dominates the move a for Player II. In this case, the value of the game and the unique 0-optimal mixed move of Player I is as in the previous case. For Player II, all mixed moves are 0-optimal.

When $z < 1$, there is no domination relation between moves of any player. We will calculate the value of the game. Identify a mixed move x_I of Player I with the probability it assigns to the move a. Similarly, identify a mixed move x_{II} of Player II with the probability it assigns to the move a. Then,

$$
\begin{aligned}
v_O(h) &= \max_{x_I \in [0,1]} \min_{x_{II} \in [0,1]} h(x_I, x_{II}) \\
&= \max_{x_I \in [0,1]} \min\{h(x_I, a), h(x_I, b)\} \\
&= \max_{x_I \in [0,1]} \min\{x_I \cdot z + 1 - x_I, x_I\}.
\end{aligned}
$$

The functions $x_I \mapsto x_I \cdot z + 1 - x_I$ and $x_I \mapsto x_I$ are linear, hence monotone. When $x_I = 1$ we have

$$
x_I \cdot z + 1 - x_I = z \le 1 = x_I,
$$

and when $x_I = 0$ we have

$$
x_I \cdot z + 1 - x_I = 1 > 0 = x_I.
$$

It follows that the maximum is attained at some point $x_I \in (0, 1]$, where $x_I \cdot z + 1 - x_I = x_I$. The solution to this equation is

$$
x_I = \frac{1}{2 - z}.
$$

It follows that the value is

$$
v_O(h) = \frac{1}{2 - z},
$$

and a 0-optimal mixed move for Player I is $x_I^* = \frac{1}{2-z}$.

The following result is known as von Neumann's Minimax Theorem.

Theorem 5.17 (von Neumann, 1928). *All two-player zero-sum one-shot games* (X_I, X_{II}, h) *where* X_I *and* X_{II} *are finite sets have a value. Moreover, both players have 0-optimal mixed moves.*

Von Neumann's Minimax Theorem follows from the more general Nash Theorem (see Theorem 6.6 below, whose proof can be found, e.g., in Maschler, Solan, and Zamir (2020, Section 5.3); see also Exercise 6.2.

As the next example shows, when X_I and X_{II} are countably infinite, the two-player zero-sum one-shot game might not have a value.

Example 5.18 (Choosing the larger integer). *Consider a two-player zero-sum one-shot game in which each player has to select a natural number and Player I wins if she selects the larger number (and loses otherwise). The value of all mixed moves of Player I is 0. Indeed, when Player II knows Player I's mixed move, she (Player II) can respond with a sufficiently high number and guarantee winning with an arbitrarily high probability. Similarly, the value of all mixed moves of Player II is 1. In particular, the game does not have a value.*

Formally, the game is a two-player zero-sum one-shot game with $X_I = X_{II} = \mathbb{N}$ *and a payoff function* h *given by*

$$h(a_I, a_{II}) = \begin{cases} 1, & \text{if } a_I > a_{II}, \\ 0, & \text{if } a_I \leq a_{II}. \end{cases}$$

Consider an arbitrary mixed move $x_I \in \Delta(X_I)$ *of Player I. For each* $n \in \mathbb{N}$*, let* $A_n := \{k \in \mathbb{N} : k \geq n\}$*. The sets* $(A_n)_{n \in \mathbb{N}}$ *are decreasing and their intersection is empty. It follows that* $\lim_{n \to \infty} x_I(A_n) = 0$*, and therefore there exists* $n = n(x_I)$ *such that* $x_I(A_n) < \frac{1}{3}$*. Let* $a_{II} = n(x_I) \in \mathbb{N} = X_{II}$*; this is a move of Player II in the game* (X_I, X_{II}, h)*. By definition,* $\mathbf{E}_{x_I, a_{II}}[h] < \frac{1}{3}$*, and therefore* $\text{val}_O(x_I) < \frac{1}{3}$*. Since* x_I *is arbitrary,*

$$\underline{v}_O(h) = \sup_{x_I \in \Delta(X_I)} \text{val}_O(x_I) \leq \frac{1}{3}.$$

A similar argument shows that for every mixed move $x_{II} \in \Delta(X_{II})$ *there is a move* $a_I \in X_I$*, such that* $\mathbf{E}_{a_I, x_{II}}[h] > \frac{2}{3}$*, which implies that* $\overline{v}_O(h) \geq \frac{2}{3}$*. Consequently, the one-shot game* (X_I, X_{II}, h) *does not have a value.*

Example 5.18 shows that Theorem 5.17 fails when both players have infinite sets of moves. When one of the sets X_I and X_{II} is infinite and the other set is finite, the value exists. Moreover, the player whose set of moves is finite has a 0-optimal mixed move, while the other player may not have a 0-optimal mixed move, see Exercise 5.2.

Readers who are interested to learn more about one-shot games are referred to Chapters 4 and 5 in Maschler, Solan, and Zamir (2020).

5.2 PROBABILITY DISTRIBUTIONS ON TREES

In this section we define the concept of a probability function over a tree. A probability function assigns to every position a probability distribution on the set of possible moves at that position.

Definition 5.19 (Probability function). *Let T be a tree over an alphabet $\mathcal{A}$. A probability function (over T) is a function $\mu : T \to \Delta(\mathcal{A})$ with the property that*

$$(\mu(p))(\{a \in \mathcal{A} \colon \langle p, a \rangle \in T\}) = 1, \quad \forall p \in T.$$

The interpretation of a probability function μ is that at position p, a move $a \in \mathcal{A}$ is selected according to the probability distribution $\mu(p)$. In particular, for every position $p \in T$, the probability that under μ the position $p = \langle a_0, a_1, \ldots, a_n \rangle$ is reached is given by

$$P(\mu, p) := (\mu(\langle \, \rangle))(a_0) \cdot (\mu(\langle a_0 \rangle))(a_1) \cdot \ldots \cdot (\mu(\langle a_0, a_1, \ldots, a_{n-1} \rangle))(a_n).$$

Definition 5.20 ($\mathbf{P}_\mu$). *Let T be a tree over a finite alphabet $\mathcal{A}$, let $\mu : T \to \mathcal{A}$ be a probability function, and let $\mathcal{Y}_T$ be the algebra generated by the sets $[T_p]$ with $p \in T$. Define a function $\mathbf{P}_\mu : \mathcal{Y}_T \to [0,1]$ by*

$$\mathbf{P}_\mu([T_p]) := P(\mu, p), \quad \forall p \in T, \tag{5.5}$$

and for every finite union of disjoint sets $B = \bigcup_{k=0}^{K}[T_{p_k}]$, define

$$\mathbf{P}_\mu(B) := \sum_{k=0}^{K} \mathbf{P}_\mu([T_{p_k}]). \tag{5.6}$$

This definition implies that the function $\mathbf{P}_\mu$ is a finitely additive probability on $([T], \mathcal{Y}_T)$. The next lemma, whose proof is left to the reader (Exercise 5.4), states that in fact it is countably additive.

Lemma 5.21. *Let T be a tree over a finite alphabet $\mathcal{A}$, let $\mu : T \to \mathcal{A}$ be a probability function, and let $\mathcal{Y}_T$ be the algebra generated by the sets $[T_p]$ with $p \in T$. Then for every sequence $(A_k)_{k \in \mathbb{N}}$ of pairwise disjoint subsets of $\mathcal{Y}_T$ that satisfy $\bigcup_{k \in \mathbb{N}} A_k \in \mathcal{Y}_T$ we have*

$$\sum_{k \in \mathbb{N}} \mathbf{P}_\mu(A_k) = \mathbf{P}_\mu \left(\bigcup_{k \in \mathbb{N}} A_k \right).$$

Recall that $\mathcal{B}(T)$ is the Borel sigma-algebra of T (see Definition 3.14). Lemma 5.21 and Theorem 2.23 imply that every probability function μ defines a probability distribution on $([T], \mathcal{B}(T))$, which is the content of the next result.

Theorem 5.22 (Kolmogorov's Extension Theorem). *Let T be a tree over a finite alphabet $\mathcal{A}$ and let $\mu : T \to \mathcal{A}$ be a probability function. The finitely additive probability $\mathbf{P}_\mu$ can be extended in a unique way to a (countably additive) probability distribution on $\mathcal{B}(T)$.*

5.3 THE MODEL OF TWO-PLAYER ZERO-SUM BLACKWELL GAMES

In this section we formally define two-player zero-sum Blackwell games.

Definition 5.23 (Blackwell game). *A two-player zero-sum Blackwell game (over an alphabet $\mathcal{A}$) is a pair $\Gamma = (T, f)$, where*

- *T is a tree over $\mathcal{A}^2$.*

- *For every $p \in T$, there are nonempty finite sets $X_{\mathrm{I}}^p, X_{\mathrm{II}}^p \subseteq \mathcal{A}$, such that*

$$\{p' \in T \colon p \prec p', \operatorname{len}(p') = \operatorname{len}(p) + 1\}$$
$$= \{\langle p, (a_{\mathrm{I}}, a_{\mathrm{II}})\rangle \colon a_{\mathrm{I}} \in X_{\mathrm{I}}^p, a_{\mathrm{II}} \in X_{\mathrm{II}}^p\}. \tag{5.7}$$

- *$f : [T] \to \mathbb{R}$ is a bounded and Borel-measurable payoff function.*

The interpretation of Definition 5.23 is as follows. In a Blackwell game, the players select their moves simultaneously. Hence, the tree T is over $\mathcal{A}^2$: the first (resp., second) coordinate represents the set of all possible moves of Player I (resp., Player II). At each position $p \in T$, the sets of available moves to the two players are X_{I}^p and X_{II}^p. Therefore the set of positions that extend p by one is the set on the right-hand side of Equation (5.7).

Note that a position in T is a finite sequence of pairs of moves, and not a finite sequence of single moves, as in the games we studied in Chapters 3 and 4.

Notation 5.24 ($\vec{a}$). *Let $\Gamma = (T, f)$ be a two-player zero-sum Blackwell game. In accordance with Notation 5.4, $\vec{a} = (a_{\mathrm{I}}, a_{\mathrm{II}})$ represents a pair of moves, one for each player. When p is a position of length n, we will denote it by $p = \langle \vec{a}_0, \vec{a}_1, \ldots, \vec{a}_{n-1}\rangle$.*

Notation 5.25 (X^p). *Let $\Gamma = (T, f)$ be a two-player zero-sum Blackwell game. For every position $p \in T$, denote*

$$X^p := \prod_{i \in \{\mathrm{I}, \mathrm{II}\}} X_i^p.$$

This is the finite set of all vectors of possible moves at position p.

Remark 5.26 (Alternating-move game). *To emphasize the difference between games as studied in Chapters 3 and 4, where the players move sequentially, and Blackwell games, where players select their moves simultaneously, we call the former alternating-move games. For the same reason, we denote Blackwell games by Γ, while alternating-move games were denoted by G.*

Remark 5.27 (Delayed observation). *We can view a Blackwell game as a game where the players select their moves alternately, as was done in Chapters 3 and 4, but Player II obtains her information with delay: she observes the move that was selected by Player I in stage $2n$ at the beginning of stage $2n + 3$ (and not at the beginning of stage $2n + 1$). Games where players obtain information with delay will be studied in Chapter 7.*

We can now define the concept of *behavior strategies*, which are strategies in Blackwell games that allow randomization.

Definition 5.28 (Behavior strategy). *Let $\Gamma = (T, f)$ be a two-player zero-sum Blackwell game and let $i \in \{I, II\}$. A behavior strategy for player i is a function σ_i whose domain is T, such that $\sigma_i(p)$ is a probability distribution on X_i^p, for every position $p \in T$. The set of all behavior strategies for each player $i \in \{I, II\}$ is denoted $\Sigma_i(T)$.*

Remark 5.29 (Pure strategies, $S_i(T)$). *Behavior strategies that do not use randomization are called* pure *strategies. When we want to emphasize that a strategy of player i is pure, we will denote it by s_i. The set of all pure strategies of player i will be denoted by $S_i(T)$.*

Definition 5.30 (The standard filtration $(\mathcal{F}^n)_{n \in \mathbb{N}}$). *Let $\Gamma = (T, f)$ be a two-player zero-sum Blackwell game. For each $n \in \mathbb{N}$, denote by $\mathcal{F}^n$ the algebra on $[T]$ generated by the positions of length n, that is, the algebra generated by the sets $[T_p]$, for all positions p with length $\text{len}(p) = n$. The* standard filtration *over $([T], \mathcal{B}(T))$ is the filtration $(\mathcal{F}^n)_{n \in \mathbb{N}}$.*

Remark 5.31 (A behavior strategy as a stochastic process). *Let $\Gamma = (T, f)$ be a two-player zero-sum Blackwell game. For each player $i \in \{I, II\}$, a behavior strategy $\sigma_i \in \Sigma_i(T)$ can be viewed as a stochastic process adapted to the filtration $(\mathcal{F}^n)_{n \in \mathbb{N}}$, with values in $\bigcup_{p \in T} \Delta(X_i^p)$.*

Remark 5.32 (Behavior strategies as probability distributions on pure strategies). *Let $\Gamma = (G, f)$ be a Blackwell game, and let $i \in \{I, II\}$. The set of pure strategies $S_i(T)$ of player i is equivalent to the set $\prod_{p \in T} X_i^p$. If X_i^p is equipped with the discrete topology for each $p \in T$ and $\prod_{p \in T} X_i^p$ with the product topology, then the space $\prod_{p \in T} X_i^p$ becomes a topological space.*

Every behavior strategy $\sigma_i \in \Sigma_i(T)$ induces a probability distribution on $S_i(T)$. Indeed, a basic open set in $S_i(T)$ is determined by the moves $(a_i^{p_0}, a_i^{p_1}, \ldots, a_i^{p_{k-1}})$ selected by player i in some finite set $\{p_0, p_1, \ldots, p_{k-1}\}$ of positions in T. The probability of this set is the product $\prod_{l=0}^{k-1} (\sigma_i(p_l))(a_i^{p_l})$. By Theorem 5.22, σ_i induces a probability distribution on $S_i(T)$.

In fact, every probability distribution on $S_i(T)$ is induced by some behavior strategy in $\Sigma_i(T)$. Let us show that for $i = I$. Suppose μ is a probability distribution on $S_I(T)$. For each position $p = \langle \vec{a}_0, \vec{a}_1, \ldots, \vec{a}_n \rangle \in T$ we can calculate the probability $\mu(p)$ that under μ Player I selects her moves along p:

$$\mu(\{s_I \in S_I(T) : s_I(\langle \ \rangle) = a_{0,I}, s_I(\langle \vec{a}_0 \rangle) = a_{1,I}, \ldots, s_I(\langle \vec{a}_0, \vec{a}_1, \ldots, \vec{a}_{n-1} \rangle) = a_{n,I}\}).$$

One can now define a behavior strategy σ_I^μ as follows: For every position $p = \langle \vec{a}_0, \vec{a}_1, \ldots, \vec{a}_{n-1} \rangle \in T$ such that $\mu(p) > 0$, and every move $a_{n,I} \in X_I^p$, set

$$(\sigma_I^\mu(p))(a_{n,I}) := \frac{\mu(\langle p, (a_{n,I}, a_{n,II}) \rangle)}{\mu(p)},$$

which is independent of $a_{n,II} \in X_{II}^p$. At all positions $p \in T$ for which $\mu(p) = 0$, *the strategy σ_I^μ can be defined arbitrarily. The definition implies that the probability distribution that σ_I^μ induces on $S_I(T)$ is μ.*

A probability distribution on pure strategies is called a mixed strategy. *The equivalence between behavior strategies and mixed strategies is called* Kuhn's Theorem *(Kuhn (1957); see also Aumann (1964) and Chapter 6 in Maschler, Solan, and Zamir (2020)).*

Notation 5.33 ($\sigma, \mathbf{P}_\sigma, \mathbf{E}_\sigma[f]$). *A pair of behavior strategies in a two-player zero-sum Blackwell game will be denoted $\sigma = (\sigma_I, \sigma_{II})$. Every such pair of behavior strategies $\sigma = (\sigma_I, \sigma_{II})$ induces a conditional probability $\mu_\sigma(\cdot \mid p) = \sigma_I(p) \otimes \sigma_{II}(p)$ on X^p, for every $p \in T$, which, by Theorem 5.22, induce a probability measure $\mathbf{P}_\sigma$ on $[T]$. The payoff induced by σ is the expectation of f with respect to the probability measure $\mathbf{P}_\sigma$, denoted by $\mathbf{E}_\sigma[f]$. In Blackwell games, the payoff function f is bounded and Borel measurable, hence the payoff $\mathbf{E}_\sigma[f]$ is well-defined for every vector of behavior strategies σ.*

Example 5.34. *Consider the tree $T = (\{0,1\} \times \{0,1\})^{<\infty}$. That is, at every position $p \in T$ each player has two possible moves: $X_I^p = X_{II}^p = \{0,1\}$. A possible behavior strategy of Player I is the constant strategy $\sigma_I \equiv (\frac{1}{2}, \frac{1}{2})$ defined by $\sigma_I(p)(a_I) = \frac{1}{2}$, for every $p \in T$ and every $a_I \in X_I^p = \{0,1\}$. Let $\sigma_{II} \equiv (\frac{1}{3}, \frac{2}{3})$ be Player II's behavior strategy, defined similarly. The law $\mathbf{P}_{\sigma_I,\sigma_{II}}$ of the play induced by (σ_I, σ_{II}) is the countable-fold product of the measure $(\frac{1}{2}, \frac{1}{2}) \otimes (\frac{1}{3}, \frac{2}{3})$.*

As an example of a payoff function take the function $f : [T] \to \mathbb{R}$ defined by

$$f(\langle \vec{a}_0, \vec{a}_1, \ldots \rangle) = \limsup_{n \to \infty} \frac{1}{n} \sum_{k<n} (a_{k,I} - a_{k,II}).$$

Then $\mathbf{E}_{\sigma_I,\sigma_{II}}[f] = \frac{1}{2} - \frac{2}{3} = -\frac{1}{6}$.

The concepts of the value guaranteed by a strategy, the maxmin value, the minmax value, the value, and ε-optimal strategies were defined for zero-sum games in Definitions 4.9, 4.10, 4.12, and 4.18. We adapt these concepts to two-player zero-sum Blackwell games.

Definition 5.35 (Value of a strategy). *Let $\Gamma = (T, f)$ be a two-player zero-sum Blackwell game. The* value *of a behavior strategy $\sigma_I \in \Sigma_I(T)$ is the highest amount that σ_I guarantees against any behavior strategy of Player II:*

$$\mathrm{val}(\sigma_I) := \inf_{\sigma_{II} \in \Sigma_{II}(T)} \mathbf{E}_{\sigma_I,\sigma_{II}}[f].$$

The value *of a behavior strategy $\sigma_{II} \in \Sigma_{II}(T)$ is the lowest amount that σ_{II} guarantees against any behavior strategy of Player I:*

$$\mathrm{val}(\sigma_{II}) := \sup_{\sigma_I \in \Sigma_I(T)} \mathbf{E}_{\sigma_I,\sigma_{II}}[f].$$

Remark 5.36 (Pure strategies as extreme points of behavior strategies). *Recall that a mixed strategy of player i in a two-player zero-sum Blackwell game is a probability* *distributions on $S_i(T)$. The space of probability distributions on $S_i(T)$ is convex, as one can calculate the weighted average of every two probability distributions, and its set of extreme points is $S_i(T)$.*

As explained in Remark 5.32, every mixed strategy is equivalent to (possibly some) behavior strategies. When identifying behavior strategies that are equivalent to the same mixed strategy, we obtain a quotient space that is isomorphic to the space of mixed strategies. In this sense, one can view the set of pure strategies as extreme points of the set of behavior strategies.

Since the expectation operator is linear in the underlying probability distribution, it follows that for every behavior strategy $\sigma_{\mathrm{I}} \in \Sigma_{\mathrm{I}}(T)$ we have

$$\inf_{\sigma_{\mathrm{II}} \in \Sigma_{\mathrm{II}}(T)} \mathbf{E}_{\sigma_{\mathrm{I}}, \sigma_{\mathrm{II}}}[f] = \inf_{s_{\mathrm{II}} \in S_{\mathrm{II}}(T)} \mathbf{E}_{\sigma_{\mathrm{I}}, s_{\mathrm{II}}}[f]. \tag{5.8}$$

Definition 5.37 (Maxmin value, minmax value). *Let $\Gamma = (T, f)$ be a two-player zero-sum Blackwell game. The* maxmin value *is*

$$\underline{v}(\Gamma) := \sup_{\sigma_{\mathrm{I}} \in \Sigma_{\mathrm{I}}(T)} \mathrm{val}(\sigma_{\mathrm{I}}) = \sup_{\sigma_{\mathrm{I}} \in \Sigma_{\mathrm{I}}(T)} \inf_{\sigma_{\mathrm{II}} \in \Sigma_{\mathrm{II}}(T)} \mathbf{E}_{\sigma_{\mathrm{I}}, \sigma_{\mathrm{II}}}[f]. \tag{5.9}$$

The minmax value *is*

$$\overline{v}(\Gamma) := \inf_{\sigma_{\mathrm{II}} \in \Sigma_{\mathrm{II}}(T)} \mathrm{val}(\sigma_{\mathrm{II}}) = \inf_{\sigma_{\mathrm{II}} \in \Sigma_{\mathrm{II}}(T)} \sup_{\sigma_{\mathrm{I}} \in \Sigma_{\mathrm{I}}(T)} \mathbf{E}_{\sigma_{\mathrm{I}}, \sigma_{\mathrm{II}}}[f]. \tag{5.10}$$

The minmax value of a player is always larger than or equal to her maxmin value. We proved this result for zero-sum games (Theorem 4.11), and the same argument is valid for Blackwell games.

Definition 5.38 (Value). *Let Γ be a two-player zero-sum Blackwell game. The real number $v = v(\Gamma)$ is the* value *of Γ if*

$$v(\Gamma) = \underline{v}(\Gamma) = \overline{v}(\Gamma).$$

Definition 5.39 (ε-optimal strategies). *Let $\Gamma = (T, f)$ be a two-player zero-sum Blackwell game that has a value. For every $\varepsilon \geq 0$, a behavior strategy $\sigma_{\mathrm{I}} \in \Sigma_{\mathrm{I}}(T)$ is called ε-optimal if*

$$\mathrm{val}(\sigma_{\mathrm{I}}) \geq v(\Gamma) - \varepsilon,$$

and a behavior strategy $\sigma_{\mathrm{II}} \in \Sigma_{\mathrm{II}}(T)$ is called ε-optimal if

$$\mathrm{val}(\sigma_{\mathrm{II}}) \leq v(\Gamma) + \varepsilon.$$

Remark 5.40 (One-shot games as Blackwell games). *Every two-player zero-sum one-shot game can be viewed as a two-player zero-sum Blackwell game in which the moves of the players after stage 0 are singletons. Thus, for such games we defined two concepts of value, one in Definition 5.8 and one in Definition 5.38. Not surprisingly, the two concepts coincide.*

Remark 5.41 (Alternating-move games as Blackwell games). *Every alternating-move game can be viewed as a two-player zero-sum Blackwell game, in which in each even (resp., odd) stage Player II (resp., Player I) has a single possible move.*

Remark 5.42 (Behavior strategies in alternating-move games). *We could define the concept of behavior strategies for alternating-move games, and ask whether the value in pure strategies (that do not use randomization) coincides with the value when randomization is allowed. This is indeed the case, see Exercise 5.5.*

Example 5.43 (A simultaneous move stopping game). *In a simultaneous move stopping game, each player selects at every stage one of two possible moves,* continue *or* stop, *and the play effectively terminates once at least one player stops; that is, the payoff of both players is determined by the stage in which a player first stops, and by the identity of the player(s) who stopped at that stage.*

In this example, we consider a stopping game where the payoff function is as follows. Denote by $n \in \mathbb{N} \cup \{\infty\}$ the first stage in which some player stops.

- *If $n = \infty$ (no player ever stops), the payoff is 0.*

- *If $n < \infty$ and at stage n both players stop, the payoff is 0.*

- *Otherwise, the payoff is 1.*

Thus, in this game, Player I wants the play to terminate by a single player, while Player II wants the play to continue forever or to terminate by both players simultaneously.

We will show that the value of this game is 1. Since the maximal payoff is 1, it is sufficient to show that for every $\varepsilon > 0$ Player I has a behavior strategy σ_I^ε such that $\mathrm{val}\,(\sigma_I^\varepsilon) \geq 1 - \varepsilon$.

Fix then $\varepsilon > 0$ and consider the following strategy $\sigma_I^\varepsilon \in \Sigma_I(T)$: At every stage n, stop *with probability ε and* continue *with probability $1 - \varepsilon$. We argue that for every strategy $\sigma_{II} \in \Sigma_{II}(T)$ we have $\mathbf{E}_{\sigma_I, \sigma_{II}}[f] \geq 1 - \varepsilon$. We show this for strategies $s_{II} \in S_{II}(T)$, see Equation (5.8).*

Let then s_{II} be a pure strategy of Player II. If s_{II} always continues, then under $(\sigma_I^\varepsilon, s_{II})$ Player I stops alone with probability 1, and so $\mathbf{E}_{\sigma_I^\varepsilon, s_{II}}[f] = 1$.

Suppose next that under s_{II}, Player II stops at stage n, if the game has not terminated before that stage. Under σ_I^ε, with probability $1 - (1 - \varepsilon)^n$ Player I stops before stage n, with probability $\varepsilon(1-\varepsilon)^n$ she stops at stage n, and with the remaining probability $(1-\varepsilon)^{n+1}$ she continues in the first $n+1$ stages. The outcome is 0 only in the second case, where both players stop at stage n. It follows that $\mathbf{E}_{\sigma_I^\varepsilon, s_{II}}[f] = 1 - \varepsilon(1-\varepsilon)^n \geq 1 - \varepsilon$.

The following example is the Big Match game, which was introduced by Gillette (1957) and analyzed by Blackwell and Ferguson (1968).

Example 5.44 ("Big Match"). *The Big Match is a two-player zero-sum Blackwell game played by Robber and Hakram. Every day Hakram selects L or R, and Robber tries to predict Hakram's choice; If Hakram selects L and Robber is correct, then Robber wins a point. If Hakram selects R and Robber is correct, then Robber wins c points, where $c > 0$. If Robber's prediction is incorrect, Robber wins nothing. This continues as long as Robber predicts R. Yet if Robber ever predicts L, all future choices for both players are required to be the same as that day's choices. Thus, if Robber is correct on that day, he wins a point every day thereafter; if he is wrong on that day,* *he wins zero every day thereafter. The payoff, which Robber tries to maximize and Hakram tries to minimize, is* $\limsup_{n\to\infty}(u_0 + \cdots + u_n)/(n+1)$, *where* u_k *is the number of points Robber wins on the* k'*th day.*

In this game, $\mathcal{A} = \{L, R\}$. *The sets of moves of the players are* $X^p_{\text{Robber}} = X^p_{\text{Hakram}} = \{L, R\}$ *for every position* p *along which Robber always selected* R, *and they are a singleton for every position* p *along which Robber selected* L *at least once.*

An alternative way to present the game is by the (2×2)-*matrix that is displayed in Figure 19. The rows of the matrix reflect the moves of Robber, its columns reflect the moves of Hakram, each entry reflects the number of points Robber gains if the corresponding moves are selected, and an asterisk in an entry indicates that once this entry is selected, the game essentially terminates because the players have no choices afterwards.*

Hakram

	L	R
R	0	c
L	1 *	0 *

Robber

Figure 19: The Big Match.

We argue that the minmax value of the Big Match is $\frac{c}{1+c}$. *We first show that* $\overline{v} \leq \frac{c}{1+c}$. *To this end we construct a strategy* σ^*_{Hakram} *for Hakram such that* $\text{val}\,(\sigma^*_{\text{Hakram}}) \leq \frac{c}{1+c}$. *Suppose that at every stage Hakram selects* L *with probability* $\frac{c}{1+c}$, *and* R *with probability* $\frac{1}{1+c}$. *If at some stage Robber selects* L, *then with probability* $\frac{c}{1+c}$ *he wins* 1 *in every subsequent stage, and with probability* $\frac{1}{1+c}$ *he wins* 0 *in every subsequent stage, hence the expected payoff is* $\frac{c}{1+c}$. *If Robber always selects* R, *then by the strong law of large numbers, with probability* 1 *the limit* $\lim_{n\to\infty}(u_0 + \cdots + u_n)/(n+1)$ *exists and is equal to* $\frac{c}{1+c}$. *It follows that* $\text{val}\,(\sigma^*_{\text{Hakram}}) \leq \frac{c}{1+c}$, *as claimed.*

We next show that $\overline{v} \geq \frac{c}{1+c}$. *To this end, for every strategy* σ_{Hakram} *of Hakram we construct a strategy* σ_{Robber} *of Robber such that*

$$\mathbf{E}_{\sigma_{\text{Hakram}},\sigma_{\text{Robber}}}\left[\limsup_{n\to\infty}\frac{u_0 + \cdots + u_n}{n+1}\right] \geq \frac{c}{1+c}.$$

Define σ_{Robber} *as follows. For every* $p \in T$, *denote by* $\sigma_{\text{Hakram}}(p)$ *the probability by which Hakram selects the move* L *at position* p.

- *If* $\sigma_{\text{Hakram}}(p) \geq \frac{c}{1+c}$, *Robber selects* L *at position* p.

- *If* $\sigma_{\text{Hakram}}(p) < \frac{c}{1+c}$, *Robber selects* R *at position* p.

Denote by A *the event that Robber ever selects* L:

$$A := \{a_{n,\text{Robber}} = L \text{ for some } n \in \mathbb{N}\},$$

and by μ *the probability of this event under* $(\sigma_{\text{Hakram}}, \sigma_{\text{Robber}})$:

$$\mu := \mathbf{P}_{\sigma_{\text{Hakram}},\sigma_{\text{Robber}}}(A).$$

We have

$$\mathbf{E}_{\sigma_{\text{Hakram}}, \sigma_{\text{Robber}}} \left[\limsup_{n \to \infty} \frac{u_0 + \cdots + u_n}{n+1} \right]$$

$$= \mu \cdot \mathbf{E}_{\sigma_{\text{Hakram}}, \sigma_{\text{Robber}}} \left[\limsup_{n \to \infty} \frac{u_0 + \cdots + u_n}{n+1} \mid A \right]$$

$$+ (1 - \mu) \cdot \mathbf{E}_{\sigma_{\text{Hakram}}, \sigma_{\text{Robber}}} \left[\limsup_{n \to \infty} \frac{u_0 + \cdots + u_n}{n+1} \mid A^{\mathrm{c}} \right].$$

By the definition of σ_{Robber}, *if* $\mathbf{P}_{\sigma_{\text{Hakram}}, \sigma_{\text{Robber}}}(A) > 0$, *then*

$$\mathbf{E}_{\sigma_{\text{Hakram}}, \sigma_{\text{Robber}}} \left[\limsup_{n \to \infty} \frac{u_0 + \cdots + u_n}{n+1} \mid A \right] \geq \frac{c}{1+c},$$

and since on A^{c} *Robber never selects* L, *if* $\mathbf{P}_{\sigma_{\text{Hakram}}, \sigma_{\text{Robber}}}(A^{\mathrm{c}}) > 0$, *then by the strong law of large numbers,*

$$\mathbf{E}_{\sigma_{\text{Hakram}}, \sigma_{\text{Robber}}} \left[\limsup_{t \to \infty} \frac{u_0 + \cdots + u_n}{n+1} \mid A^{\mathrm{c}} \right] \geq \frac{c}{1+c}.$$

The claim follows.

Remark 5.45 (The minmax value process as a sub-martingale). *Let* $\Gamma = (T, f)$ *be a two-player zero-sum Blackwell game. Recall that* $\overline{v}(\Gamma_p)$ *is the minmax value in the subgame that starts at position* p. *For each position* $p \in T$, *let* $x^p \in \Delta(X_{\mathrm{I}}^p) \times \Delta(X_{\mathrm{II}}^p)$ *be a pair of mixed moves that satisfies*

$$\overline{v}(\Gamma_p) \leq \sum_{\vec{a} = (a_{\mathrm{I}}, a_{\mathrm{II}}) \in X^p} x_{\mathrm{I}}^p(a_{\mathrm{I}}) \cdot x_{\mathrm{II}}^p(a_{\mathrm{II}}) \cdot \overline{v}(\Gamma_{\langle p, (a_{\mathrm{I}}, a_{\mathrm{II}}) \rangle}).$$

In Exercise 5.6 we establish the existence of such a pair of mixed moves. For each $i = \mathrm{I}, \mathrm{II}$, *let* σ_i *be the strategy of player* i *where she selects at each position* $p \in T$ *the mixed move* x_i^p, *and let* $\sigma = (\sigma_{\mathrm{I}}, \sigma_{\mathrm{II}})$. *Define a sequence* $(X_n)_{n \in \mathbb{N}}$ *by*

$$X_n(\langle \vec{a}_0, \vec{a}_1, \dots \rangle) := \overline{v}(\langle \vec{a}_0, \vec{a}_1, \dots, \vec{a}_n \rangle), \quad \forall \langle \vec{a}_0, \vec{a}_1, \dots \rangle \in [T], \forall n \in \mathbb{N}.$$

Then $(X_n)_{n \in \mathbb{N}}$ *is a sub-martingale on the probability space* $([T], \mathcal{B}, \mathbf{P}_{\sigma})$.

5.4 EXISTENCE OF THE VALUE OF BLACKWELL GAMES

This section is devoted to the proof of the following result, which is the analog of Theorems 3.37 and 5.17 for Blackwell games.

Theorem 5.46 (Martin, 1998). *All two-player zero-sum Blackwell games have a value.*

Recall that Example 5.18 shows that Theorem 5.46 would not hold if we had allowed players to have countably infinite sets of moves in Blackwell games; our definition only allows for finitely many moves.

Before proving Theorem 5.46, we will show that without loss of generality, we can assume that the payoffs are between 0 and 1.

Remark 5.47 (The game $(T, cf + d)$). *Given two real numbers $c > 0$ and $d \in \mathbb{R}$, and a two-player zero-sum Blackwell game $\Gamma = (T, f)$, consider the two-player zero-sum Blackwell game $\Gamma' = (T, cf + d)$. The games Γ and Γ' have the same sets of behavior strategies, and for all pairs of behavior strategies $(\sigma_{\mathrm{I}}, \sigma_{\mathrm{II}}) \in \Sigma_{\mathrm{I}}(T) \times \Sigma_{\mathrm{II}}(T)$ we have*

$$\mathbf{E}_{\sigma_{\mathrm{I}}, \sigma_{\mathrm{II}}}[cf + d] = c\mathbf{E}_{\sigma_{\mathrm{I}}, \sigma_{\mathrm{II}}}[f] + d.$$

Therefore, if $c > 0$, then $\overline{v}(\Gamma') = c\overline{v}(\Gamma) + d$ and $\underline{v}(\Gamma') = c\underline{v}(\Gamma) + d$.

As a result of Remark 5.47 we deduce the following.

Lemma 5.48. *For every two-player zero-sum Blackwell game $\Gamma = (T, f)$ and all real numbers $c > 0$ and $d \in \mathbb{R}$, the game Γ has a value if and only if the game $\Gamma' = (T, cf + d)$ has a value. Moreover, if the two games have a value, than $v(\Gamma') = cv(\Gamma) + d$.*

Proof of Theorem 5.46. Throughout the proof, let $\Gamma = (T, f)$ be a fixed two-player zero-sum Blackwell game. By Lemma 5.48, we can assume without loss of generality that $0 \leq f \leq 1$.

The structure of the proof is as follows. For every $u \in (0, 1]$ we will define an auxiliary alternating-move game G_u (Step 1), and we will prove that this game is well-defined (Steps 2 and 3). We will then prove that there is $u^* \in (0, 1]$ such that Player I wins G_u if $u < u^*$, and Player II wins G_u if $u > u^*$ (Step 4). We will finally prove that u^* is the value of the original game Γ (Steps 5 and 6). Though the structure of the proof is similar to the structure of the proof of Theorem 4.15, the auxiliary game G_u that we will define here is much more complicated than the one defined in the proof of Theorem 4.15.

Step 1: Definition of an alternating-move game $G_u = (T_u, W_u)$.

We now define for every $u \in (0, 1]$ an auxiliary alternating-move game $G_u = (T_u, W_u)$. To distinguish between the players in the original game Γ, and the players in the auxiliary game G_u, we will denote the players in G_u by Player 1 and Player 2.

In G_u, Player 2 selects at each stage $2n + 1$ a pair of moves $\vec{a}_n$, such that $\vec{a}_n \in X^{\langle \vec{a}_0, \dots, \vec{a}_{n-1} \rangle}$. That is, Player 2 constructs a play in the tree T. Player 1 selects in each stage $2n$ a payoff matrix h_n for the one-shot game with sets of moves $X_{\mathrm{I}}^{\langle \vec{a}_0, \dots, \vec{a}_{n-1} \rangle}$ and $X_{\mathrm{II}}^{\langle \vec{a}_0, \dots, \vec{a}_{n-1} \rangle}$. Thus, a play in G_u is a sequence of the form $\langle h_0, \vec{a}_0, h_1, \vec{a}_1, h_2, \dots \rangle$.

The choices of the players are not arbitrary; they have to satisfy some constraints, as follows:

- At stage 0, Player 1 selects a function $h_0 \colon X_{\mathrm{I}}^{\langle \ \rangle} \times X_{\mathrm{II}}^{\langle \ \rangle} \to [0, 1]$. The function h_0 is interpreted as a two-player zero-sum one-shot game with sets of moves $X_{\mathrm{I}}^{\langle \ \rangle}$ and $X_{\mathrm{II}}^{\langle \ \rangle}$. Denote by $v_O(h_0)$ the value of this game, which exists by Theorem 5.17. We require Player 1 to select a function h_0 that satisfies

$$v_O(h_0) \geq u.$$

- At stage 1, Player 2 selects a pair of moves $\vec{a}_0 \in X_{\mathrm{I}}^{\langle\,\rangle} \times X_{\mathrm{II}}^{\langle\,\rangle}$ such that $h_0(\vec{a}_0) > 0$. Denote
$$u_1 := h_0(\vec{a}_0).$$

- At stage 2, Player 1 selects a function $h_1 \colon X_{\mathrm{I}}^{\langle \vec{a}_0 \rangle} \times X_{\mathrm{II}}^{\langle \vec{a}_0 \rangle} \to [0,1]$ such that
$$v_O(h_1) \geq u_1.$$

- At stage 3, Player 2 selects a pair of moves $\vec{a}_1 \in X_{\mathrm{I}}^{\langle \vec{a}_0 \rangle} \times X_{\mathrm{II}}^{\langle \vec{a}_0 \rangle}$ such that $h_1(\vec{a}_1) > 0$. Denote
$$u_2 := h_1(\vec{a}_1).$$

- Similarly, for each $n \in \mathbb{N}$, at stage $2n$, Player 1 selects a function $h_n \colon X_{\mathrm{I}}^{\langle \vec{a}_0, \vec{a}_1, \ldots, \vec{a}_{n-1} \rangle} \times X_{\mathrm{II}}^{\langle \vec{a}_0, \vec{a}_1, \ldots, \vec{a}_{n-1} \rangle} \to [0,1]$ such that
$$v_O(h_n) \geq u_n, \tag{5.11}$$

and at stage $2n+1$, Player 2 selects a pair of moves $\vec{a}_n \in X_{\mathrm{I}}^{\langle \vec{a}_0, \vec{a}_1, \ldots, \vec{a}_{n-1} \rangle} \times X_{\mathrm{II}}^{\langle \vec{a}_0, \vec{a}_1, \ldots, \vec{a}_{n-1} \rangle}$ such that
$$u_{n+1} := h_n(\vec{a}_n) > 0. \tag{5.12}$$

- The winning set of Player 1 in G_u is the set W_u consisting of all the plays $\langle h_0, \vec{a}_0, h_1, \vec{a}_1, \ldots \rangle$ such that
$$\limsup_{n \to \infty} h_n(\vec{a}_n) \leq f(\langle \vec{a}_0, \vec{a}_1, \ldots \rangle).$$

We next prove that the game G_u is well-defined, that is, T_u is a tree according to Definition 3.2 (Step 2). We will then prove that W_u is Borel (Step 3).

Step 2: T_u is a tree.

Let us prove that for every $u \in (0,1]$ the set T_u of positions in G_u is a tree. To do this, we show that at every position in T_u, the player who moves at that position has a valid move.

We start with Player 1. Since $u \leq 1$ and since h_n is $[0,1]$-valued, if Player 2 has a valid move in stage $2n + 1$, then at stage $2n + 2$ Player 1 can select the function $h_{n+1} \equiv 1$, whose value is 1.

We turn to show that Player 2 also always has a valid move. Consider stage $2n+1$, and let h_n be the last move of Player 1. Since the move of Player 1 is valid, $v(h_n) > 0$. If all entries of h_n were 0, its value would be 0. Since $v(h_n) > 0$, there is at least one entry $\vec{a}_n$ such that $h_n(\vec{a}_n) > 0$, and hence Player 2 has a valid move.

Step 3: The winning set W_u is Borel.

The set W_u is given by

$$W_u = \left\{ (h_0, \vec{a}_0, h_1, \vec{a}_1, \ldots) \in [T_u] \colon \limsup_{n \to \infty} h_n(\vec{a}_n) \leq f(\langle \vec{a}_0, \vec{a}_1, \ldots \rangle) \right\}.$$

The function $(h_0, \vec{a}_0, h_1, \vec{a}_1, \dots) \mapsto h_n(\vec{a}_n)$ is continuous, hence the function $(h_0, \vec{a}_0, h_1, \vec{a}_1, \dots) \mapsto \limsup_{n \to \infty} h_n(\vec{a}_n)$ is Borel measurable (Lemma 2.14).

The function $(h_0, \vec{a}_0, h_1, \vec{a}_1, \dots) \mapsto f(\langle \vec{a}_0, \vec{a}_1, \dots \rangle)$ is the composition of the Borel-measurable function f and the projection $\varphi : (h_0, \vec{a}_0, h_1, \vec{a}_1, \dots) \mapsto \langle \vec{a}_0, \vec{a}_1, \dots \rangle$. The projection φ is continuous, hence the function $(h_0, \vec{a}_0, h_1, \vec{a}_1, \dots) \mapsto f(\langle \vec{a}_0, \vec{a}_1, \dots \rangle)$ is Borel measurable, as the composition of two Borel-measurable functions. It follows that the set W_u is Borel.

Discussion: The idea behind the definition of G_u.

To prove Theorem 5.46, we use the determinacy of alternating-move win/lose games. We therefore would like to define for every $u \in (0,1]$ an auxiliary alternating-move game G_u that satisfies the following statements:

- If Player 1 has a winning strategy in G_u, then Player I can guarantee u in Γ.

- If Player 2 has a winning strategy in G_u, then Player II can guarantee[1] u in Γ.

Let $u^* \in [0,1]$ denote the cut-off such that Player 1 wins G_u if $u < u^*$, and Player 2 wins G_u if $u > u^*$. The definition of u^* implies that it is the value of Γ.

How should we define G_u? Player I can guarantee u in Γ, if and only if we can assign a quantity $h_0(\vec{a}_0)$ to each position $\langle \vec{a}_0 \rangle$ of length 1, such that the following two properties hold:

(i) Player I can guarantee $h_0(\vec{a}_0)$ in the subgame $\Gamma_{\langle \vec{a}_0 \rangle}$.

(ii) In the two-player zero-sum one-shot game with payoff function h_0, Player I can guarantee u. In other words, $v_O(h_0) \geq u$.

Indeed, if (i) and (ii) hold, then Player I can guarantee u in Γ, by the following strategy:

- In stage 0, select an optimal strategy in the two-player zero-sum one-shot game with payoff function h_0.

- From stage 1 and on, follow a strategy that guarantees $h_0(\langle \vec{a}_0 \rangle)$ in the subgame $\Gamma_{\langle \vec{a}_0 \rangle}$.

Conversely, suppose that Player I can guarantee u in Γ, and denote by σ_I a strategy of Player I that guarantees u in Γ. Denote by $h_0(\vec{a}_0)$ the maximal amount that Player I can guarantee in $\Gamma_{\langle \vec{a}_0 \rangle}$, for every $a_0 \in X^{\langle \rangle}$. Then the mixed move $\sigma_I(\langle \rangle)$ guarantees u in the one-shot game with payoffs h_0, and hence (i) and (ii) hold.

The definition of G_u reflects the intuition provided above. Player 1 in G_u wants to prove that Player I can guarantee u in Γ. Player 2 is a quality controller, who verifies that the proof is valid. So, at every stage n, when the play is at some position p of length n, Player 1 needs to show that Player I can guarantee u_n in the subgame Γ_p. To do this, Player 1 selects quantities $(h_n(\vec{a}_n))_{\vec{a}_n \in X^p}$, which she claims that Player I

---

[1] In fact, we will be able to prove a slightly weaker claim: If Player 2 has a winning strategy in G_u, then for every $\delta > 0$ Player II can guarantee $u + \delta$ in Γ.

can guarantee in the subgames $(\Gamma_{\langle p, \vec{a}_n \rangle})_{\vec{a}_n \in X^p}$, such that the value of the two-player zero-sum one-shot game with payoff function h_n is at least u_n. Player 2 then selects a position $\langle p, \vec{a}_n \rangle$ that extends p, and asks Player 1 to continue the proof from the new position, with $u_{n+1} = h_n(\vec{a}_n)$.

When is the proof valid? When the actual payoff along the play that Player 2 selected, namely, $f(\langle \vec{a}_0, \vec{a}_1, \ldots \rangle)$, is at least the values of the subgames that Player 1 selected,[2] namely, $\limsup_{n \to \infty} u_n$.

Step 4: There is $u^* \in [0, 1]$ such that Player 1 wins G_u for all $u < u^*$, and Player 2 wins G_u for all $u > u^*$.

By Step 3 and Theorem 3.37, for every u the game G_u is determined. We argue that if Player 1 has a winning strategy in G_u, then she has a winning strategy in $G_{u'}$ for every $u' < u$. Indeed, the winning strategy in G_u is a valid strategy in $G_{u'}$, and it is a winning strategy in $G_{u'}$ (for an alternative argument of this derivation, see Exercise 5.9).

Denote

$$u^* := \sup\{u \in (0, 1] : \text{Player 1 has a winning strategy in } G_u\}. \tag{5.13}$$

This u^* satisfies that Player 1 wins in G_u, for every $u < u^*$, and Player 2 wins G_u for every $u > u^*$.

Our last goal is to show that u^* is the value of Γ. This will follow from the following two lemmas, which will be proved respectively in Steps 5 and 6.

Lemma 5.49. *If Player 1 has a winning strategy in G_u, then Player I can guarantee u in Γ: there is a behavior strategy $\sigma_I \in \Sigma_I(T)$ such that $\inf_{\sigma_{II} \in \Sigma_{II}(T)} \mathbf{E}_{\sigma_I, \sigma_{II}}[f] \geq u$.*

Lemma 5.50. *If Player 2 has a winning strategy in G_u, then, for every $\delta > 0$, Player II can guarantee $u + \delta$ in Γ: there is a behavior strategy $\sigma_{II} \in \Sigma_{II}(T)$ such that $\sup_{\sigma_I \in \Sigma_I(T)} \mathbf{E}_{\sigma_I, \sigma_{II}}[f] \leq u + \delta$.*

Step 5: Proof of Lemma 5.49.

Fix a winning strategy $s_1 \in S_1(T_u)$ for Player 1 in G_u.

Definition 5.51 (Position compatible with s_1). *For $p = \langle \vec{a}_0, \ldots, \vec{a}_{n-1} \rangle \in T$, we say that p is compatible with s_1 if there are $h_0, \ldots, h_{n-1}$, such that $\langle h_0, \vec{a}_0, \ldots, \vec{a}_{n-1}, h_{n-1} \rangle$ is consistent with s_1. Namely, $h_0 = s_1(\langle \ \rangle)$, $h_{k+1} = s_1(\langle h_0, \vec{a}_0, \ldots, h_k, \vec{a}_k \rangle)$ for each $k = 0, \ldots, n - 2$, and $h_k(\vec{a}_k) > 0$ for each $k = 0, \ldots, n - 1$.*

Remark 5.52 (Compatibility and consistency). *The concept of compatibility resembles that of consistency, yet the two are different. Indeed, so that a position can be consistent with a strategy, the two need to relate to the same game. However, the position $p = \langle \vec{a}_0, \ldots, \vec{a}_{n-1} \rangle$ is in T, while s_1 is a strategy in the game $G_u = (T_u, W_u)$.*

---

[2]The limit superior could be replaced by limit inferior as shown in Exercise 5.7.

Step 5.1: Definition of a function $H_{s_1} \colon T \to [0, 1]$.

For every stage $n \in \mathbb{N}$ and every position $p = \langle \vec{a}_0, \ldots, \vec{a}_{n-1} \rangle \in T$ that is compatible with s_1, the strategy s_1 selects a function $h_n(s_1, p) \colon X_{\mathrm{I}}^p \times X_{\mathrm{II}}^p \to [0, 1]$. We denote this function $h_n(s_1, p)$, to emphasize that it is the move selected by s_1 at p. Equivalently, s_1 defines a function $H_{s_1} \colon T \to [0, 1]$:

- $H_{s_1}(\langle \, \rangle) := u$.

- $H_{s_1}(\langle \vec{a}_0, \vec{a}_1, \ldots, \vec{a}_n \rangle) := h_n(s_1, \langle \vec{a}_0, \vec{a}_1, \ldots, \vec{a}_{n-1} \rangle)(\vec{a}_n)$, for every position $\langle \vec{a}_0, \vec{a}_1, \ldots, \vec{a}_{n-1}, \vec{a}_n \rangle \in T$ that is compatible with s_1.

- $H_{s_1}(\langle \vec{a}_0, \vec{a}_1, \ldots, \vec{a}_n \rangle) := 0$, for every position $\langle \vec{a}_0, \vec{a}_1, \ldots, \vec{a}_{n-1}, a_n \rangle \in T$ that is not compatible with s_1.

Step 5.2: Definition of a behavior strategy $\sigma_{\mathrm{I}} \in \Sigma_{\mathrm{I}}(T)$.

Define a behavior strategy $\sigma_{\mathrm{I}} \in \Sigma_{\mathrm{I}}(T)$ as follows:

- If $p \in T$ is compatible with s_1, then $\sigma_{\mathrm{I}}(p)$ is a 0-optimal mixed move for Player I in the two-player zero-sum one-shot game $h_n(s_1, p) \colon X^p \to [0, 1]$.

- If $p \in T$ is not compatible with s_1, then $\sigma_{\mathrm{I}}(p)$ is defined arbitrarily.

Step 5.3: Definition of a sequence of random variables $(X_n)_{n \in \mathbb{N}}$.

For every $n \in \mathbb{N}$, define a function $X_n \colon [T] \to [0, 1]$ as

$$X_n(\langle \vec{a}_0, \vec{a}_1, \ldots \rangle) := H_{s_1}(\langle \vec{a}_0, \ldots, \vec{a}_{n-1} \rangle). \tag{5.14}$$

For every $n \in \mathbb{N}$, the function X_n is determined by the first n coordinates of $\langle \vec{a}_0, \vec{a}_1, \ldots \rangle$. It follows that the stochastic process $(X_n)_{n \in \mathbb{N}}$ is adapted to the standard filtration $(\mathcal{F}^n)_{n \in \mathbb{N}}$.

Step 5.4: The process $(X_n)_{n \in \mathbb{N}}$ is a sub-martingale under $\mathbf{P}_{\sigma_{\mathrm{I}}, \sigma_{\mathrm{II}}}$, for every $\sigma_{\mathrm{II}} \in \Sigma_{\mathrm{II}}(T)$.

By Equation (5.14), we need to prove that

$$H_{s_1}(p) \leq \mathbf{E}_{\sigma_{\mathrm{I}}(p), \sigma_{\mathrm{II}}(p)}[H_{s_1}(\langle p, \cdot \rangle)], \quad \forall p \in T. \tag{5.15}$$

Indeed, Equation (5.15) implies that for every play $x \in [T]$,

$$X_n(x) = H_{s_1}(x_{n-1}) \leq \mathbf{E}_{\sigma_{\mathrm{I}}(x_{n-1}), \sigma_{\mathrm{II}}(x_{n-1})}[H_{s_1}(x_n)] = \mathbf{E}_{\sigma_{\mathrm{I}}, \sigma_{\mathrm{II}}}[X_{n+1} \mid \mathcal{F}^n](x).$$

It remains to show that Equation (5.15) holds for every position $p \in T$ and every mixed move $x_{\mathrm{II}} \in \Delta(X_{\mathrm{II}}^p)$. If p is not compatible with s_1, then $H_{s_1}(p) = 0$ and $H_{s_1}(\langle p, \cdot \rangle) \equiv 0$, and therefore both sides of Equation (5.15) are 0. Suppose now that $p \in T$ is compatible with s_1, and denote its length by n. Then,

$$\begin{aligned}
H_{s_1}(p) &= u_n & (5.16) \\
&\leq \operatorname{val}_O(h_n(s_1, p)) & (5.17) \\
&\leq \mathbf{E}_{\sigma_{\mathrm{I}}(p), \sigma_{\mathrm{II}}(p)}[h_n(s_1, p)(\cdot)] & (5.18) \\
&= \mathbf{E}_{\sigma_{\mathrm{I}}(p), \sigma_{\mathrm{II}}(p)}[H_{s_1}(\langle p, \cdot \rangle)], & (5.19)
\end{aligned}$$

where Equation (5.16) follows from the definitions of H_{s_1} and u_n (see Equation (5.12)), Equation (5.17) follows from Equation (5.11), Equation (5.18) holds since $\sigma_I(p)$ is optimal for Player I in the one-shot game with payoff function $h_n(s_1, p)$, and Equation (5.19) follows from the definition of H_{s_1}.

Step 5.5: $\text{val}(\sigma_I) \geq u$.

Let $X := \limsup_{n \to \infty} X_n$.[3] By the chain rule of conditional expectations, $\mathbf{E}_{\sigma_I, \sigma_{II}}[X_n] \geq \mathbf{E}_{\sigma_I, \sigma_{II}}[X_0] = u$, for every $n \in \mathbb{N}$. It follows by Fatou's lemma (Theorem 2.41) that

$$\mathbf{E}_{\sigma_I, \sigma_{II}}[X] = \mathbf{E}_{\sigma_I, \sigma_{II}}\left[\limsup_{n \to \infty} X_n\right] \geq \limsup_{n \to \infty} \mathbf{E}_{\sigma_I, \sigma_{II}}[X_n] \geq u. \tag{5.20}$$

To complete the proof we show that $X(x) \leq f(x)$, for every $x \in [T]$, which will imply, together with Equation (5.20), that

$$u \leq \mathbf{E}_{\sigma_I, \sigma_{II}}[X] \leq \mathbf{E}_{\sigma_I, \sigma_{II}}[f]. \tag{5.21}$$

If all finite prefixes of x are compatible with s_1, then $X_n = u_n$, for all $n \in \mathbb{N}$, and since s_1 is a winning strategy of Player 1 in G_u,

$$X(x) = \limsup_{n \to \infty} X_n(x) = \limsup_{n \to \infty} u_n(x) \leq f(x). \tag{5.22}$$

If some prefix of x is incompatible with s_1, then all the extensions of that prefix are incompatible with s_1. By definition, this implies that $X_n(x) = 0$ for all n sufficiently large. Since $f \geq 0$, it follows that

$$X(x) = 0 \leq f(x). \tag{5.23}$$

The claim follows by Equations (5.20)–(5.23).

Step 6: Proof of Lemma 5.50.

To prove Lemma 5.49, we fixed a winning strategy s_1 of Player 1 in G_u, and defined a strategy σ_I for Player I in Γ that satisfies two properties:

- For every strategy σ_{II} of Player II in Γ, the process H_{s_1} is a sub-martingale under (σ_I, σ_{II}).

- The payoff is at least the limit of H_{s_1}.

Unfortunately, it is not clear how to define a similar function using a winning strategy of Player 2 in G_u, because in the game Γ the information available to Player II at stage n is the sequence of past moves of the two players, while to apply the winning strategy s_2 in G_u one needs to know the sequence $h_0, h_1, \ldots, h_{n-1}$ that Player 1 selected in the past. The challenge in the proof of Lemma 5.50 is the definition of a function $H : T \to [0, 1]$ that will take the role of H_{s_1}.

---

[3]The limit superior is, in fact, a limit, by Theorem 2.40 applied to the super-martingale $(1 - X_n)_{n=0}^{\infty}$.

Let $s_2 \in S_2(T_u)$ be a winning strategy of Player 2 in the auxiliary game G_u, and let $\delta > 0$.

Step 6.1: Acceptable positions.

We here define the following objects recursively:

(AC.1) a subtree $T' \subset T$ of *acceptable* positions;[4]

(AC.2) for every acceptable position $\langle \vec{a}_0, \ldots, \vec{a}_{n-1} \rangle \in T'$, a position

$$\psi(\langle \vec{a}_0, \ldots, \vec{a}_{n-1} \rangle) = \langle h_0, \vec{a}_0, \ldots, h_{n-1}, \vec{a}_{n-1} \rangle \in T_u$$

consistent with s_1, such that the function $\psi \colon T' \to T_u$ is monotone;

(AC.3) a function $H \colon T \to (0, 1]$.

What is the role of these quantities? Roughly, we will construct a strategy s_1 of Player 1 in the game G_u, which in a sense is optimal from the point of view of Player 2; that is, among the strategies of Player 1 in G_u, it minimizes the expected payoff. The function ψ describes this strategy, as the functions $h_0, h_1, \ldots$ that are added to acceptable positions are the moves selected by s_1. A position in T is acceptable if it is the projection of some position in G_u consistent with s_2. The quantity $H(p)$ will approximate the value of Γ_p, the subgame that starts at p.

We turn to the formal definition.

Definition 5.53 (Acceptable positions). *Define the set of acceptable positions for s_2, a monotone function ψ that assigns a position in T_u to each acceptable position for s_2, and a function $H \colon T \to (0, 1]$, as follows.*

- *The empty position is acceptable for s_2. Set*

$$\psi(\langle \, \rangle) := \langle \, \rangle$$

   and

$$H(\langle \, \rangle) := u.$$

- *If $p \in T$ is unacceptable for s_2, then any extension of p is unacceptable for s_2. In this case set*

$$H(p) := 1.$$

- *Let $p = \langle \vec{a}_0, \ldots, \vec{a}_{n-1} \rangle$ be an acceptable position for s_2, and denote $\psi(\langle \vec{a}_0, \ldots, \vec{a}_{n-1} \rangle) = \langle h_0, \vec{a}_0, \ldots, h_{n-1}, \vec{a}_{n-1} \rangle$. An extension $q = \langle \vec{a}_0, \ldots, \vec{a}_{n-1}, \vec{a}_n \rangle$ of p is acceptable for s_2 if and only if there exists a payoff matrix $h \colon X_{\mathrm{I}}^p \times X_{\mathrm{II}}^p \to [0, 1]$ such that*

   AD.1) *h is a legal move of Player 1 at the position $\psi(p)$ in G_u.*

   AD.2) *$s_2(\langle \psi(p), h \rangle) = \vec{a}_n$.*

----

[4]We use the term "acceptable" (rather than "compatible"), as in the proof of Lemma 5.49, to differentiate the two concepts.

Denote by $\mathcal{H}(\langle p, \vec{a}_n \rangle)$ the set of all payoff matrices h that satisfy (AD.1)–(AD.2). *Choose $h_n \in \mathcal{H}(\langle p, \vec{a}_n \rangle)$ such that*

$$h_n(\vec{a}_n) \leq \inf_{h \in \mathcal{H}(\langle p, \vec{a}_n \rangle)} h(\vec{a}_n) + \delta \cdot 2^{-(n+1)}. \tag{5.24}$$

Note that h_n depends on s_2, p (through $\psi(p)$), $\vec{a}_n$, and δ. Define

$$H(\langle p, \vec{a}_n \rangle) := h_n(\vec{a}_n). \tag{5.25}$$

The fact that $h_n \in \mathcal{H}(\langle p, \vec{a}_n \rangle)$ implies that h_n is a legal move of Player 1 at $\psi(p)$, in which the response of Player 2 is $\vec{a}_n$. Therefore, $H(\langle p, \vec{a}_n \rangle) > 0$. Moreover, since by induction $\psi(p) \in T_u$, since h_n is a valid move of Player 1 at the position $\psi(p)$ in G_u, and since $h_n(\vec{a}_n) > 0$, it follows that $\vec{a}_n$ is a valid move of Player 2 at the position $\langle \psi(p), h_n \rangle$ in G_u, and therefore $\langle \psi(p), h_n, \vec{a}_n \rangle \in T_u$. Thus, the position $\langle \psi(p), h_n, \vec{a}_n \rangle$ is consistent with s_2 and extends $\psi(p)$. Set

$$\psi(\langle p, \vec{a}_n \rangle) := \langle \psi(p), h_n, \vec{a}_n \rangle.$$

It follows that ψ satisfies (AC.2).

The set of acceptable positions for s_2 is a subtree (Exercise 5.17).

Step 6.2: Definition of a strategy $\sigma_{II} \in \Sigma_{II}(T)$.

We here define a strategy σ_{II} of Player II in T. We will then show that when Player II adopts this strategy, the payoff in Γ is at most $u + \delta$, whatever strategy Player I adopts, thereby proving Lemma 5.50.

If $p \in T$ is unacceptable for s_2, then $\sigma_{II}(p)$ is defined arbitrarily. If $p \in T$ is acceptable for s_2, then $\sigma_{II}(p)$ is an optimal strategy of Player II in the two-player zero-sum one-shot game $h^p \colon X_I^p \times X_{II}^p \to [0,1]$, defined by

$$h^p(\vec{a}) := H(\langle p, \vec{a} \rangle), \quad \forall \vec{a} \in X^p. \tag{5.26}$$

Note that we do not claim that h^p is a valid move of Player 1 at the position $\psi(p)$ in G_u, that is, we do not claim that $v_O(h^p) \geq H(p)$. Even though we do not provide a lower bound to $v_O(h^p)$, in the next step we do provide an upper bound.

Step 6.3: For every acceptable position $p \in T$ for s_2,

$$v_O(h^p) \leq H(p) + \delta \cdot 2^{-(n+1)}.$$

Suppose to the contrary that $v_O(h^p) = H(p) + \epsilon$, for some $\epsilon > \delta \cdot 2^{-(n+1)}$. Consider the two-player zero-sum one-shot game $h \colon X_I^p \times X_{II}^p \to [0,1]$ defined by $h := \max\{0, h^p - \epsilon\}$. Since $h \geq h^p - \epsilon$, Theorem 5.12 implies that $v_O(h) \geq v_O(h^p) - \epsilon = H(p) > 0$. Therefore, h is a legal move for Player I in G_u at $\langle h_0, \vec{a}_0, \dots, h_{n-1}, \vec{a}_{n-1} \rangle$.

Denote $\vec{a}_n := s_2(\langle h_0, \vec{a}_0, \dots, h_{n-1}, \vec{a}_{n-1}, h \rangle)$. Since in G_u the pair of moves $\vec{a}_n$ is selected by s_2 at the position $\langle h_0, \vec{a}_0, \dots, h_{n-1}, \vec{a}_{n-1}, h \rangle$, it follows that $\vec{a}_n$ is a legal move for Player 2 in G_u at that position, and therefore $h(\vec{a}_n) > 0$ and $h \in \mathcal{H}(p, \vec{a}_n)$. Hence,

$$
\begin{align}
h(\vec{a}_n) &= h^p(\vec{a}_n) - \epsilon \tag{5.27}\\
&< h^p(\vec{a}_n) - \delta \cdot 2^{-(n+1)} \tag{5.28}\\
&= H(\langle p, \vec{a}_n \rangle) - \delta \cdot 2^{-(n+1)} \tag{5.29}\\
&\leq \inf_{\widehat{h} \in \mathcal{H}(p, \vec{a}_n)} \widehat{h}(\vec{a}_n), \tag{5.30}
\end{align}
$$

where Equation (5.27) follows by the definition of h, Equation (5.28) holds by the choice of ε, Equation (5.29) holds by the definition of h^p (Equation (5.26)), and Equation (5.30) holds by Equations (5.24) and (5.25). However, the list of equations (5.27)–(5.30) cannot hold because $h \in \mathcal{H}(p, \vec{a}_n)$.

We next define a stochastic process $(Y_n)_{n \in \mathbb{N}}$, which will be related to the expected payoff.

Step 6.4: Definition of a stochastic process $(Y_n)_{n \in \mathbb{N}}$.

For every $n \in \mathbb{N}$ define a function $Y_n : [T] \to \mathbb{R}$ by

$$
Y_n(\langle \vec{a}_0, \vec{a}_1, \ldots \rangle) := H(\langle \vec{a}_0, \vec{a}_1, \ldots \vec{a}_{n-1} \rangle) + \delta \cdot 2^{-n}, \quad \forall \langle \vec{a}_0, \vec{a}_1, \ldots \rangle \in [T].
$$

The function Y_n is measurable with respect to $\mathcal{F}^n$. In particular, the sequence $(Y_n)_{n \in \mathbb{N}}$ is a stochastic process, adapted to the standard filtration $\mathcal{F}^n$.

Step 6.5: The process $(Y_n)_{n \in \mathbb{N}}$ is a super-martingale under $\mathbf{P}_{\sigma_{\mathrm{I}}, \sigma_{\mathrm{II}}}$, for every $\sigma_{\mathrm{I}} \in \Sigma_{\mathrm{I}}(T)$.

Recall that for every play $x = \langle \vec{a}_0, \vec{a}_1, \ldots \rangle$, we denoted its prefix of length $n + 1$ by $x_n = \langle \vec{a}_0, \vec{a}_1, \ldots \vec{a}_n \rangle$.

Fix a strategy $\sigma_{\mathrm{I}} \in \Sigma_{\mathrm{I}}(T)$. The fact that $(Y_n)_{n \in \mathbb{N}}$ is a super-martingale follows from the following chain of equalities and inequalities, which hold for every $x \in [T]$:

$$
\begin{align}
Y_n(x) &= H(x_n) + \delta \cdot 2^{-n} \tag{5.31}\\
&\geq v_O(h^{x_n}) + \delta \cdot 2^{-(n+1)} \tag{5.32}\\
&\geq \mathbf{E}_{\sigma_{\mathrm{I}}, \sigma_{\mathrm{II}}}\left[h^{x_n}(\vec{a}_{n+1}) \mid \mathcal{F}^n\right](x) + \delta \cdot 2^{-(n+1)} \tag{5.33}\\
&= \mathbf{E}_{\sigma_{\mathrm{I}}, \sigma_{\mathrm{II}}}\left[H(x_{n+1}) \mid \mathcal{F}^n\right](x) + \delta \cdot 2^{-(n+1)} \tag{5.34}\\
&= \mathbf{E}_{\sigma_{\mathrm{I}}, \sigma_{\mathrm{II}}}\left[Y_{n+1} \mid \mathcal{F}^n\right](x), \tag{5.35}
\end{align}
$$

where Equation (5.31) follows from the definition of Y_n, Equation (5.32) follows from Step 6.3, Equation (5.33) holds since $\sigma_{\mathrm{II}}(x_n)$ is a 0-optimal strategy in the one-shot game with payoff function h^{x_n}, and Equations (5.34) and (5.35) follow from the definitions of H and Y_{n+1}, respectively.

By the Martingale Convergence Theorem (Theorem 2.40), the process $\{Y_n\}_{n \in \mathbb{N}}$ converges $\mathbf{P}_{\sigma_{\mathrm{I}}, \sigma_{\mathrm{II}}}$-a.s. to a random variable Y, and therefore the stochastic process $\{H(x_n)\}_{n \in \mathbb{N}}$ converges a.s. to Y as well.

Equations (5.31)–(5.34) also prove that

$$\mathbf{E}_{\sigma_{\mathrm{I}},\sigma_{\mathrm{II}}}[H(x_{n+1}) \mid \mathcal{F}^n] \leq H(x_n) + \delta \cdot 2^{-(n+1)}. \tag{5.36}$$

We finally prove that σ_{II} guarantees $u + \delta$ in G_u.

Step 6.6: $\mathbf{E}_{\sigma_{\mathrm{I}},\sigma_{\mathrm{II}}}[f] \leq \mathbf{E}_{\sigma_{\mathrm{I}},\sigma_{\mathrm{II}}}[Y] \leq u + \delta$ for every $\sigma_{\mathrm{I}} \in \Sigma_{\mathrm{I}}(T)$.

Fix again $\sigma_{\mathrm{I}} \in \Sigma_{\mathrm{I}}(T)$. We first argue that $Y(x) \geq f(x)$ with probability 1 under $\mathbf{P}_{\sigma_{\mathrm{I}},\sigma_{\mathrm{II}}}$. Fix for the moment a play $x = \langle \vec{a}_0, \vec{a}_1, \ldots \rangle$. If all the prefixes of x are acceptable for s_2, then $\langle h_0, \vec{a}_0, h_1, \vec{a}_1, \ldots \rangle$ is consistent with s_2 in G_u, and since s_2 is a winning strategy of Player 2 in G_u, we have $\limsup_{n\to\infty} H(x_n) \geq f(x)$. If x has a prefix that is unacceptable for s_2, then $\lim_{n\to\infty} H(x_n) = 1 \geq f(x)$. Since $Y = \lim_{n\to\infty} H(x_n)$ with probability 1 under $\mathbf{P}_{\sigma_{\mathrm{I}},\sigma_{\mathrm{II}}}$, it follows that $Y(x) \geq f(x)$ with probability 1 under $\mathbf{P}_{\sigma_{\mathrm{I}},\sigma_{\mathrm{II}}}$.

By Fatou's lemma (Theorem 2.41),

$$\mathbf{E}_{\sigma_{\mathrm{I}},\sigma_{\mathrm{II}}}[Y] \leq \liminf_{n\to\infty} \mathbf{E}_{\sigma_{\mathrm{I}},\sigma_{\mathrm{II}}}[H(x_n)].$$

In view of the law of iterated expectation, applying Equation (5.36) repeatedly, for every $n \in \mathbb{N}$ we have

$$\mathbf{E}_{\sigma_{\mathrm{I}},\sigma_{\mathrm{II}}}[H(x_{n+1})] = \mathbf{E}_{\sigma_{\mathrm{I}},\sigma_{\mathrm{II}}}\big[\mathbf{E}_{\sigma_{\mathrm{I}},\sigma_{\mathrm{II}}}[H(x_{n+1}) \mid \mathcal{F}^{n+1}]\big] \leq \mathbf{E}_{\sigma_{\mathrm{I}},\sigma_{\mathrm{II}}}[H(x_0)] + \delta = u + \delta.$$

It follows that

$$\mathbf{E}_{\sigma_{\mathrm{I}},\sigma_{\mathrm{II}}}[f] \leq \mathbf{E}_{\sigma_{\mathrm{I}},\sigma_{\mathrm{II}}}[Y] \leq u + \delta,$$

which completes the proof of Lemma 5.50, and with it the proof of Theorem 5.46. □

We now provide a result that follows from the proof of Lemma 5.49. To state this result, we need a new notation, which is borrowed from the proof of Lemma 5.50.

Notation 5.54 $(H(\langle p, * \rangle))$. *Let $\Gamma = (T, f)$ be a Blackwell game. Whenever $H : T \to \mathbb{R}$ and $p \in T$, we denote by $H(\langle p, * \rangle)$ the collection $(H(\langle p, \vec{a} \rangle))_{\vec{a} \in X^p}$, which is viewed as a payoff function of a two-player zero-sum one-shot game with sets of moves X_{I}^p and X_{II}^p.*

Theorem 5.55. *Let $\Gamma = (T, f)$ be a two-player zero-sum Blackwell game, and let $u < v(\Gamma)$. There is a function $H : T \to \mathbb{R}$ such that:*

- $H(\langle \, \rangle) = u$.

- $H(p) \leq v_O(H(\langle p, * \rangle))$, *for every $p \in T$.*

- $f(x) \geq \limsup_{n\to\infty} H(x_n)$, *for every play $x \in [T]$.*

- *For every pair of behavior strategies $(\sigma_{\mathrm{I}}, \sigma_{\mathrm{II}}) \in \Sigma_{\mathrm{I}}(T) \times \Sigma_{\mathrm{II}}(T)$ that satisfies*

$$\mathbf{E}_{\sigma_{\mathrm{I}}(p),\sigma_{\mathrm{II}}(p)}[H(\langle p, * \rangle)] \geq H(p), \quad \forall p \in T$$

we have $\mathbf{E}_{\sigma_{\mathrm{I}},\sigma_{\mathrm{II}}}[f] \geq u$.

Example 5.56 (Stopping game). *In Example 5.43 we studied a simultaneous move two-player zero-sum stopping game, and we saw that its value is 1. In the terminology of the proof of Theorem 5.46, Player 1 has a winning strategy in the game G_u, for every $u < 1$. We here define a winning strategy of Player 1 in G_u. We do so by defining a function $H : T \to [0,1]$ that satisfies (a) $H(\langle \ \rangle) = u$, (b) $v_O(H(\langle p, * \rangle)) \geq H(p)$ for every $p \in T$, and (c) $f(x) \geq \limsup_{n \to \infty} H(x_n)$ for every $x \in [T]$.*

In this example, as long as both players selected c, they can select between c and s. Once at least one player selects s, both players have a single available move – s. The payoff is 0 if both players always select c or if at the first stage in which some player selects s, it happened that both players selected s. The payoff is 1 otherwise.

In Example 5.16 we calculated that the value of the following two-player zero-sum one-shot game is $\frac{1}{2-z}$:

	c	s
c	z	1
s	1	0

Setting $u = \frac{1}{2-z}$, we find that $z = \frac{2u-1}{u}$. That is, the value of the following two-player zero-sum one-shot game is u:

	c	s
c	$\frac{2u-1}{u}$	1
s	1	0

Similarly, for every $n \geq 1$, the value of the following two-player zero-sum one-shot game is $\frac{nu-(n-1)}{u^{n-1}}$:

	c	s
c	$\frac{(n+1)u-n}{u^n}$	1
s	1	0

We use these observations to define H. Given a position $p = \langle \vec{a}_0, \vec{a}_1, \ldots, \vec{a}_{n-1} \rangle$ denote the first stage in which some player selects s by

$$k_p := \min\{k < n : a_{k,1} = s \ or \ a_{k,2} = s\}.$$

As usual, the minimum of an empty set is ∞, so that $k_p = \infty$ if along p both players always select c. Define for every position $p \in T$,

- *$H(p) = 0$ if along p the play terminates by both players stopping together, that is, if $\vec{a}_{k_p} = (s,s)$.*

- *$H(p) = 1$ if along p the play terminates by a single player, that is, if $\vec{a}_{k_p} = (s,c)$ or $\vec{a}_{k_p} = (c,s)$.*

- *$H(p) = \max\left\{ \frac{(n+1)u-n}{u^n}, 0 \right\}$, if p has length n and along p both players always select c.*

The definition implies that H satisfies (a) and (b). Since $u < 1$, for every n large enough $H(\langle \underbrace{(c,c), (c,c), \ldots, (c,c)}_{n \ times} \rangle)$ is negative, and hence H satisfies (c) as well.

5.5   SUBGAME ε-OPTIMAL STRATEGIES

In Theorem 4.29 we showed how the existence of an ε-optimal strategy in alternating-move games implies the existence of a subgame ε-optimal strategy. We will now state the analogous result for two-player zero-sum Blackwell games, Theorem 5.62.

Given a strategy $\sigma_I \in \Sigma_I(T)$ and a position $p \in T$, we denote by σ_I^p the strategy in the subgame T_p in which Player I selects the moves indicated by p until the position p is reached, and afterwards she follows σ_I. The formal definition follows.

Notation 5.57 (The restricted strategy σ_I^p). *Let $\sigma_I \in \Sigma_I(T)$ be a behavior strategy of player i in the game (T, f), and let $p = \langle \vec{a}_0, \vec{a}_1, \dots, \vec{a}_{n-1} \rangle \in T$. The restriction of σ_I to the subtree T_p is denoted $\sigma_I^p \in \Sigma_I(T_p)$. Formally,*

$$\sigma_I^p(p') = \begin{cases} \mathbf{1}_{a_{k+1,i}}, & \text{if } p' = \langle \vec{a}_0, \vec{a}_1, \dots, \vec{a}_k \rangle \text{ for some } k < n-1, \\ \sigma_I(p'), & \text{if } p \preceq p'. \end{cases}$$

Notation 5.58 (Expected payoff conditional on a position). *For every pair of behavior strategies $(\sigma_I, \sigma_{II}) \in \Sigma_I(T) \times \Sigma_{II}(T)$ and every position $p \in T$, denote by $\mathbf{E}_{\sigma_I, \sigma_{II}}[f \mid [T_p]]$ the expected payoff under (σ_I, σ_{II}) in the subgame Γ_p; that is, the expected payoff conditioned that the position p occurred. Formally,*

$$\mathbf{E}_{\sigma_I, \sigma_{II}}[f \mid [T_p]] := \mathbf{E}_{\sigma_I^p, \sigma_{II}^p}[f]. \tag{5.37}$$

Notation 5.59 (Value of a strategy in a subgame, $\mathrm{val}(\sigma_i; \Gamma_p)$). *Let $\Gamma = (T, f)$ be a two-player zero-sum Blackwell game, let $p \in T$ be a position, let $\sigma_I \in \Sigma_I(T)$ be a behavior strategy of Player I, and let $\sigma_{II} \in \Sigma_{II}(T)$ be a behavior strategy of Player II. The value of σ_I in Γ_p is*

$$\mathrm{val}(\sigma_I; \Gamma_p) := \inf_{\sigma_{II}' \in \Sigma_{II}(T)} \mathbf{E}_{\sigma_I, \sigma_{II}'}[f \mid [T_p]],$$

and the value of σ_{II} in Γ_p is

$$\mathrm{val}(\sigma_{II}; \Gamma_p) := \sup_{\sigma_I' \in \Sigma_I(T)} \mathbf{E}_{\sigma_I', \sigma_{II}}[f \mid [T_p]].$$

Remark 5.60 (The expected payoff conditional on a position and conditional expectation). *The expected payoff conditional on a position, defined in Notation 5.58, can be alternatively written using conditional expectations:*

$$\mathbf{E}_{\sigma_I, \sigma_{II}}[f \mid [T_p]] := \mathbf{E}_{\sigma_I, \sigma_{II}}[f \mid \mathcal{F}^n]([T_p]), \tag{5.38}$$

where $(\mathcal{F}^n)_{n \in \mathbb{N}}$ is the standard filtration (see Definition 5.30). Note, though, that the two definitions in Equations (5.37) and (5.38) are not the same. The two definitions coincide at positions p that are realized with positive probability under (σ_I, σ_{II}), namely, $\mathbf{P}_{\sigma_I, \sigma_{II}}([T_p]) > 0$. They do not coincide at positions p that are realized with probability 0 under (σ_I, σ_{II}). Indeed, in this case, $\mathbf{E}_{\sigma_I^p, \sigma_{II}^p}[f]$ is uniquely defined, while $\mathbf{E}_{\sigma_I, \sigma_{II}}[f \mid \mathcal{F}^n]([T_p])$ can be arbitrary.

Definition 5.61 (Subgame ε-optimal strategy). *Let $\Gamma = (T, f)$ be a two-player zero-sum blackwell game, and let $\varepsilon > 0$. A behavior strategy $\sigma_{\mathrm{I}} \in \Sigma_{\mathrm{I}}(T)$ is* subgame ε-optimal *if*

$$\mathrm{val}\,(\sigma_{\mathrm{I}}; \Gamma_p) \geq v(\Gamma_p) - \varepsilon, \quad \forall p \in T.$$

A behavior strategy $\sigma_{\mathrm{II}} \in \Sigma_{\mathrm{II}}(T)$ is subgame ε-optimal *if*

$$\mathrm{val}\,(\sigma_{\mathrm{II}}; \Gamma_p) \leq v(\Gamma_p) + \varepsilon, \quad \forall p \in T.$$

Theorem 5.62. *In every two-player zero-sum Blackwell game, both players have a subgame ε-optimal strategy, for every $\varepsilon > 0$.*

The proof of Theorem 5.62, which is analogous to the proof of Theorem 4.29, is left to the reader (Exercise 5.14). The main conceptual difference between the proofs is that since in two-player zero-sum Blackwell games the ε-optimal strategy is a behavior strategy, the quantity K defined in the proof of Theorem 4.29 is a random variable rather than a constant, and we need to show that this random variable is finite a.s.

5.6   DISCUSSION

The model of two-player zero-sum Blackwell games was first studied by Blackwell (1969), who proved that when the function f is the characteristic function of a G_δ-set (namely, a countable intersection of open sets) the game has a value. This result has been extended by Orkin (1972) to the case where f is the characteristic function of a set in the Boolean algebra generated by the G_δ-sets, by Vervoort (1996) to the case where f is the characteristic function of a $G_{\delta\sigma}$ set (namely, a countable union of G_δ sets), and finally by Martin (1998) to any bounded and Borel measurable function. Maitra and Sudderth (1998) observed that Theorem 5.46 is valid in the more general model of *stochastic games*. In this model, at every position the game is in some state, and the state changes randomly according to a probability distribution that depends on the current state and on the moves of the players. Exercise 5.15 is concerned with this extension. For a detailed treatment of stochastic games the reader is referred to the textbooks Filar and Vrieze (1997) or Solan (2022).

Though a function H as described in Theorem 5.55 is guaranteed to exist, finding it explicitly is not an easy task. In Example 5.56 we constructed H for a specific stopping game. In Exercise 5.23 we identify H when the payoff function f is the discounted payoff. Mertens and Neyman (1981) identified a function H in the context of stochastic games when f is the long-run average payoff.

5.7 EXERCISES

The solution to Exercise 5.20 requires Exercise 5.2. Exercise 5.15 is used to prove Theorem 7.27.

Exercises 5.19 and 5.21 are adapted from Martin (1998). Exercise 5.26 is inspired by Solan and Vieille (2002) and Gimbert, Renault, Sorin, Venel, and Zielonka (2016). Exercise 5.27 is adapted from Gimbert, Renault, Sorin, Venel, and Zielonka (2016).

Exercises marked with a star are more challenging than unmarked ones, while those marked with two stars are even more difficult.

1. Prove Theorem 5.12: Let (X_I, X_{II}, h) and $(X_I, X_{II}, \widehat{h})$ be two two-player zero-sum one-shot games that share the same sets of moves, and let $c \geq 0$. Prove the following assertions.

   (a) If $h(\vec{a}) \geq \widehat{h}(\vec{a})$ for every $\vec{a} \in X_I \times X_{II}$, then $v_O(h) \geq v_O(\widehat{h})$.

   (b) If $h(\vec{a}) = \widehat{h}(\vec{a}) + c$ for every $\vec{a} \in X_I \times X_{II}$, then $v_O(h) = v_O(\widehat{h}) + c$.

2. Prove the following extension of Theorem 5.17. Let X_I be a nonempty finite set, let X_{II} a countably infinite set, and let $h \colon X_I \times X_{II} \to [0,1]$ be a bounded function. Prove that the two-player zero-sum one-shot game (X_I, X_{II}, h) has a value and Player I has a 0-optimal strategy.

3. Provide an example of a two-player zero-sum one-shot game that satisfies the assumptions of Exercise 5.2 in which Player II does not have a 0-optimal strategy.

4. Prove Lemma 5.21.

5. In this exercise we compare the value of alternating-move games when randomization is allowed to their value when randomization is not allowed.

   Let $G = (T, f)$ be an alternating-move game where f is bounded and Borel measurable, and the sets X_I^p and X_{II}^p are finite for every position $p \in T$. Prove that the value v of the game G as defined in Definition 4.12 is equal to $\sup_{\sigma_I \in \Sigma_I(T)} \inf_{\sigma_{II} \in \Sigma_{II}(T)} \mathbf{E}_{\sigma_I, \sigma_{II}}[f]$.

6. Let $\Gamma = (T, f)$ be a two-player zero-sum Blackwell game. Prove that for every position $p \in T$ there is a mixed move $x_I^p \in X_I^p$ that satisfies

$$v(\Gamma_p) \leq \sum_{\vec{a} \in X^p} x_I^p(a_I) \cdot x_{II}^p(a_{II}) \cdot v(\Gamma_{\langle p, \vec{a} \rangle}), \quad \forall x_{II}^p \in X_{II}^p.$$

7. (*) Suppose we modify the definition of the game G_u in the proof of Theorem 5.46 by changing the winning condition to "$\liminf_{n \to \infty} w_n \leq f$". Prove that Lemmas 5.49 and 5.50 will still hold.

8. (*) Suppose we simplify the definition of the game G_u in the proof of Theorem 5.46 by allowing Player 2 to take moves $\vec{a}_n$ such that $h_n(\vec{a}_n) = 0$. Prove that Lemmas 5.49 and 5.50 will still hold.

9. Prove that for every $0 < u < u' \leq 1$, the game tree of G_u is a 1-imposed subtree of the game tree of $G_{u'}$. Conclude that if Player 1 wins in $G_{u'}$, then she also wins in G_u.

10. Provide an example of a two-player zero-sum Blackwell game Γ with a $\{0, 1\}$-valued payoff function, where for every $\varepsilon > 0$ there is a strategy $\sigma_I^\varepsilon \in \Sigma_I(T)$ that guarantees that the probability of a win is at least $1 - \varepsilon$, but there is no strategy $\sigma_I \in \Sigma_I(T)$ that guarantees that the probability of a win is 1.

11. (*) Provide an example of a two-player zero-sum Blackwell game in which none of the players has a 0-optimal strategy.

12. (*) Like in Exercise 5.11, provide an example of a two-player zero-sum Blackwell game in which none of the players has a 0-optimal strategy, and in addition the payoff function is the indicator of a Borel set.

13. (**) Consider the two-player zero-sum Blackwell game $\Gamma = (T, f)$, where

   - $T := (\{0, 1\} \times \{0, 1\})^{<\mathbb{N}}$.
   - $f : [T] \to \{0, 1\}$ is defined as follows: $f(\langle \vec{a}_0, \vec{a}_1, \vec{a}_2, \ldots \rangle) = 1$ if $a_{n,I} = a_{n,II}$ infinitely often, and $f(\langle \vec{a}_0, \vec{a}_1, \vec{a}_2, \ldots \rangle) = 0$ otherwise.

   (a) What is the value $v(\Gamma)$ of the game?

   (b) Do the players have 0-optimal strategies in Γ? If so, find such a strategy for each player.

   (c) For every $\varepsilon > 0$, describe a winning strategy of Player 1 in the game $G_{v(\Gamma)-\varepsilon}$ that was defined in the proof of Theorem 5.46.

14. (*) Prove Theorem 5.62: In every two-player zero-sum Blackwell game, both players have a subgame ε-optimal strategy, for every $\varepsilon > 0$.

15. (*) In this exercise we generalize Theorem 5.46 to Blackwell games that include chance moves. We model such a game as a game between three players – Player I, Player II, and Player C, who specifies the chance moves.

   A *two-player zero-sum Blackwell game with chance moves* (over an alphabet $\mathcal{A}$) is given by $\Gamma = (\{I, II, C\}, T, f, \sigma_C)$, where

   - T is a tree over $\mathcal{A}^3$.
   - For every $p \in T$, there are nonempty finite sets $X_I^p, X_{II}^p, X_C^p \subseteq \mathcal{A}$, such that

   $$\{p' \in T : p \prec p', \operatorname{len}(p') = \operatorname{len}(p) + 1\} = \{\langle p, \vec{a} \rangle : \vec{a} \in X_I^p \times X_{II}^p \times X_C^p\}.$$

   - $f : [T] \to \mathbb{R}$ is bounded and Borel-measurable payoff function.
   - σ_C is a behavior strategy of Player C.

The strategy σ_C describes the chance moves along the play: Player C is not allowed to choose her moves, but rather she follows the pre-defined strategy σ_C, which is known to Players I and II.

For example, Backgammon is an alternating-move game, where a player rolls two dice before moving her pieces. We can describe Backgammon as an alternating-move game, where Player I moves in stages $n = 1 \bmod 4$, Player II moves in stages $n = 3 \bmod 4$, and all moves in even stages are chance moves, where nature selects an element of $\{1, 2, 3, 4, 5, 6\}^2$ with the uniform distribution.

The *minmax value* is given by

$$\overline{v}(\Gamma) := \min_{\sigma_{II} \in \Sigma_{II}(T)} \max_{\sigma_I \in \Sigma_I(T)} \mathbf{E}_{\sigma_I, \sigma_{II}, \sigma_C}[f],$$

and the *maxmin value* is given by

$$\underline{v}(\Gamma) := \max_{\sigma_I \in \Sigma_I(T)} \min_{\sigma_{II} \in \Sigma_{II}(T)} \mathbf{E}_{\sigma_I, \sigma_{II}, \sigma_C}[f].$$

The *value* of the game exists if $\overline{v}(G) = \underline{v}(G)$.

Prove that every two-player zero-sum Blackwell game with chance moves, has a value.

Hint: Follow the proof of Theorem 5.46.

16. (*) In Step 4 of the proof of Theorem 5.46 we showed that there is $u^* \in [0, 1]$ such that Player I wins G_u for all $u < u^*$, and Player II wins G_u for all $u > u^*$. Which player wins the game G_{u^*}? Is it always Player I, always Player II, or maybe the identity of the winner depends on the game Γ? Justify your answers.

17. Prove that the set of acceptable positions defined in Step 6.1 of the proof of Theorem 5.46 is a subtree.

18. (*) In the proof of Theorem 5.46 we defined the concepts of position compatible with a strategy s_1 and acceptable for a strategy s_2. In this exercise, we compare these concepts.

Consider the setup of Step 6 of the proof. Say that a position $p \in T$ is compatible with s_2 if it is the projection to the $\vec{a}$-coordinates of a position in T_u that is consistent with s_2. Prove the correctness of each of the following statements, or provide a counterexample.

  (a) Every position $p \in T$ that is acceptable for s_2 is also compatible with s_2.

  (b) Every position $p \in T$ that is compatible with s_2 is also acceptable for s_2.

19. (*) In this exercise we consider games in which the payoff function is a characteristic function of a Borel set, and see how the proof of Theorem 5.46 adapts to this case.

Consider a two-player zero-sum Blackwell game $\Gamma = (T, f)$ where the payoff function is the characteristic function of some set: $f = \mathbf{1}_W$ for some Borel set $W \subseteq [T]$. Consider the variation of the auxiliary game G_u that is defined in Step 1 of the proof of Theorem 5.46, where Player 1 wins if the play that is generated by the moves of Player 2 are in W.

(a) Prove that the game G_u is determined for every $u \in \mathbb{R}$.

(b) Define u^* as in Equation (5.13). Prove that u^* is the value of Γ.

Hint: You can use a proof analogous to that of Theorem 5.46.

20. (**) In this exercise we extend Theorem 5.46 to the case in which at every position, the set of moves of at least one player is finite.

Let (T, f) be a Blackwell game where for every position p one of the sets X_{I}^p and X_{II}^p is finite, and the other is a measurable space (not necessarily finite or countably infinite). Prove that the game has a value.

Hint: Recall Exercise 5.2.

21. (**) As in Exercise 5.19, consider a two-player zero-sum Blackwell game $\Gamma = (T, f)$ where the payoff function is the characteristic function $f = \mathbf{1}_W$ of a Borel set $W \subseteq [T]$.

(a) Prove that for every $\varepsilon > 0$ there is a closed subset $W_0 \subseteq W$ such that
$v(T, \mathbf{1}_{W_0}) \geq v(T, \mathbf{1}_W) - \varepsilon$.

(b) Is the set W_0 necessarily compact?

Hint: Consider a winning strategy of Player 1 in the auxiliary game defined in the proof of Theorem 5.46, and plays x such that $H(x_n) \geq \varepsilon$ for every $n \in \mathbb{N}$.

22. (**) Consider the two-player zero-sum Blackwell game $\Gamma = (T, f)$, where $T = \{c, s\}^{<\mathbb{N}}$ and the payoff function is equal to $1 - \frac{1}{n+1}$, where n is the first stage in which Player I selects the move s. Answer the following questions.

(a) For each position $p \in T$, calculate $v(\Gamma_p)$.

(b) For every $\varepsilon > 0$, describe an ε-optimal strategy of Player I.

(c) For each $u < v(\Gamma)$, provide a function H that is a winning strategy for Player 1 in the auxiliary game G_u defined in the proof of Theorem 5.46.

(d) Is it true that at each position p, the ε-optimal strategy you identified in Part (c) selects a move that attains $\max\{v(\Gamma_{\langle p,c \rangle}), v(\Gamma_{\langle p,s \rangle})\}$?

23. (**) Let (T, f) be a two-player zero-sum Blackwell game. Suppose there exists a function $u : T \to [0, 1]$ and a real number $\lambda \in (0, 1)$ such that

$$f(x) = \sum_{n \in \mathbb{N}} (1 - \lambda)\lambda^n u(x_{n-1}), \quad \forall x \in [T].$$

The constant λ is called the *discount factor*, the function u is called the *stage payoff*, and the function f is known as the *discounted payoff*.

(a) Prove that $v(\Gamma_p) \in [0,1]$ for every position $p \in T$.

(b) Define a function $H : T \to [0,1]$ by

$$H(p) := v(\Gamma_p), \quad \forall p \in T. \tag{5.39}$$

Prove that H satisfies the conclusion of Theorem 5.55 with $u = v(\Gamma)$.

24. (*) In Step 1 of the proof of Theorem 5.46 we defined an auxiliary game G_u. A strategy of Player 1 in G_u is equivalent to a function $H : T \to [0,1]$ that satisfies (a) $H(\langle\,\rangle) = u$ and (b) $H(p) \leq v_O(H(\langle p, *\rangle))$ for every $p \in T$. The strategy H is winning in G_u if $\limsup_{n\to\infty} H(x_n) \leq f(x)$ for every play $x \in [T]$.

   (a) Let $u \in [0,1)$, and let $H^0, H^1 : T \to [0,1]$ be two winning strategies of Player 1 in G_u. Define a function $H : T \to [0,1]$ by

   $$H(p) := \max\{H^0(p), H^1(p)\}, \quad \forall p \in T.$$

   Is H a valid strategy of Player 1 in G_u? Is it a winning strategy of Player 1 in Γ_u?

   (b) Let $u \in [0,1)$, and suppose that for each $k \in \mathbb{N}$ the function $H^k : T \to [0,1]$ is a winning strategy of Player 1 in Γ_u. Define a function $H : T \to [0,1]$ by

   $$H(p) := \sup\{H^k(p), k \in \mathbb{N}\}, \quad \forall p \in T,$$

   Is H a valid strategy of Player 1 in G_u? If yes, is H a winning strategy?

   Justify your answers.

25. (*) In this exercise we extend Exercise 5.21 to the setup of Exercise 5.15. In the latter setup, suppose that $f = \mathbf{1}_W$ for a Borel set W, and denote $v(W) := \sup_{\sigma_I \in \Sigma_I(T_p)} \inf_{\sigma_{II} \in \Sigma_{II}(T_p)} \mathbf{E}_{\sigma_I, \sigma_{II}, \sigma_{III}}[\mathbf{1}_W]$. Prove that for every $\varepsilon > 0$ there exists a compact set $W_0 \subseteq W$ such that $v(W_0) \geq v(W) - \varepsilon$.

26. (*) In this exercise we provide a class of Blackwell games where the existence of the value can be deduced without the use of Theorem 5.46.

   Let $\Gamma = (T, f)$ be a Blackwell game, where f satisfies the following condition: there is a subset $T_0 \subset T$ and a bounded and non-negative function $g : T_0 \to \mathbb{R}_+$ such that

   - No position in T_0 is a prefix of another position in T_0: if $p, p' \in T_0$, then $p \not\prec p'$.
   -

   $$f(x) = \begin{cases} g(p), & \text{if } p \in T_0 \text{ and } p \prec x, \\ 0, & \text{otherwise.} \end{cases}$$

That is, if the play ever reaches a position in T_0, the payoff is given by g; otherwise, the payoff is 0.

In this exercise, we will show without using Theorem 5.46 that the game has a value.

For every $n \in \mathbb{N}$, let $f^n : T \to \mathbb{R}_+$ be the function defined as follows:

$$f^n(x) = \begin{cases} g(p), & \text{if } p \in T_0, p \prec x, \text{and } \text{len}(p) \leq n, \\ 0, & \text{otherwise.} \end{cases}$$

(a) Without using Theorem 5.46, prove that for every $n \in \mathbb{N}$, the game $\Gamma^n = (T, f^n)$ admits a value v^n, and that both players have 0-optimal strategies, denoted σ_I^n and σ_{II}^n.

(b) Prove that $v_{n+1} \geq v_n$ for every $n \in \mathbb{N}$.

Denote $v^\infty := \lim_{n \to \infty} v^n$. Let σ_{II}^∞ be an accumulation point of the strategies $(\sigma_{II}^n)_{n \in \mathbb{N}}$; that is, there is a sequence $(n_k)_{k \in \mathbb{N}}$ of increasing natural numbers such that $\sigma_{II}^\infty(p) = \lim_{k \to \infty} \sigma_{II}^{n_k}(p)$, for every position $p \in T$.

(c) Prove that for every $\varepsilon > 0$ there is $n_\varepsilon \in \mathbb{N}$ such that σ_I^n guarantees $v^\infty - \varepsilon$ in Γ, for every $n \geq n_\varepsilon$.

(d) Prove that σ_{II}^∞ guarantees v^∞ in Γ.

27. (*) In this exercise we consider a variant of the Big Match game presented in Example 5.44, where players do not observe the moves selected by their opponent. Therefore, a strategy for each player $i \in \{\text{Robber}, \text{Hakram}\}$ is a behavior function $\sigma_i : \bigcup_{n \in \mathbb{N}} \{L, R\}^n \to [0, 1]$, where $\sigma_i(a_{0,i}, a_{1,i}, \ldots, a_{n-1,i})$ is the probability by which player i selects the move L at stage n if player i's moves at the first n stages were $(a_{0,i}, a_{1,i}, \ldots, a_{n-1,i})$.

(a) Prove that if $c \leq 0$, then the value of the game is 0.

(b) Prove that if $c > 0$, then the value does not exist.

CHAPTER 6

Multiplayer Blackwell Games

In this chapter we study multiplayer Blackwell games; that is, Blackwell games that are played by several players, in which the goals of the players are not necessarily antagonistic.

We start by proving that in every multiplayer Blackwell game there is a vector of behavior strategies that yields to all players a payoff larger than or equal to their minmax value, up to ε, see Section 6.3.

We then prove that given a vector of behavior strategies and a target set of plays, if a single player deviates from her prescribed strategy in such a way that the realized play is not in the target set, then an outside observer who observes only the play can identify the deviator, see Section 6.4.

We will then adapt the solution concept of ε-equilibrium to multiplayer Blackwell games. The question of whether every multiplayer Blackwell game admits an ε-equilibrium for every $\varepsilon > 0$ is still open. We will exhibit two classes of games in which an ε-equilibrium exists. Both of these results will be for games where the set of moves of each player remains fixed throughout the game. Such games are called *repeated games*. These results will hold when the payoff functions of the players satisfy a certain property, called *tail measurability*. Roughly, this property holds when the payoff of a player does not depend on the play in the first n stages of the game, for every $n \in \mathbb{N}$, as was the case with the payoff functions in Example 5.34. We will present tail-measurable functions in Section 6.6.

In Section 6.7 we will show that multiplayer repeated games with tail-measurable payoffs admit an ε-equilibrium, for every $\varepsilon > 0$. In Section 6.8 we will present another class of repeated games – two-player Big Match games – and prove that they also admit ε-equilibria for every $\varepsilon > 0$.

In Section 6.9 we will show that in every multiplayer Blackwell game there is a subgame in which an ε-equilibrium exists. This will be the most involved proof in the chapter.

To prove these results, we will adapt the proof technique of Theorem 5.46 to multiplayer Blackwell games, and we will introduce new proof techniques, like Lévy's 0-1 law and inner regularity of measures.

DOI: 10.1201/9781003582106-6

Before turning to study multiplayer Blackwell games, we will recall the concept of multiplayer one-shot game, which extends the notion of two-player zero-sum one-shot game.

6.1 MULTIPLAYER ONE-SHOT GAMES

In Section 5.1 we presented two-player zero-sum one-shot games. In this section, we extend this model to any finite number of players. The following notation will be used throughout the discussion on multiplayer games.

Notation 6.1 (z_{-i}). *When $z = (z_i)_{i \in N}$ is a vector and $i \in N$, we denote by $z_{-i} := (z_j)_{j \neq i}$ the vector that includes all coordinates of z, except for the i'th.*

Definition 6.2 (Multiplayer one-shot game). *A multiplayer one-shot game is a triplet $(N, (X_i)_{i \in N}, (h_i)_{i \in N})$, where N is a finite set of players, and for each player $i \in N$, X_i is a nonempty finite or countably infinite set of moves and $h_i \colon X \to \mathbb{R}$ is her payoff function, where $X := \prod_{i \in N} X_i$.*

Definition 6.2 differs from Definition 5.1 of two-player zero-sum one-shot games in three respects:

- In Definition 5.1 the game is played by two players, while in Definition 6.2 the set of players is any finite set.

- In Definition 5.1 the goals of the two players are antagonistic: Player I's goal is to maximize the expected payoff, while Player II's goal is to minimize this quantity. In Definition 6.2 each player has her own payoff function.

Notation 6.3 ($\mathbf{E}_z$). *Every vector of mixed moves $z = (z_i)_{i \in N} \in \prod_{i \in N} \Delta(X_i)$ induces a product probability measure on $\Delta\left(\prod_{i \in N} X_i\right)$. Denote the corresponding expectation operator by $\mathbf{E}_z[\cdot]$.*

Notation 6.4 ($h(z)$). *Let $(N, (X_i)_{i \in N}, (h_i)_{i \in N})$ be a multiplayer one-shot game, and let $z = (z_i)_{i \in N} \in \prod_{i \in N} \Delta(X_i)$ be a vector of mixed moves. The expected payoff of each player $i \in N$ is*

$$h_i(z) := \mathbf{E}_z[h_i].$$

In Definition 4.33 we defined the concept of ε-equilibrium for alternating-move games. We here provide the analogous concept for multiplayer one-shot games.

Definition 6.5 (ε-equilibrium). *Let $\varepsilon \geq 0$. A vector of mixed moves $z^* = (z_i^*)_{i \in N} \in \prod_{i \in N} \Delta(X_i)$ is an ε-equilibrium if*

$$h_i(z^*) \geq h_i(z_i, z_{-i}^*) - \varepsilon, \quad \forall i \in N, \forall z_i \in \Delta(X_i).$$

The following result asserts that a 0-equilibrium exists whenever the sets of moves of all players are finite. For a guided proof, see Exercise 6.2.

Theorem 6.6 (Nash, 1950). *Every multiplayer one-shot game $(N, (X_i)_{i \in N}, (h_i)_{i \in N})$ where the sets $(X_i)_{i \in N}$ are finite admits a 0-equilibrium.*

As Example 5.18 shows, when at least two of the sets $(X_i)_{i\in\mathbb{N}}$ are countably infinite, Theorem 6.6 does not necessarily hold: there may not exist an ε-equilibrium for every sufficiently small $\varepsilon > 0$. In Exercise 6.3 we will prove that two-player one-shot games where one player has a finite set of moves and the other has a countably infinite set of moves admit an ε-equilibrium for every $\varepsilon > 0$.

Remark 6.7 (Countably infinite set of players). *Definition 6.2 of multiplayer one-shot games naturally extends to games with countably many players. In this book we will not study such games, and will present a game with a countably infinite set of players only in the following example, to show that Nash's Theorem (Theorem 6.6) does not extend to this setup.*

The following example, due to Voorneveld (2010), shows that when the number of players is countably infinite, an ε-equilibrium need not exist for small $\varepsilon > 0$ even when the sets of moves are finite.

Example 6.8 (Minority game). *Let $(N, (X_i)_{i\in N}, (h_i)_{i\in N})$ be the following multiplayer one-shot game:*

- *The set of players is $N = \mathbb{N}$.*

- *The set of moves of each player $i \in N$ is $X_i = \{0, 1\}$.*

- *The payoff of each player $i \in N$ is as follows. For every vector $\vec{a} = (a_j)_{j\in N} \in X = \prod_{j\in N} X_j$,*

$$h_i(\vec{a}) = \begin{cases} 1, & \text{if } x_i = 0 \text{ and } \limsup_{n\to\infty} \frac{1}{n}\sum_{j=0}^{n-1} a_j > \frac{1}{2}, \\ 1, & \text{if } x_i = 1 \text{ and } \limsup_{n\to\infty} \frac{1}{n}\sum_{j=0}^{n-1} a_j \le \frac{1}{2}, \\ 0, & \text{otherwise.} \end{cases} \quad (6.1)$$

In this game, each player would like to select the move that is chosen by the minority of the players.

We will show that this multiplayer one-shot game has no ε-equilibrium when $\varepsilon \in [0, 1/2)$. Suppose, by contradiction, that z^* is an ε-equilibrium.

The sigma-algebra $\mathcal{F}$ over X is the product sigma-algebra. For each $n \in \mathbb{N}$, let $\mathcal{F}_n$ be the sub-sigma-algebra of $\mathcal{F}$ defined by the moves of player n. This sub-sigma-algebra contains four sets:

$$\mathcal{F}_n = \Big\{\emptyset, (\{0,1\})^{n-1} \times \{0\} \times (\{0,1\})^\infty,$$
$$(\{0,1\})^{n-1} \times \{1\} \times (\{0,1\})^\infty, (\{0,1\})^\infty\Big\}.$$

The sub-sigma-algebras $(\mathcal{F}_n)_{n\in\mathbb{N}}$ are independent because the distribution on X induced by z^* is a product distribution.

The set of vectors of moves

$$B = \Big\{\vec{a} \in X: \limsup_{n\to\infty} \frac{1}{n}\sum_{j<n} a_j > \frac{1}{2}\Big\} \subseteq X$$

lies in the tail sigma-algebra $\bigcap_{n\in\mathbb{N}}\sigma\left(\bigcup_{m\geq n}\mathcal{F}_m\right)$. Note that tail-measurability is with respect to the set of players, which is countably infinite.

By Kolmogorov's zero-one law (Theorem 2.53), either $\mathbf{P}_{z^*}(B)=0$ or $\mathbf{P}_{z^*}(B)=1$. Suppose first that $\mathbf{P}_{z^*}(B)=0$. In this case, a player who chooses the move 0 gets the payoff 0, while a player who chooses the move 1 gets the payoff 1. Therefore, the strategy z_i^* of each player i selects the move 1 with probability at least $1-\varepsilon > 1/2$, which contradicts the fact that $\mathbf{P}_{z^*}(B)=0$. The argument is analogous for the case $\mathbf{P}_{z^*}(B)=1$. Hence, z^* cannot be an ε-equilibrium.

Remark 6.9 (Strong equilibrium). *A vector of mixed moves z^* is an ε-equilibrium if no player can profit by unilaterally deviating from z_i^*, assuming the other players follow z^*. It might happen that by deviating simultaneously, several players can profit. For example, in the game displayed in Figure 20 there are two 0-equilibria: (T,R) and (B,L). Though (B,L) is a 0-equilibrium, both players benefit if both of them change their moves, thereby switching to the vector of moves (T,R).*

Player II

		L	R
	T	$-1,-1$	$1,1$
Player I	B	$0,0$	$-1,-1$

Figure 20: A game with a 0-equilibrium that is not strong.

A vector of mixed moves with the property that no subset of players can all profit by changing their moves is called a strong 0-equilibrium. As the game in Figure 21 shows, a strong 0-equilibrium may not exist. Indeed, in this game, (B,L) is the unique 0-equilibrium, yet this 0-equilibrium is not strong, because both players benefit when both of them switch to their other move, that is, they switch to the vector of moves (T,R).

Player II

		L	R
	T	$0,4$	$3,3$
Player I	B	$1,1$	$4,0$

Figure 21: A game with no strong equilibrium.

We next adapt the definitions of the value of a strategy, the maxmin value, and the minmax value, to multiplayer one-shot games.

Definition 6.10. *(value of strategy, maxmin value, minmax value) Let $(N,(X_i)_{i\in N},(h_i)_{i\in N})$ be a multiplayer one-shot game, and let $i \in N$ be a player. The value of the mixed move $z_i \in \Delta(X_i)$ is*

$$\mathrm{val}_O(z_i) := \inf_{z_{-i}\in\prod_{j\neq i}\Delta(X_j)} h(z_i, z_{-i}) = \inf_{a_{-i}\in\prod_{j\neq i}X_j} h(z_i, a_{-i}). \tag{6.2}$$

The maxmin value *of player i is*

$$\underline{v}_{i,O}(h) := \sup_{z_i \in \Delta(X_i)} \text{val}(z_i) = \sup_{z_i \in \Delta(X_i)} \inf_{a_{-i} \in \prod_{j \neq i} X_j} h(z_i, a_{-i}). \tag{6.3}$$

The minmax value *of player i is*

$$\overline{v}_{i,O}(h) := \inf_{x_{-i} \in \prod_{j \neq i} \Delta(X_j)} \sup_{a_i \in X_i} h(a_i, z_{-i}). \tag{6.4}$$

Remark 6.11 (Comparison between the minmax value and the maxmin value). *As we have seen in Theorem 4.11, the minmax value of a player is always larger than or equal to her maxmin value. By von Neumann's Minmax Theorem (Theorem 5.17), in two-player games in which the sets of moves X_I and X_II are finite, these two quantities coincide. Indeed, the minmax (resp., maxmin) value of the two-player one-shot game $(\{\text{I}, \text{II}\}, X_\text{I}, X_\text{II}, h_\text{I}, h_\text{II})$ coincides with the minmax (resp., maxmin) value of the two-player zero-sum one-shot game $(X_\text{I}, X_\text{II}, h_\text{I})$, which are equal by von Neumann's Minmax Theorem. In the presence of more than two players, the minmax value of a player can be strictly higher than her maxmin value, see Exercise 6.4.*

The next result states that for every ε-equilibrium, the corresponding payoff of each player i is at least her minmax value minus ε.

Theorem 6.12. *Let $(N, (X_i)_{i \in N}, (h_i)_{i \in N})$ be a multiplayer one-shot game, let $\varepsilon \geq 0$, and let $z^* \in \prod_{i \in N} \Delta(X_i)$ be an ε-equilibrium. Then for every $i \in N$ we have*

$$h_i(z^*) \geq \overline{v}_{i,O}(h) - \varepsilon. \tag{6.5}$$

Proof. Equation (6.5) follows from the chain of equations

$$h_i(z^*) \geq \sup_{z_i \in \Delta(X_i)} h_i(z_i, z^*_{-i}) - \varepsilon \tag{6.6}$$

$$\geq \inf_{z_{-i} \in \prod_{j \neq i} \Delta(X_j)} \sup_{z_i \in \Delta(X_i)} h_i(z_i, z_{-i}) - \varepsilon$$

$$= \overline{v}_{i,O}(h) - \varepsilon, \tag{6.7}$$

where Equation (6.6) follows from the definition of an ε-equilibrium, and Equation (6.7) follows from the definition of the minmax value. □

The analogy between ε-equilibria and ε-optimal strategies is expressed by the following result. The proof is left to the reader (Exercise 6.1).

Theorem 6.13. *Let $(N, (X_i)_{i \in N}, (h_i)_{i \in N})$ be a multiplayer one-shot game where $N = \{\text{I}, \text{II}\}$ and $h_\text{I} + h_\text{II} = 0$, and let $\varepsilon \geq 0$. The pair of mixed moves $z^* = (z^*_\text{I}, z^*_\text{II})$ is an ε-equilibrium if and only if z^*_I and z^*_II are ε-optimal strategies of the two players according to Definition 5.13.*

6.2 THE MODEL OF MULTIPLAYER BLACKWELL GAMES

In this section, we formally define multiplayer Blackwell games, which are similar to two-player zero-sum Blackwell games, except that possibly more than two players participate in the game, and each player has her own payoff function.

Definition 6.14 (Multiplayer Blackwell game). *A* multiplayer *Blackwell game (over an alphabet $\mathcal{A}$) is a triplet $\Gamma = (N, T, (f_i)_{i \in N})$, where*

- *N is a finite set of players.*

- *T is a tree over $\mathcal{A}^N$.*

- *For every position $p \in T$ and every player $i \in N$ there is a nonempty finite set $X_i^p \subseteq \mathcal{A}$, such that*

$$\{p' \in T \colon p \prec p', \ \text{len}(p') = \text{len}(p) + 1\} = \left\{ \langle p, \vec{a} \rangle \colon \vec{a} \in \prod_{i \in N} X_i^p \right\}.$$

- *$f_i : [T] \to \mathbb{R}$ is the bounded and Borel-measurable payoff function of player i, for each $i \in N$.*

As in Definition 5.23, X_i^p is the set of moves available to player i at position p Denote by $X^p := \prod_{i \in N} X_i^p$ the set of vectors of moves available at position p. The game is played similarly to two-player zero-sum Blackwell games, except that all players in N simultaneously select moves in each stage, and, as in Section 4.2, each player maximizes her own payoff.

Strategies in multiplayer Blackwell games are defined analogously to their counterparts in two-player zero-sum Blackwell games, see Definition 5.28 and Remark 5.29. Recall that $\Sigma_i(T)$ is the set of player i's behavior strategies in the Blackwell game $\Gamma = (N, T, (f_i)_{i \in N})$.

Notation 6.15 ($\Sigma(T)$, $\Sigma_{-i}(T)$)**.** *Let $\Gamma = (N, T, (f_i)_{i \in N})$ be a multiplayer Blackwell game and let $i \in N$. Denote by $\Sigma(T) := \prod_{i \in N} \Sigma_i(T)$ the set of all vectors of behavior strategies of all players. Denote by $\Sigma_{-i}(T) := \prod_{j \neq i} \Sigma_j(T)$ the set of all vectors of behavior strategies of all players except player i.*

We next extend the definition of the probability distribution over plays induced by a vector of behavior strategies and of the expected payoff from two-player zero-sum Blackwell games (see Notation 5.33) to multiplayer Blackwell games.

Notation 6.16 (σ, $\mathbf{P}_\sigma$, $\mathbf{E}_\sigma[f]$)**.** *A vector of behavior strategies in a multiplayer Blackwell game will be denoted $\sigma = (\sigma_i)_{i \in N}$. Every such vector σ induces a conditional probability $\mu_\sigma(\cdot \mid p) = \bigotimes_{i \in N} \sigma_i(p)$ on X^p, for every $p \in T$, which, by Theorem 5.22, induce a probability measure $\mathbf{P}_\sigma$ on $[T]$. The payoff induced by σ is the expectation of f with respect to the probability measure $\mathbf{P}_\sigma$, denoted by $\mathbf{E}_\sigma[f]$. In Blackwell games, the payoff function f is bounded and Borel measurable, hence the payoff $\mathbf{E}_\sigma[f]$ is well-defined for every vector of behavior strategies σ.*

Definitions 5.35 and 5.37 introduced the concepts of value of strategy, maxmin value, and minmax value for two-player zero-sum Blackwell games, and Definition 6.10 introduced these concepts for multiplayer one-shot games. We now adapt these definitions to multiplayer Blackwell games.

Definition 6.17 (Value of strategy). *Let* $\Gamma = (N, T, (f_i)_{i \in N})$ *be a multiplayer Blackwell game, and let* $i \in N$ *be a player. The* value *of a behavior strategy* $\sigma_i \in \Sigma_i(T)$ *is*

$$\mathrm{val}(\sigma_i) := \inf_{\sigma_{-i} \in \Sigma_{-i}(T)} \mathbf{E}_{\sigma_i, \sigma_{-i}}[f_i].$$

Definition 6.18 (Maxmin value, minmax value, ε-minmax strategy). *Let* $\Gamma = (N, T, (f_i)_{i \in N})$ *be a multiplayer Blackwell game, and let* $i \in N$. *The* maxmin value *for player* i *is*

$$\underline{v}_i(\Gamma) := \sup_{\sigma_i \in \Sigma_i(T)} \mathrm{val}(\sigma_i) = \sup_{\sigma_i \in \Sigma_i(T)} \inf_{\sigma_{-i} \in \Sigma_{-i}(T)} \mathbf{E}_{\sigma_i, \sigma_{-i}}[f_i]. \tag{6.8}$$

The minmax value *for player* i *is*

$$\overline{v}_i(\Gamma) := \inf_{\sigma_{-i} \in \Sigma_{-i}(T)} \sup_{\sigma_i \in \Sigma_i(T)} \mathbf{E}_{\sigma_i, \sigma_{-i}}[f_i]. \tag{6.9}$$

A vector of behavior strategies $\sigma_{-i} \in \Sigma_{-i}(T)$ *that attains the infimum in Equation* (6.9) *up to* ε *is called an* ε-minmax strategy *of the players* $N \setminus \{i\}$ *against player* i.

Remark 6.19 (Comparison between the minmax value and the maxmin value). *As in Theorem 4.11, the minmax value of a player is always larger than or equal to her maxmin value. Since one-shot games are a special case of Blackwell games, as discussed in Remark 6.11, in the presence of more than two players, the minmax value of a player can be strictly higher than her maxmin value. As for one-shot games, the existence of the value in two-player zero-sum Blackwell games ensures that in two-player (not necessarily zero-sum) Blackwell games the minmax value of each player coincides with her maxmin value.*

We next adapt the concepts of subgames, strategies restricted to subgames, and expected payoff in a subgame, to multiplayer Blackwell games.

Notation 6.20 (Subgame, Γ_p). *Let* $\Gamma = (N, T, (f_i)_{i \in N})$ *be a multiplayer Blackwell game. For every position* $p \in T$, *denote by* $\Gamma_p = (N, T_p, (f_i|_{[T_p]})_{i \in N})$ *the subgame that starts at* p, *where* $f_i|_{[T_p]}$ *is the restriction of* f_i *to* $[T_p]$.

Notation 6.21 (Restricted strategy, σ_i^p). *For every behavior strategy* $\sigma_i \in \Sigma_i(T)$ *and every position* $p \in T$, *the restriction of* σ_i *to* Γ_p, *denoted* σ_i^p, *is defined as in Notation 5.57.*

Notation 6.22 (Expected payoff in a subgame). *For every vector* $\sigma = (\sigma_i)_{i \in I}$ *of behavior strategies we denote by* $\mathbf{E}_\sigma[f_i \mid [T_p]] := \mathbf{E}_{(\sigma_i^p)_{i \in N}}[f_i]$ *the expected payoff of player* i *under* σ *in the subgame* Γ_p. *To simplify the notations, we will sometimes use this notation when some of the behavior strategies in the vector* σ *are restricted to* Γ_p, *while others are strategies in* Γ.

Notation 6.23 ($\overline{v}_{p,i}$). *Let* $\Gamma = (N, T, (f_i)_{i \in N})$ *be a multiplayer Blackwell game, and let* $i \in N$. *The minmax value of player* i *in the subgame* Γ_p *is denoted* $\overline{v}_{p,i}$.

The following result strengthens Theorem 5.55 in two respects. First, it extends Theorem 5.55 from two players to any number of players. Second, it requires that the conclusions of the theorem hold for all positions $p \in T$, and not just for the initial position $\langle \ \rangle$.

Theorem 6.24 (Ashkenazi-Golan, Flesch, Predtechinski, and Solan, 2022). *Let* $\Gamma = (N, T, f)$ *be a multiplayer Blackwell game, let* $i \in N$, *and let* $\varepsilon > 0$. *There is a function* $H_i : T \to \mathbb{R}$ *that satisfies the following three properties, for every* $p \in T$:

M.1) $\overline{v}_i(\Gamma_p) - \varepsilon \leq H_i(p) \leq \overline{v}_i(\Gamma_p)$.

M.2) $H_i(p) \leq \overline{v}_{i,O}(H_i(\langle p, * \rangle))$.

M.3) *For every vector of behavior strategies* $\sigma \in \Sigma(T)$ *such that*

$$\mathbf{E}_{\sigma(p)}[H_i(\langle p, * \rangle)] \geq \overline{v}_{i,O}(H_i(\langle p, * \rangle)), \qquad \forall p \in T, \tag{6.10}$$

we have

$$\mathbf{E}_\sigma[f_i \mid [T_p]] \geq \overline{v}_{i,O}(H_i(\langle p, * \rangle)) - \varepsilon, \qquad \forall p \in T. \tag{6.11}$$

Proof. **Step 1:** Adapting the proof of Theorem 5.46 to the multiplayer setup.

As in the proof of Theorem 5.46, assume that the payoffs satisfy $0 \leq f_i \leq 1$. For every $u \in (0, 1]$ define an auxiliary two-player alternating-move win/lose game $G_u = (T_u, W_u)$, played by Players 1 and 2. In this game, for every $n \in \mathbb{N}$, in stage $2n$ Player 1 selects a function $h_n : X^p \to [0, 1]$, and in stage $2n + 1$ Player 2 selects a vector of moves $\vec{a}_n \in X^p$. Hence, a typical position in stage $2n$ (which Player 1 controls) has the form $\langle h_0, \vec{a}_0, h_1, \vec{a}_1, \dots, \vec{a}_{n-1} \rangle$, and a typical position in stage $2n+1$ (which Player 2 controls) has the form $\langle h_0, \vec{a}_0, h_1, \vec{a}_1, \dots, \vec{a}_{n-1}, h_n \rangle$.

Set $u_0 = u$. For every $n \in \mathbb{N}$, in stage $2n$ Player 1 selects a function $h_n : X^p \to \mathbb{R}$ such that $\overline{v}_{i,O}(h_n) \geq u_n$, and in stage $2n + 1$ Player 2 selects a vector of moves $\vec{a}_n \in X^p$ such that $u_{n+1} := h_n(\vec{a}_n) > 0$. Player 1 wins at the play $\langle h_0, \vec{a}_0, h_1, \vec{a}_1, \dots \rangle$ if $\limsup_{n \to \infty} h_n(\vec{a}_n) \geq f(\vec{a}_0, \vec{a}_1, \dots)$:

$$W_u := \{ \langle h_0, \vec{a}_0, h_1, \vec{a}_1, \dots \rangle \in T_u : \limsup_{n \to \infty} h_n(\vec{a}_n) \geq f(\vec{a}_0, \vec{a}_1, \dots) \}.$$

The arguments in the proof of Lemma 5.49 imply that if Player 1 wins the game G_u, then $\overline{v}_i(\Gamma) \geq u$: for every vector of behavior strategies of players $N \setminus \{i\}$ in Γ, player i has a response that ensures that her payoff in Γ is at least u. Similarly, the arguments in the proof of Lemma 5.50 imply that if Player 2 wins the game G_u, then $\overline{v}_i(\Gamma) \leq u$: For every $\delta > 0$ players $N \setminus \{i\}$ have a vector of behavior strategies σ_{-i} in Γ that guarantees that player i's payoff is at most $u + \delta$.

Fix $\varepsilon > 0$. By the above discussion, Player 1 wins the game $G_{\overline{v}_i(\Gamma) - \varepsilon/2}$. As we have seen in the proof of Theorems 5.46, Player 1's winning strategy in $G_{\overline{v}_i(\Gamma) - \varepsilon/2}$ defines a function $H_i : T \to [0, 1]$ which satisfies the following:

- $H_i(\langle \, \rangle) \geq \overline{v}_i(\Gamma) - \frac{\varepsilon}{2}$.

- H_i satisfies (M.2) and the right-hand side inequality in (M.1).

- For every play x,

$$\limsup_{n \to \infty} H_i(x_n) \leq f_i(x). \tag{6.12}$$

Step 2: Definition of an increasing sequence of stopping times.

For every position $p \in T$, let H_i^p be the function constructed in Part 1 for the subgame Γ_p.

Define recursively an increasing sequence of stopping times $(\theta_k)_{k \in \mathbb{N}}$ as follows:

$$\theta_0(x) := 0, \quad \forall x \in [T],$$

and for every $k \in \mathbb{N}$,

$$\theta_{k+1}(x) := \min \left\{ n > \theta_k(x) \colon H_i^{\theta_k(x)}(x_n) < \overline{v}_i(\Gamma_{x_n}) - \varepsilon \right\},$$

where the minimum of an empty set is ∞. Thus,

- $\theta_1(x)$ is the length of the first prefix p of x where $H_i^{\langle \, \rangle}(p) < \overline{v}_i(\Gamma_p) - \varepsilon$.

- $\theta_2(x)$ is the length of the first prefix p of x longer than $\theta_1(x)$ where $H_i^{x_{\theta_1(x)}}(p) < \overline{v}_i(\Gamma_p) - \varepsilon$.

- Etc.

In other words, suppose we start at $\langle \, \rangle$ and proceed along the tree, and for each position we reach, we check whether the left-hand side inequality in (M.1) holds for $H_i^{\langle \, \rangle}$. The stage in which this inequality fails for the first time is θ_1. We then continue along the tree, and for each position we reach, we check whether the left-hand side inequality in (M.1) holds for $H_i^{x_{\theta_1}}$. The stage in which this inequality fails for the first time is θ_2, etc.

It will be convenient to view θ_k as a function defined on T (instead of on $[T]$): if there is an extension $x \in [T]$ of p such that $\theta_k(x) \leq \mathrm{len}(p)$, we set $\theta_k(p) := \theta_k(x)$; otherwise we set $\theta_k(p) := \infty$. The stages $(\theta_k(x))_{k \in \mathbb{N}}$ are called *bad stages for x*, and the stages $(\theta_k(p))_{k \in \mathbb{N}}$ are called *bad stages for p*. A position p such that $\theta_k(p) = \mathrm{len}(p)$ for some $k \in \mathbb{N}$ is called *bad*.

Step 3: Definition of a sequence of functions $(H_i^k)_{k \in \mathbb{N}}$ and their limit H_i.

For every $k \in \mathbb{N}$, let $H_i^k : T \to [0,1]$ be the function that coincides with $H_i^{\langle \, \rangle}$ between stages 0 and $\theta_1 - 1$, with $H_i^{x_{\theta_1}}$ between stages $\theta_1 + 1$ and $\theta_2 - 1$, and so on, and after stage θ_k it coincides with $H_i^{\theta_k}$; And at bad positions p it coincides with $\overline{v}_i(\Gamma_p) - \frac{\varepsilon}{2}$. Formally, the functions $(H_i^k)_{k \in \mathbb{N}}$ are defined recursively by

$$H_i^0(p) = H_i^{\langle \, \rangle}(p), \quad \forall p \in T,$$

and

$$H_i^k(p) := \begin{cases} H_i^{k-1}(p), & \text{if } \theta_k(p) = \infty, \\ \overline{v}_i(\Gamma_p) - \frac{\varepsilon}{2}, & \text{if } \text{len}(p) = \theta_k(p), \\ H_i^{\theta_k(p)}(p), & \text{if } \text{len}(p) < \theta_k(p). \end{cases}$$

It follows by induction on k that for each $k \in \mathbb{N}$, the function H_i^k satisfies (M.2), (M.3), and the right-hand side inequality in (M.1). Moreover, it satisfies the left-hand side inequality in (M.1) for all positions p shorter than θ_k, that is, whenever $\theta_k(p) = \infty$, and for all positions p such that $\text{len}(p) \in (\theta_l(p))_{l \in \mathbb{N}}$.

The function H_i^k agrees with H_i^{k+1} on positions with length at most θ_k. It follows that for each position p, the sequence $(H_i^k(p))_{k \in \mathbb{N}}$ is eventually constant. In particular, the limit

$$H_i(p) := \lim_{k \to \infty} H_i^k(p)$$

exists for every $p \in T$.

Step 4: The function H_i satisfies (M.1), (M.2), and (M.3).

Conditions (M.1) and (M.2) are local conditions. Since for each given position p they hold with respect to H_i^k for each $k \in \mathbb{N}$ sufficiently large, they hold also for H_i.

It remains to show that Condition (M.3) holds for H_i as well. Let then $\sigma \in \Sigma(T)$ be a strategy profile that satisfies Equation (6.10). We will show that Equation (6.11) holds for $p = \langle \rangle$ with 2ε instead of ε. The proof that it holds for all positions is analogous.

By (M.2) and Equation (6.10), the process $(Y_n)_{n \in \mathbb{N}}$ that is defined by

$$Y_n(x) := H_i(x_n), \quad \forall x \in [T]$$

is a submartingale under σ, and hence by the Martingale Convergence Theorem (Theorem 2.40), $\lim_{n \to \infty} Y_n$ exists $\mathbf{P}_\sigma$-a.s. Moreover,

$$Y_n(x) \leq \mathbf{E}_\sigma[Y_{n+1} \mid [T_{x_n}]] - \frac{\varepsilon}{2} \cdot \mathbf{P}_\sigma(\text{stage } n+1 \text{ is bad} \mid [T_{x_n}]).$$

Since the process $(Y_n)_{n \in \mathbb{N}}$ is bounded between 0 and 1, it follows that

$$1 \geq \mathbf{E}\left[\limsup_{n \to \infty} Y_n(x)\right] \geq \frac{\varepsilon}{2} \cdot \sum_{n \in \mathbb{N}} \mathbf{P}_\sigma(\text{stage } n+1 \text{ is bad}),$$

and hence by the Borel-Cantelli Lemma the number of bad stages along the play is finite $\mathbf{P}_\sigma$-a.s.

Let $n_0 \in \mathbb{N}$ be sufficiently large such that

$$\mathbf{P}_\sigma(\text{there is no bad stage after stage } n_0) \geq 1 - \varepsilon,$$

and let X_0 be the set of all plays along which there are no bad stages after stage n_0, so that $\mathbf{P}_\sigma(X_0) \geq 1 - \varepsilon$. Let

$$\alpha(x) := \max\{k \in \mathbb{N} : \theta_k(x) \leq n_0\}, \quad \forall x \in [T],$$

be the last time among $(\theta_k(x))_{k\in\mathbb{N}}$ that is smaller or equal to n_0. Then,

$$\mathbf{P}_\sigma\left(H_i(x_n) = H_i^{\alpha(x)}(x_n), \quad \forall n \geq n_0\right) \geq 1 - \varepsilon.$$

Moreover, for every $x \in X_0$,

$$f_i(x) \geq \limsup_{n\to\infty} H_i^{\alpha(x)}(x_n) = \limsup_{n\to\infty} H_i(x_n). \tag{6.13}$$

We finally have,

$$\mathbf{E}_\sigma[f_i(x)] \geq \mathbf{E}_\sigma\left[\limsup_{n\to\infty} H_i(x_n)\right] - \varepsilon \tag{6.14}$$

$$\geq H_i(\langle\,\rangle) - \varepsilon \tag{6.15}$$

$$\geq \overline{v}_i(\Gamma) - 2\varepsilon, \tag{6.16}$$

where Equation (6.14) holds since $\mathbf{P}_\sigma(X_0) \geq 1 - \varepsilon$ and since $0 \leq f_i \leq 1$, Equation (6.15) holds since $(Y_n)_{n\in\mathbb{N}}$ is a submartingale under σ, and Equation (6.16) holds by (M.1). □

6.3 CONSTRUCTING A GOOD VECTOR OF BEHAVIOR STRATEGIES

In the previous sections we studied the concepts of value and ε-equilibrium, which capture the intuition that players try to obtain as high a payoff as they can. In this section we take a different approach, and look for a vector of behavior strategies σ^* with the following property: under σ^*, the expected payoff of each player i in each subgame Γ_p is at least $\overline{v}_i(\Gamma_p)$. Such a vector of behavior strategies yields an adequate payoff to each player, yet players can profit by deviating from it.

Definition 6.25 (Acceptable vector of strategies). *Let $\Gamma = (N, T, (f_i)_{i\in N})$ be a multiplayer Blackwell game and let $\varepsilon > 0$. A vector of strategies σ^* is ε-acceptable[1] if for every position $p \in T$ and every player $i \in N$ we have*

$$\mathbf{E}_{\sigma^*}[f_i \mid [T_p]] \geq \overline{v}_i(\Gamma_p) - \varepsilon.$$

Theorem 6.26 (Flesch and Solan (2024)). *Let $\Gamma = (N, T, (f_i)_{i\in N})$ be a multiplayer Blackwell game. For every $\varepsilon > 0$ there is an ε-acceptable vector of strategies.*

Proof. For each player $i \in N$, let H_i be the function described in Theorem 6.24 for player i.

For every position $p \in T$ consider the multiplayer one-shot game $(N, (X_i^p)_{i\in N}, (H_i(\langle p, *\rangle))_{i\in N})$. By Theorem 6.6, this game has an equilibrium in mixed moves $z_p^* = (z_{p,i}^*)_{i\in N}$.

For each player $i \in N$ consider the behavior strategy $\sigma_i^* \in \Sigma_i(T)$, which selects at each position $p \in T$ the mixed move $z_{p,i}^*$:

$$\sigma_i^*(p) := z_{p,i}^*, \quad \forall p \in T.$$

---

[1]The reader should note that ε-acceptable vectors of strategies are unrelated to acceptable positions, which were defined in Step 6.1 of the proof of Lemma 5.50.

Let $\sigma^* := (\sigma_i^*)_{i \in N}$. Thus, under σ^*, at every position $p \in T$ the players' mixed moves form a 0-equilibrium in the game $(N, (X_i^p)_{i \in N}, (H_i(\langle p, * \rangle))_{i \in N})$. In particular, by Theorem 6.12, for each position $p \in T$ and every player $i \in N$ we have

$$\mathbf{E}_{\sigma_p^*}[H_i(\langle p, * \rangle)] \geq \overline{v}_{i,O}(H_i(\langle p, * \rangle)).$$

It follows that σ^* satisfies Equation (6.10), and therefore it also satisfies Equation (6.11). By Theorem 6.24,

$$\mathbf{E}_{\sigma^*}[f_i \mid [T_p]] \geq \overline{v}_{i,O}(H_i(\langle p, * \rangle)) \geq \overline{v}_i(\Gamma_p) - \varepsilon, \qquad (6.17)$$

as claimed. □

Remark 6.27 (ε-acceptable vectors of strategies and ε-equilibria). *In the proof of Theorem 4.37 we constructed an ε-acceptable strategy pair, and supplemented it with a statistical test and punishment strategies to create a 3ε-equilibrium. A natural question is whether the same approach would work for simultaneous-move games, as we study in this chapter. The answer is, unfortunately, negative. When the game is alternating-move, the ε-acceptable strategy pair that we constructed in the proof of Theorem 4.37 was pure: the players did not use randomization. Hence, identifying a deviation was an easy task: a player is declared a deviator once she selects a move that she is not supposed to select. When a player uses randomization along the ε-acceptable vector of strategies, and, say, instead of implementing the mixed move $[\frac{1}{2}(a_1), \frac{1}{2}(a_1')]$ she selects the move a_1 with probability 1, there is no way we can identify the deviation.*

In the next section we develop a general way to detect deviations from behavior strategies, which will be used in Section 6.9 to construct ε-equilibria in multiplayer Blackwell games.

6.4   IDENTIFYING THE DEVIATOR

In this section we study a situation in which the players need to cooperate and reach a certain goal, yet some of them may wish to deviate. Such a case arises, for example, when a team of workers must jointly meet a production quota, while each would prefer to shirk. Taking the point of view of Nash equilibrium, we will assume that at most one player may deviate. If the goal is reached, then whether any player deviated is not important. However, if the goal is not met, then we would like to be able to identify the deviator. We will say that a goal is δ-testable if there is an identification method such that if some player deviates, then the probability that the goal is not reached *and* an honest player is erroneously identified as the deviator is at most δ. In this section we will present a formal model for this situation, and study which goals are testable.

Definition 6.28 (Goal). *Let N be a finite set of players, let $\mathcal{A}$ be a finite alphabet, and let $T = (\mathcal{A}^N)^{<\mathbb{N}}$. A goal is a pair (σ^*, D) where $\sigma^* \in \Sigma(T)$ is a vector of behavior strategies and $D \subseteq [T]$ is a Borel set of plays, which is termed the target set.*

Remark 6.29 (On the sets of moves of the players). *In Definition 6.28 we assumed that the set of moves of each player at all positions is $\mathcal{A}$. The model and results can be extended to the case where this set varies among positions and players, and also to the case where these sets are countably infinite.*

The vector of behavior strategies σ^* is a prescribed way for the players to select their mixed moves. The target set D is a set of plays that the players are supposed to reach if they follow their prescribed strategies. We are interested in cases in which the probability $\mathbf{P}_{\sigma^*}(D)$ that the prescribed vector of behavior strategies attains the target set is 1 or close to 1.

Remark 6.30 (Alternate play). *In our model, players make their choices simultaneously. Yet, the model can accommodate an alternate play. Indeed, if $N = \{0, 1, \ldots, |N| - 1\}$, and if for each player i, the strategy σ_i^* randomizes only in stages n such that $n \bmod |N| = i$, then in effect the players play alternately.*

For the rest of the section, we fix the finite set of players N and the finite alphabet $\mathcal{A}$, and we set $T = (\mathcal{A}^N)^{<\mathbb{N}}$.

Definition 6.31 (Blame function). *Let (σ^*, D) be a goal. A* blame function *(for D) is a Borel measurable function $g : D^c \to N$.*

The interpretation of a blame function is that if the players generate a play $x \in [T]$ that misses the target set, then player $g(x)$ is blamed as the player who did not follow her part of σ^*.

Definition 6.32 (δ-testability). *Let (σ^*, D) be a goal, and let $\delta \geq 0$. The goal (σ^*, D) is* δ-testable *if there exists a blame function g such that for every player $i \in N$ and every strategy $\sigma_i \in \Sigma_i(T)$ we have $\mathbf{P}_{\sigma_i, \sigma_{-i}^*}(D^c \text{ and } \{g \neq i\}) \leq \delta$.*

The interpretation of testability is that if some player i deviates, then the probability that a different player j is blamed is at most δ. Thus, with probability $\mathbf{P}_{\sigma_i, \sigma_{-i}^*}(D)$ no player is blamed, and with probability at least $\mathbf{P}_{\sigma_i, \sigma_{-i}^*}(D^c) - \delta$ the blame function correctly identifies the player who deviated from σ^*.

Remark 6.33 (Testability and the requirement of Borel measurability). *In Definition 6.28 we required that the target set D is Borel, and in Definition 6.31 we required that the blame function g is Borel measurable. These assumptions ensure that the probability $\mathbf{P}_{\sigma^*}(D^c \text{ and } \{g \neq i\})$ is well-defined.*

Remark 6.34 (Pure goal). *Suppose that σ^* is a vector of pure strategies and let $D \subseteq [T]$ be a Borel set that contains $x(\sigma^*)$, the play induced by σ^*. In particular, $\mathbf{P}_{\sigma^*}(D) = 1$. The goal (σ^*, D) is 0-testable. A blame function that correctly identifies the deviator is the function that blames the player who is the first to deviate from $x(\sigma^*)$.*

The next result states simple properties of δ-testability.

Theorem 6.35. *Suppose that (σ^*, D) is δ-testable.*

1. If $D' \supseteq D$ is Borel, then (σ^*, D') is δ-testable.

2. $\mathbf{P}_{\sigma^*}(D) \geq 1 - \frac{|N|}{|N|-1}\delta$.

Proof. The first claim follows from the definition of δ-testability. The second claim follows from the following chain of equalities and inequality:

$$
\begin{aligned}
\mathbf{P}_{\sigma^*}(D) &= 1 - \mathbf{P}_{\sigma^*}(D^c) \\
&= 1 - \sum_{i \in N} \mathbf{P}_{\sigma^*}(D^c \text{ and } \{g = i\}) \\
&= 1 - \frac{1}{|N|-1} \sum_{i \in N} \mathbf{P}_{\sigma^*}(D^c \text{ and } \{g \neq i\}) \\
&\geq 1 - \frac{|N|}{|N|-1}\delta,
\end{aligned}
$$

$\square$

Our first main result states that a goal is 0-testable whenever $\mathbf{P}_{\sigma^*}(D) = 1$.

Theorem 6.36 (Alon, Gunby, He, Shmaya, and Solan (2024)). *Let (σ^*, D) be a goal such that $\mathbf{P}_{\sigma^*}(D) = 1$. Then (σ^*, D) is 0-testable.*

Our second main result is an analogous statement when $\mathbf{P}_{\sigma^*}(D)$ is smaller than 1.

Theorem 6.37 (Alon, Gunby, He, Shmaya, and Solan (2024)). *Let $\varepsilon > 0$, and let (σ^*, D) be a goal such that $\mathbf{P}_{\sigma^*}(D) > 1 - \varepsilon$. Then (σ^*, D) is $2\sqrt{(|N|-1)\varepsilon}$-testable.*

We start by showing that Theorem 6.36 is a consequence of Theorem 6.37 and the Borel-Cantelli Lemma.

Proof of Theorem 6.36 using Theorem 6.37. Fix a goal (σ^*, D) with $\mathbf{P}_{\sigma^*}(D) = 1$. Let $(\delta_k)_{k \in \mathbb{N}}$ be a sequence of positive reals such that $\sum_{k \in \mathbb{N}} \delta_k < \infty$. By Theorem 6.37, the goal (σ^*, D) is δ_k-testable for every $k \in \mathbb{N}$. Let $g_k \colon D^c \to I$ be a blame function such that $\mathbf{P}_{\sigma_i, \sigma^*_{-i}}(D^c \text{ and } \{g_k \neq i\}) < \delta_k$, for every player $i \in N$ and every strategy $\sigma_i \in \Sigma_i(T)$.

By the Borel-Cantelli Lemma, for every player $i \in N$ and every strategy $\sigma_i \in \Sigma_i(T)$,

$$\mathbf{P}_{\sigma_i, \sigma^*_{-i}}(D^c \text{ and } \{g_k \neq i \text{ for infinitely many } k\text{'s}\}) = 0.$$

Let $g \colon D^c \to I$ be a blame function such that $g(x) = i$ only if $g_k(x) = i$ for infinitely many k's. Then $\mathbf{P}_{\sigma_i, \sigma^*_{-i}}(D^c \text{ and } \{g \neq i\}) = 0$, for every player $i \in N$ and every strategy $\sigma_i \in \Sigma_i(T)$, which implies that the goal (σ^*, D) is 0-testable (with the blame function g). $\square$

We will provide a remark and two examples before turning to the proof of Theorem 6.37.

Remark 6.38 (Tightness of Theorem 6.37). *A natural question is whether the bound* $2\sqrt{(|N|-1)\varepsilon}$ *in Theorem 6.37 is tight. In Exercise 6.13 we ask the reader to improve this bound to* $2\lceil\ln_2(|N|)\rceil\sqrt{\varepsilon}$. *In Exercise 6.9 we ask the reader to show that if we allow for random blame functions, that is, blame functions that select a player at random, then we could get rid of the constant 2 in the bound. Apart from these results, we do not know whether this bound can be improved.*

While the bound in Theorem 6.37 is not tight, the next example shows that it is tight up to a multiplicative constant.

Example 6.39. *The set of player is* $N = \{\mathrm{I}, \mathrm{II}\}$, *the alphabet is* $\mathcal{A} = \{0,1\}$, *and the tree is* $T = (\mathcal{A} \times \mathcal{A})^{<\mathbb{N}}$. *Let* $\mu \in (0,1]$ *be given, and let* σ^* *be the vector of strategies defined by*

$$\sigma_{\mathrm{I}}^*(p) = \sigma_{\mathrm{II}}^*(p) = [\mu(1), (1-\mu)(0)], \quad \forall p \in T.$$

That is, at all positions, both players are supposed to select the move 1 with probability μ *and the move 0 with probability* $1 - \mu$. *The target set* D *is given by*

$$D = T_{\langle(0,0)\rangle} \cup T_{\langle(0,1)\rangle} \cup T_{\langle(1,0)\rangle}.$$

In words, the goal is that at least one player does not select the move 1 in the first stage. Note that

$$\mathbf{P}_{\sigma^*}(D) = 1 - \mu^2. \tag{6.18}$$

We argue that for every blame function g, *there is a player* $i \in N$ *and a strategy* $\sigma_i \in S_i(T)$ *such that*

$$\mathbf{P}_{\sigma_i, \sigma_{-i}^*}(D^c \text{ and } \{g \neq i\}) \geq \frac{\mu}{2}. \tag{6.19}$$

Indeed, let g *be a blame function, and denote the probability that Player* I *is blamed when both players are honest by* q:

$$q := \mathbf{P}_{\sigma^*}(g = \mathrm{I} \mid D^c).$$

Assume without loss of generality that $q \geq \frac{1}{2}$. *Let* σ_{II} *be the following strategy of Player* II:

- *In the first stage select the move 1 with probability 1.*

- *In all other stages select the mixed move* $[\mu(1), (1-\mu)(0)]$.

Then

$$\mathbf{P}_{\sigma_i, \sigma_{-i}^*}(D^c) = \mu, \tag{6.20}$$

and since σ_{II} *and* σ_{II}^* *coincide after stage 1,*

$$\mathbf{P}_{\sigma_{\mathrm{I}}^*, \sigma_{\mathrm{II}}}(g = \mathrm{I} \mid D^c) = \mathbf{P}_{\sigma^*}(g = \mathrm{I} \mid D^c) \geq \frac{1}{2}. \tag{6.21}$$

Equations (6.20) and (6.21) imply that

$$\mathbf{P}_{\sigma_{\mathrm{I}}^*, \sigma_{\mathrm{II}}}(D^c \text{ and } \{g = \mathrm{I}\}) \geq \frac{\mu}{2}. \tag{6.22}$$

Equations (6.18) and (6.22) imply that the bound in Theorem 6.37 is indeed tight up to a multiplicative constant.

We next provide an example where a blame function can be identified.

Example 6.40. *In this example we consider a problem where the players move alternately. Two players generate an infinite sequence in $\{0,1\}^{\mathbb{N}}$, where Player I selects the even bits and Player II selects the odd bits. Thus, $N = \{I, II\}$ and $T = \{0,1\}^{<\mathbb{N}}$. Let $\mu \in (0,1]$ be given. In stage n, the player whose turn it is to select the bit is supposed to select 1 with probability $\mu/(n+1)$, and 0 with probability $1 - \frac{\mu}{n+1}$. Formally, the strategy pair $\sigma^* = (\sigma_I^*, \sigma_{II}^*)$ is given by $(\sigma_I^*(p))(1) = \frac{\mu}{n+1}$ for every $p \in T$ of even length n, and $(\sigma_{II}^*(p))(1) = \frac{\mu}{n+1}$ for every $p \in T$ of odd length n.*

We next define the target set D. A play $x = \langle a_0, a_1, \dots \rangle \in [T]$ is in D^c if there is $n \in \mathbb{N}$ such that $a_{2n} = 1$ and $a_{2n+1} = 1$; that is, Player II selects 1 exactly one stage after Player I selects 1. Thus, D is the set of all plays $x = \langle a_0, a_1, \dots \rangle \in [T]$ such that $a_{2n} = 0$ or $a_{2n+1} = 0$, for every $n \in \mathbb{N}$.

Recall that

$$\sum_{n=0}^{\infty} \frac{1}{(2n+1)(2n+2)} = \frac{1}{1 \cdot 2} + \frac{1}{3 \cdot 4} + \frac{1}{5 \cdot 6} + \cdots$$

$$= \frac{1}{1} - \frac{1}{2} + \frac{1}{3} - \frac{1}{4} + \frac{1}{5} - \frac{1}{6} + \cdots < 1.$$

By the union bound we therefore obtain that the probability of D^c under σ^* is at most

$$\sum_{n=0}^{\infty} \frac{\mu^2}{(2n+1)(2n+2)} < \mu^2,$$

so that $\mathbf{P}_{\sigma^*}(D) > 1 - \mu^2$.

By Theorem 6.37, the goal (σ^*, D) described above is 2μ-testable. We will now show that in fact (σ^*, D) is μ-testable, by constructing a specific blame function. Intuitively, if Player I selects the move 1 too often, she should be proclaimed the deviator, and otherwise it should be Player II. Formally, let $x = \langle a_0, a_1, a_2, \dots \rangle$ be a sequence in D^c, that is, a sequence containing the number 1 in two consecutive positions $2n$ and $2n+1$, for some $n \geq 0$. Define a blame function $g : [T] \to \{I, II\}$ as follows:

$$g(x) := \begin{cases} I, & \text{if } \sum_{\{n \geq 0 : a_{2n} = 1\}} \frac{\mu}{2n+2} > \mu, \\ II, & \text{otherwise.} \end{cases}$$

We first show that if Player I follows σ_I^*, then the probability that she is blamed is smaller than μ. Let then σ_{II} be any strategy of Player II. Then

$$\mathbf{E}_{\sigma_I^*, \sigma_{II}} \left[\sum_{\{n \geq 0 : a_{2n} = 1\}} \frac{\mu}{2n+2} \right] \leq \sum_{n \geq 0} \frac{\mu}{2n+1} \cdot \frac{\mu}{2n+2} < \mu^2.$$

By Markov's Inequality,

$$\mathbf{P}_{\sigma_I^*, \sigma_{II}}(D^c \cap \{g = I\}) \leq \mathbf{P}_{\sigma_I^*, \sigma_{II}} \left(\sum_{\{n \geq 0 : a_{2n} = 1\}} \frac{\mu}{2n+2} > \mu \right) < \frac{\mu^2}{\mu} = \mu. \qquad (6.23)$$

We next show that if Player II follows σ_{II}^*, then the probability that she is blamed is smaller than μ. Let then σ_I be any strategy of Player I. Note that $x \in D^c$ and $g(x) = II$ only if $\sum_{\{n \geq 0: a_{2n}=1\}} \frac{\mu}{2n+2} \leq \mu$ and there is an index $n \geq 0$ such that $a_{2n} = a_{2n+1} = 1$.

For every $x \in D^c$, denote

$$k(x) := \min \left\{ k \in \mathbb{N}: \sum_{\{n \geq 0:\, n \leq k, a_{2n}=1\}} \frac{\mu}{2n+2} > \mu \right\},$$

where as usual the minimum of an empty set is ∞. Then, $2k(x)$ is the first stage at which it is known that Player I will be blamed. In particular, for every $k < k(x)$ we have

$$\sum_{\{n \geq 0:\, n \leq k, a_{2n}=1\}} \frac{\mu}{2n+2} \leq \mu. \tag{6.24}$$

Therefore,

$$\mathbf{P}_{\sigma_I, \sigma_{II}^*}(D^c \cap \{g = II\}) \leq \sum_{n \in \mathbb{N}} \mathbf{P}_{\sigma_I, \sigma_{II}^*}(\{a_{2n} = 1: n < k(x)\}) \cdot \frac{\mu}{2n+2} \tag{6.25}$$

$$= \int_{x \in [T]} \sum_{\{n < k(x):\, a_{2n}=1\}} \frac{\mu}{2n+2} \, d\mathbf{P}_{\sigma_I, \sigma_{II}^*}(x) \tag{6.26}$$

$$\leq \int_{x \in [T]} \mu \, d\mathbf{P}_{\sigma_I, \sigma_{II}^*}(x) = \mu, \tag{6.27}$$

where Equation (6.25) holds by the definition of $k(x)$, Equation (6.26) is a change of the order of summation, and Equation (6.27) holds by Equation (6.24).

Equation (6.23) and Equations (6.25)–(6.27) imply that (σ^*, D) is μ-testable, and g is the corresponding blame function.

In the proof of Theorem 6.37 we will use the Cauchy-Schwarz inequality, which we recall now.

Lemma 6.41 (The Cauchy-Schwarz Inequality). *For every two vectors $x, y \in \mathbb{R}^n$ we have*

$$\left(\sum_{i=1}^n x_i y_i \right)^2 \leq \sum_{i=1}^n (x_i)^2 \cdot \sum_{i=1}^n (y_i)^2.$$

Proof of Theorem 6.37. The structure of the proof is as follows. We consider a zero-sum game between an adversary (minimizer) and a statistician (maximizer), in which

- the adversary selects a player $i \in N$ and a deviation for that player,

- then Nature selects a play x as if player i uses the deviation chosen for her, while all other players use σ^*, and

- if $x \notin D$, the statistician observes the play x and has to guess the deviator.

- The payoff is 1 if either $x \in D$ or the statistician correctly guessed the identity of the deviator, and 0 otherwise.

A strategy of the statistician in this game is a blame function. We use von Neumann's Minimax Theorem (Theorem 5.17) to establish that the statistician has a strategy that guarantees a high payoff. For the Minimax Theorem to apply, the set of moves of one of the players needs to be finite. We will consider finite-stage versions of the game to ensure that this is indeed the case.

We turn to the formal proof. We may assume without loss of generality that the target set D is closed. Indeed, by Theorem 2.20 the probability measure $\mathbf{P}_{\sigma^*}$ is regular, and hence every Borel set D with $\mathbf{P}_{\sigma^*}(D) > 1 - \varepsilon$ contains a closed subset F with $\mathbf{P}_{\sigma^*}(F) > 1 - \varepsilon$. If D is not closed, we can replace it by the closed subset F. By Theorem 6.35(1), once we prove that F is $(2\sqrt{(|N| - 1)\varepsilon})$-testable, so is D.

Since the set D is closed, its complement D^c is open, and therefore there is a set $Z \subseteq (\mathcal{A}^N)^{<\mathbb{N}}$ with the following properties:

- No element of Z is a prefix of another element of Z.

- $D^c = \bigcup_{p \in Z} [T_p]$.

Since $(\mathcal{A}^N)^{<\mathbb{N}}$ is countably infinite, the set Z is finite or countably infinite.

We now consider the following auxiliary zero-sum game $\Gamma(D)$ between an adversary and a statistician.

- The adversary selects an element $i \in N$ (a player in the original problem) and a pure strategy $s_i \in S_i(T)$ for that player. These choices are *not* revealed to the statistician.

- A play $x \in (\mathcal{A}^N)^{\mathbb{N}}$ is selected according to $\mathbf{P}_{s_i, \sigma^*_{-i}}$.

- If $x \notin D$, the statistician is told the element $p \in Z$ such that $x \in [T_p]$ (and she is not told the rest of x). The statistician has then to select an element $j \in N$.

- The statistician wins if $x \in D$ or $i = j$.

- The adversary wins otherwise, that is, if $x \notin D$ and $i \neq j$.

The interpretation of the game is as follows. The statistician has to detect which player deviated, and the adversary tries to cause the statistician to blame an innocent player. Thus, the adversary's strategy is to select the identity of the deviator $i \in N$ and a strategy of that deviator. Then a play x is selected according to the strategy that player i deviated to and the prescribed strategies of the other players. If $x \in D$, then the statistician wins. If $x \notin D$, then the statistician learns the minimal prefix p of the play at which it is clear that the play will be outside D, and she wins only if she correctly guesses the identity of the deviator.

Remark 6.42 (Games with imperfect monitoring). *In the auxiliary game $\Gamma(D)$, the statistician does not observe the choice (i, σ_i) of the adversary; all she observes is the prefix p of the realized play x that lies in Z, in case $x \notin D$. This is in contrast to all* *games we studied so far, in which players observed the choices of the other players that were made in the past. Games in which players do not necessarily observe past choices of the other players are called* games with imperfect monitoring. *In Chapter 7 we will study alternating-move games with imperfect monitoring, and ask whether they admit a value.*

Remark 6.43 (Mixed strategies for the adversary in the auxiliary game). *A pure strategy of the adversary is a pair (i, s_i). When the adversary uses randomization, she picks such a pair randomly. Recall that a probability distribution on pure strategies is called a* mixed strategy, *see Remark 5.32. In the game $\Gamma(D)$, a mixed strategy of the adversary is a probability distribution on $\bigcup_{i \in N}(\{i\} \times S_i(T))$. It follows that an equivalent way to present a mixed strategy of the adversary in $\Gamma(D)$ is as a pair $(q, \sigma) \in \Delta(N) \times \prod_{i \in N} \Sigma_i(T)$, where $\sigma = (\sigma_i)_{i \in N}$: the adversary selects a player $i \in N$ according to q, and the deviation of that player is σ_i.*

The following result states that the minmax value of the auxiliary game is at least $1 - \sqrt{(|N| - 1)\varepsilon}$.

Lemma 6.44. *Let D be a closed set such that $\mathbf{P}_{\sigma^*}(D) > 1 - \varepsilon$. Then for every mixed strategy of the adversary, the statistician has a response that wins in $\Gamma(D)$ with probability at least $1 - \sqrt{(|N| - 1)\varepsilon}$.*

Proof. Fix a mixed strategy $(q, \sigma) \in \Delta(N) \times \prod_{i \in N} \Sigma_i(T)$ of the adversary in $\Gamma(D)$.

Recall that $D^c = \bigcup_{p \in Z}[T_p]$ for some finite or countably infinite set Z of positions with the property that no position in Z is a prefix of a different position in Z.

For a position $p = (\vec{a}_0, \vec{a}_1, \ldots, \vec{a}_{n-1}) \in Z$ let

$$\ell_i(p) := \prod_{m=0}^{n-1} \frac{\sigma_i(a_{m,i} \mid \vec{a}_0, \vec{a}_1, \ldots, \vec{a}_{m-1})}{\sigma_i^*(a_{m,i} \mid \vec{a}_0, \vec{a}_1, \ldots, \vec{a}_{m-1})}, \quad \forall i \in N,$$

where $\frac{0}{0} = 1$ and $\frac{c}{0} = \infty$ for $c > 0$. The quantity $\ell_i(p)$ is the likelihood ratio of σ_i (the deviation strategy of player i) over σ_i^* (the goal strategy of player i) at the position p.

For every position $p = (\vec{a}_0, \vec{a}_1, \ldots, \vec{a}_{n-1}) \in T$ and every strategy profile σ', denote by

$$\mathbf{P}_{\sigma'}(p) := \mathbf{P}_{\sigma'}([T_p]) = \prod_{m=0}^{n-1} \prod_{i \in N} \sigma_i'(a_{m,i} \mid \vec{a}_0, \vec{a}_1, \ldots, \vec{a}_{m-1})$$

the probability that the position p is realized under σ'. Note that $\mathbf{P}_{\sigma_i, \sigma_{-i}^*}(p) = \ell_i(p)\mathbf{P}_{\sigma^*}(p)$, provided $\ell_i(p) < \infty$. Similarly, $\mathbf{P}_{\sigma_i, \sigma_j, \sigma_{-i,j}^*}(p) = \ell_i(p)\ell_j(p)\mathbf{P}_{\sigma^*}(p)$, provided $\ell_i(p), \ell_j(p) < \infty$, where $\sigma_{-i,j}^* = (\sigma_k^*)_{k \in N \setminus \{i,j\}}$.

Consider a pure strategy of the statistician that, after observing a position $p \in Z$, blames a player j whose likelihood ratio is maximal. When there are several players who maximize the likelihood ratio, the way the statistician selects whom among them to blame is not important. For each $j \in N$, denote by E_j the set of all positions in Z where the statistician blames j. Then

$$E_j \subseteq \bigcup\{[T_p]: p \in Z, \ell_j(p) \geq \ell_i(p) \text{ for every } i \neq j\},$$

where the inclusion may be strict when for some $p \in Z$ the maximum of $\{\ell_i(p), i \in N\}$ is attained at j together with some other index. The set of positions where player j is blamed is, therefore,

$$Z_j := \{p \in Z \colon [T_p] \subseteq E_j\}.$$

Observe that

$$
\begin{aligned}
\left(\mathbf{P}_{\sigma_i, \sigma^*_{-i}}(E_j)\right)^2 &= \left(\sum_{p \in Z_j} \ell_i(p) \mathbf{P}_{\sigma^*}(p)\right)^2 \\
&\le \left(\sum_{p \in Z_j} \ell_i(p)^2 \mathbf{P}_{\sigma^*}(p)\right) \cdot \left(\sum_{p \in Z_j} \mathbf{P}_{\sigma^*}(p)\right) \quad (6.28) \\
&\le \left(\sum_{p \in Z_j} \ell_i(p) \ell_j(p) \mathbf{P}_{\sigma^*}(p)\right) \cdot \left(\sum_{p \in Z_j} \mathbf{P}_{\sigma^*}(p)\right) \quad (6.29) \\
&= \mathbf{P}_{\sigma_i, \sigma_j, \sigma^*_{-i,j}}(E_j) \cdot \mathbf{P}_{\sigma^*}(E_j) \le \mathbf{P}_{\sigma^*}(E_j), \quad (6.30)
\end{aligned}
$$

where Equation (6.28) holds by the Cauchy-Schwarz Inequality (Lemma 6.41), Equation (6.29) holds since $\ell_i(p) \le \ell_j(p)$ on E_j, and Equation (6.30) follows from the definitions. By the Cauchy-Schwarz Inequality once again, it follows that

$$
\begin{aligned}
\left(\sum_{j \ne i} \mathbf{P}_{\sigma_i, \sigma^*_{-i}}(E_j)\right)^2 &\le (|N| - 1) \cdot \sum_{j \ne i} \mathbf{P}_{\sigma_i, \sigma^*_{-i}}(E_j)^2 \\
&\le (|N| - 1) \cdot \sum_{j \ne i} \mathbf{P}_{\sigma^*}(E_j) \\
&= (|N| - 1) \cdot \mathbf{P}_{\sigma^*}(Z) \le (|N| - 1) \cdot \varepsilon,
\end{aligned}
$$

and the claim follows. □

We now conclude the proof of Theorem 6.37. For every $n \in \mathbb{N}$, let $Z_n = \{p \in Z \colon \text{length of } p < n\}$, and let $D_n \subseteq A$ be the set whose complement is given by

$$D_n^c = \bigcup \{[T_p] \colon p \in Z_n\}.$$

For every $n \in \mathbb{N}$, we can view the game $\Gamma(D_n)$ as a finite-stage game, because in stage n it is known whether the play will be in D_n. In particular, the set of pure strategies of both players is finite. By Theorem 5.17, this game admits a value in mixed strategies, and the statistician has a 0-optimal strategy $\xi_n : Z_n \to \Delta(I)$, which depends only on the play until a position in Z_n is reached.

The sequence of sets $(Z_n)_{n \in \mathbb{N}}$ is non-decreasing, and consequently $(D_n)_{n \in \mathbb{N}}$ is a non-increasing sequence of closed sets that contain D. Since D is closed, $D = \bigcap_{n \in \mathbb{N}} D_n$. In particular, $\mathbf{P}_{\sigma^*}(D_n) \ge \mathbf{P}_{\sigma^*}(D) > 1 - \varepsilon$ for every $n \in \mathbb{N}$.

By Lemma 6.44, for every $n \in \mathbb{N}$, the value of the game $\Gamma(D_n)$ is at least $1 - \sqrt{(|N| - 1)\varepsilon}$. Let $g_n \colon Z_n \to N$ be a blame function such that $g_n(p) \in$

$\mathrm{argmax}_{i \in N} \, \xi_n(p)$, where p is the prefix of x that lies in Z_n. It follows that if $g_n(p) \neq i$, then $(\xi_n(p))(N \setminus \{i\}) \geq \frac{1}{2}$, and hence

$$\mathbf{P}_{\sigma_i, \sigma_{-i}^*}(D_n^{\mathrm{c}} \text{ and } \{g_n \neq i\}) \leq 2\mathbf{P}_{\sigma_i, \sigma_{-i}^*, \xi_n}(D_n^{\mathrm{c}} \text{ and } \{j \neq i\}) \leq 2\sqrt{(|N|-1)\varepsilon},$$

for every $i \in N$ and every $\sigma_i \in \Sigma_i(T)$. Abusing notations, we view g_n as a function from D_n^{c} to I, such that for every $p \in Z_n$ and every $x \in [T_p]$, we set $g_n(x) := g_n(p)$.

For every $n \in \mathbb{N}$, the domain of g_n is the finite set Z_n. By a diagonal argument, there is a function $g : D^{\mathrm{c}} \to I$ that is an accumulation point of the sequence $(g_n)_{n \in \mathbb{N}}$: there is a subsequence $(n_k)_{k \in \mathbb{N}}$ such that for every $p \in Z$ and every $x \in [T_p]$, $g(x)$ is equal to $g_{n_k}(x)$, for all sufficiently large $k \in \mathbb{N}$.

We claim that g guarantees to the statistician at least $1 - 2\sqrt{(|N|-1)\varepsilon}$ in $\Gamma(D)$. Indeed, let $i \in N$ and $\sigma_i \in \Sigma_i(T)$ be arbitrary. Since $D^{\mathrm{c}} = \bigcup_{k \in \mathbb{N}} D_{n_k}^{\mathrm{c}}$,

$$\begin{aligned}\mathbf{P}_{\sigma_i, \sigma^* -i}(D^{\mathrm{c}} \text{ and } \{g \neq i\}) &= \lim_{k \to \infty} \mathbf{P}_{\sigma_i, \sigma^* -i}(D_{n_k}^{\mathrm{c}} \text{ and } \{g \neq i\}) \\ &\leq 2\sqrt{(|N|-1)\varepsilon},\end{aligned}$$

and the result follows. □

Remark 6.45 (The case of closed D). *When the target set D is closed, the set D^{c} is a union of open sets: $D^{\mathrm{c}} = \bigcup_{p \in Z} [T_p]$, for some finite or countably infinite set $Z \subseteq T$. The proof of Theorem 6.37 shows that there is a blame function $g : Z \to N$ such that for every player $i \in N$ and every strategy σ_i of player i,*

$$\mathbf{P}_{\sigma_i, \sigma_{-i}^*}(D^{\mathrm{c}} \text{ and } \{g \neq i\}) \leq 2\sqrt{(|N|-1)\varepsilon}.$$

That is, the blame function identifies the deviator in the first stage in which it is clear that a deviation occurred.

Remark 6.46 (The value of the auxiliary game $\Gamma(D)$). *In the proof of Theorem 6.37 we did not show that the auxiliary game $\Gamma(D)$ has a value. We argued that for each $n \in \mathbb{N}$, the auxiliary game $\Gamma(D_n)$ has a value, because both players have finitely many pure strategies in this game, and then used a sequence of 0-optimal strategies of the statistician in these games to construct a blame function that guarantees that $\mathbf{P}_{\sigma_i, \sigma_{-i}^*}(D) \geq 1 - 2\sqrt{(|N|-1)\varepsilon}$ for every $i \in N$ and every $\sigma_i \in \Sigma_i(T)$.*

Using the fact that the payoff function in $\Gamma(D)$ is upper semi-continuous, one can prove that $\Gamma(D)$ has a value, see Theorem 7.33 below.

6.5 ε-EQUILIBRIUM IN MULTIPLAYER BLACKWELL GAMES

The next concept that we will study is that of ε-equilibrium, which we discussed in Chapter 4 in the context of two-player alternating-move games and in Section 6.1 in the context of multiplayer one-shot games (see Definition 6.5). We here adapt this concept to multiplayer Blackwell games.

Definition 6.47 (ε-equilibrium). *Let $\Gamma = (N, T, (f_i)_{i \in N})$ be a multiplayer Blackwell game, and let $\varepsilon \geq 0$. A vector of behavior strategies $\sigma^* = (\sigma_i^*)_{i \in N} \in \Sigma(T)$ is an ε-equilibrium if for every player $i \in N$,*

$$\mathbf{E}_{\sigma_i, \sigma_{-i}^*}[f_i] \ \leq \ \mathbf{E}_{\sigma^*}[f_i] + \varepsilon, \quad \forall \sigma_i \in \sigma_i(T).$$

Whether every multiplayer Blackwell game admits an ε-equilibrium for every $\varepsilon > 0$ is still an open problem. In Section 6.7 we will prove that for every $\varepsilon > 0$ an ε-equilibrium exists in multiplayer Blackwell games where the set of moves of each player is independent of the position, and the payoff functions satisfy a certain property, called tail measurability. In Section 6.8 we will prove the same result for a second class of multiplayer Blackwell games, called Big Match games with tail-measurable payoffs. In Section 6.9 we will prove that for every multiplayer Blackwell game there is a subgame that admits an ε-equilibrium for every $\varepsilon > 0$. The construction of the ε-equilibrium in Section 6.7 is simple and ensuring that players cannot gain more than ε by deviating is not difficult. In Section 6.8 we will see several constructions of ε-equilibria, which will reveal additional ways to identify deviations. The construction in Section 6.9 will be the most challenging and will require using a wide variety of tools.

Remark 6.48 (Deviations from an ε-equilibrium). *According to Definition 6.47, a vector of behavior strategies σ^* is an ε-equilibrium if no player can profit more than ε by deviating to another behavior strategy, provided the other players keep following σ^*. For each player $i \in N$, when we fix the vector of strategies σ_{-i} of the other players, the expectation operator $\mathbf{E}_{\sigma_i, \sigma_{-i}}$ is linear in σ_i. Therefore, to verify that a vector of behavior strategies σ^* is an ε-equilibrium, it is enough to verify that no player i can profit more than ε by deviating to a strategy which is an extreme point of $\Sigma_i(T)$. Kuhn's Theorem (see Remark 5.32) shows that the set of extreme points of $\Sigma_i(T)$ is $S_i(T)$ – the set of pure strategies of player i. Therefore, σ^* is an ε-equilibrium if*

$$\mathbf{E}_{s_i, \sigma_{-i}^*}[f_i] \ \leq \ \mathbf{E}_{\sigma^*}[f_i] + \varepsilon, \quad \forall s_i \in S_i(T).$$

Theorem 6.12 states that in multiplayer one-shot games, the payoff of player i in an ε-equilibrium cannot be much smaller than the minmax value of player i. The analogous result for multiplayer Blackwell games, which is stated without a proof, is the following.

Theorem 6.49. *Let $\Gamma = (N, T, (f_i)_{i \in N})$ be a multiplayer Blackwell game, let $\varepsilon > 0$, let σ^* be an ε-equilibrium, and let $i \in N$. Then*

$$\mathbf{E}_{\sigma^*}[f_i] \geq \overline{v}_i(\Gamma) - \varepsilon.$$

6.6 TAIL-MEASURABLE FUNCTIONS

In this section we present the concepts of tail sets and tail-measurable functions. In Sections 6.7 and 6.8 we will study multiplayer Blackwell games where the payoff functions are tail measurable.

In Definition 2.50 we introduced the tail sigma-algebra. We here adapt this definition to trees.

Definition 6.50 (Tail sigma-algebra on a tree). *Let T be a tree over an alphabet $\mathcal{A}$, and let $\mathcal{F}$ be the tree sigma-algebra (see Definition 3.14). For every $n \in \mathbb{N}$, let $\mathcal{F}_n$ be the sub-sigma-algebra of $\mathcal{F}$ defined by the move at the n'th stage; that is, the sets in $\mathcal{F}_n$ are all the sets $B \in \mathcal{F}$ that can be presented as*

$$B = \{x = \langle a_0, a_1, \ldots \rangle \in [T] \colon a_n \in C\},$$

for some set $C \subseteq \mathcal{A}$. The tail sigma-algebra of T is the sigma-algebra $\mathcal{F}^{\mathrm{tail}}(T) := \bigcap_{n \in \mathbb{N}} \sigma\left(\bigcup_{m \geq n} \mathcal{F}_m\right)$.

An alternative way to define the tail sigma-algebra on a tree is using the concept of tail sets. A set $B \subseteq [T]$ is a *tail set* if to answer the question whether a certain play is in B one need not know the first moves along the play. The formal definition is as follows.

Definition 6.51 (Tail set). *Let T be a tree over an alphabet $\mathcal{A}$. We say that two plays $x = \langle a_0, a_1, \ldots \rangle$ and $x' = \langle a'_0, a'_1, \ldots \rangle$ in $[T]$ have the same tail if there exists $n \in \mathbb{N}$, such that $(a_n, a_{n+1}, \ldots) = (a'_n, a'_{n+1}, \ldots)$. A set $B \subseteq [T]$ is a tail set if for any two plays $x, x' \in [T]$ with the same tail, $x \in B$ if and only if $x' \in B$.*

Remark 6.52 (On the definition of tail sets). *Note that in Definition 6.51 we can replace the condition "$x \in B$ if and only if $x' \in B$" by the condition "if $x \in B$, then $x' \in B$".*

The collection of all tail sets in $[T]$ is a sigma-algebra, which coincides with the tail sigma-algebra $\mathcal{F}^{\mathrm{tail}}(T)$ (Exercise 6.18).

Definition 6.53 (Tail measurability). *Let T be a tree over an alphabet $\mathcal{A}$. A function $f \colon [T] \to \mathbb{R}$ is tail measurable if $f(x) = f(x')$ whenever $x, x' \in [T]$ have the same tail. In other words, f is measurable with respect to the tail sigma-algebra.*

Example 6.54 (Büchi winning condition). *Let T be a tree over an alphabet $\mathcal{A}$, and let $A \subseteq \mathcal{A}$. Define the function*

$$f(\langle a_0, a_1, \ldots \rangle) := \begin{cases} 1, & \text{if } a_t \in A \text{ infinitely often,} \\ 0, & \text{otherwise.} \end{cases}$$

The function f, which is termed Büchi *in the computer science literature, is tail measurable.*

Example 6.55 (Limsup of average payoffs). *Let $u \colon T \to [0,1]$. Define the function*

$$f(\langle a_0, a_1, \ldots \rangle) := \limsup_{N \to \infty} \frac{1}{N+1} \sum_{n=0}^{N} u(\langle a_0, a_1, \ldots, a_{n-1} \rangle).$$

Then f is tail measurable.

Remark 6.56 (Tail sets and the Borel hierarchy). *Let $\mathcal{A}$ be an alphabet, let $T = \mathcal{A}^{<\mathbb{N}}$, let $A \subseteq \mathcal{A}$, and let f be the function defined in Example 6.54. For every $n \in \mathbb{N}$ and every $a \in A$, denote the set of all positions of length n whose last element is a by $T_{n,a}$:*

$$T_{n,a} = \{\langle a_0, a_1, \ldots, a_{n-1}\rangle \in T \colon a_{n-1} = a\}.$$

We then have

$$\{x \in A^{\mathbb{N}} \colon f(x) = 1\} = \bigcap_{k \in \mathbb{N}} \bigcup_{n \geq k} \bigcup_{p \in T_{n,a}} [T_p].$$

Since the set $[T_p]$ is open for every $p \in T$, so is the set $\bigcup_{n \geq k} \bigcup_{p \in T_{n,a}} [T_p]$, and therefore the set $\{x \in A^{\mathbb{N}} \colon f(x) = 1\}$ is of rank 2 in the Borel hierarchy of T. One may wonder if all tail sets belong to low ranks in the Borel hierarchy. It turns out that this is not the case: for every countable ordinal α, there is a tail set of rank α in the Borel hierarchy, see Exercise 6.20. Thus, tail sets can be quite complicated.

6.7 REPEATED GAMES WITH TAIL-MEASURABLE PAYOFFS

A repeated game is a multiplayer Blackwell game where the set of moves of each player is the same at all positions. To define a repeated game, it is therefore not necessary to provide the tree, but rather one can provide the sets of moves of the players.

Definition 6.57 (Repeated game). *A repeated game is a multiplayer Blackwell game where the sets of available moves of the players are independent of the position. We denote a repeated game as $\Gamma = (N, (X_i)_{i \in N}, (f_i)_{i \in N})$, where X_i is the finite set of moves of player i, for each $i \in N$.*

Notation 6.58 (Set X of vector of moves). *Denote by $X := \prod_{i \in N} X_i$ the set of vectors of moves, one for each player.*

With this notation, the tree that underlies the repeated game is $T = X^{<\mathbb{N}}$.

As mentioned before, whether every multiplayer Blackwell game, or even every repeated game, admits an ε-equilibrium for every $\varepsilon > 0$, is still an open problem. In this section we study this problem in the setup of repeated games with tail-measurable payoffs.

Remark 6.59 (Repeated games and tail measurability). *The reader may wonder why we restrict attention to repeated games when studying tail-measurable payoffs. The reason is that without this restriction, by changing the names of the moves, all payoff functions can be presented as tail measurable. Indeed, consider a multiplayer Blackwell game $\Gamma = (N, T, (f_i)_{i \in N})$, and define an auxiliary game Γ', which is similar to Γ, except that we change the names of the moves: the move a_i at position p in Γ is called move (p, a_i) in Γ'. With the new names, the payoff function in Γ' is tail measurable, because the suffixes of two plays coincide only if the two plays coincide.*

Since the set of positions is countable, renaming moves as described in the previous paragraph requires a countable set of moves. The restriction that each player has the same finite set of moves at all positions does not allow us to rename moves, and makes the concept of tail-measurable payoff function meaningful.

Recall that the maxmin value and the minmax value where defined in Definition 6.18 for all Blackwell games, and in particular for repeated games. When the payoff function of a repeated game is tail measurable, for every fixed $n \in \mathbb{N}$ the payoff is independent of the moves in the first n stages. This implies that for every two positions p, p' that have the same length, that is, $\text{len}(p) = \text{len}(p')$, the minmax value of a player in the subgame Γ_p is equal to her minmax value in the subgame $\Gamma_{p'}$, and an analogous property holds for the maxmin value. As the next result states, this property holds for *every* two positions, whether or not they have the same length.

Theorem 6.60. *Let* $\Gamma = (N, (X_i)_{i \in N}, (f_i)_{i \in N})$ *be a repeated game where the payoff functions* $(f_i)_{i \in N}$ *are tail measurable. Then* $\overline{v}_i(\Gamma_p) = \overline{v}_i(\Gamma)$ *and* $\underline{v}_i(\Gamma_p) = \underline{v}_i(\Gamma)$ *for each* $i \in N$ *and every position* $p \in T$.

Remark 6.61 (Theorem 6.60 for Blackwell games). *One of the conditions of repeated games is that the set of moves of each player is independent of the position. This assumption is crucial for Theorem 6.60 to hold. Indeed, if the set of moves of some player i depends on the position, then it might be that in all positions that extend some position p_1, the only move available to player i is $a_{i,1}$, while in all positions that extend some position p_2, the only move available to player i is $a_{i,2}$. If the payoff to player i when she selects the move $a_{i,1}$ after position p_1 differs from her payoff when she selects the move $a_{i,2}$ after position p_2, then the payoff function may be tail measurable yet the minmax value of player i may differ at the two positions.*

Proof of Theorem 6.60. We will prove the claim for the minmax value. The proof for the maxmin value is analogous.

Step 1: $\overline{v}_i(\Gamma) \leq \overline{v}_i(\Gamma_p)$, for every $p \in T$.

We will prove that $\overline{v}_i(\Gamma) \leq \overline{v}_i(\Gamma_p) + \varepsilon$, for every position $p \in T$ and every $\varepsilon > 0$. Fix then $p \in T$ and $\varepsilon > 0$, and denote $n = \text{len}(p)$. Since f_i is tail measurable, $\overline{v}_i(\Gamma_p) = \overline{v}_i(\Gamma_{p'})$ for every position p' of length n. For each such position p', let $\sigma_{-i}^{p'} \in \Sigma_{-i}(T_{p'})$ be an ε-minmax strategy against player i in the subgame $\Gamma_{p'}$, so that

$$\mathbf{E}_{\sigma_i^{p'}, \sigma_{-i}^{p'}} \left[f_i \mid [T_{p'}] \right] \leq \overline{v}_i(\Gamma_{p'}) + \varepsilon = \overline{v}_i(\Gamma_p) + \varepsilon, \quad \forall \sigma_i^{p'} \in \Sigma_i(T_{p'}). \tag{6.31}$$

For every $j \neq i$, let $\sigma_j^* \in \Sigma_j(T)$ be a behavior strategy that coincides with $\sigma_j^{p'}$ on the subgame $\Gamma_{p'}$, for each position p' of length n. The definition of σ_j^* for the first n stages can be arbitrary. By Equation (6.31),

$$\mathbf{E}_{\sigma_i, \sigma_{-i}^*}[f_i] = \mathbf{E}_{\sigma_i, \sigma_{-i}^*} \left[\mathbf{E}_{\sigma_i^{x_{n-1}}, \sigma_{-i}^{x_{n-1}}} \left[f_i \mid [T_{x_{n-1}}] \right] \right]$$

$$\leq \overline{v}_i(\Gamma_p) + \varepsilon, \quad \forall \sigma_i \in \Sigma_i(T_p),$$

where $\sigma_{-i}^* = (\sigma_j^*)_{j \neq i}$. Hence,

$$\overline{v}_i(\Gamma) = \inf_{\sigma_{-i} \in \Sigma_{-i}(T)} \sup_{\sigma_i \in \Sigma_i(T)} \mathbf{E}_{\sigma_i, \sigma_{-i}}[f_i] \leq \sup_{\sigma_i \in \Sigma_i(T)} \mathbf{E}_{\sigma_i, \sigma_{-i}^*}[f_i] \leq \overline{v}_i(\Gamma_p) + \varepsilon.$$

Step 2: $\overline{v}_i(\Gamma) \geq \overline{v}_i(\Gamma_p)$, for every $p \in T$.

We will prove that $\overline{v}_i(\Gamma) \geq \overline{v}_i(\Gamma_p) - \varepsilon$, for every position $p \in T$ and every $\varepsilon > 0$. Fix then $p \in T$ and $\varepsilon > 0$, and denote $n = \text{len}(p)$. For any vector of behavior strategies $\sigma_{-i} \in \Sigma_{-i}(T)$ and any position p' of length n, denote by $\sigma_{-i}^{p'}$ the restriction of σ_{-i} to $T_{p'}$. By the definition of the minmax value, and since $v_i(\Gamma_{p'}) = v_i(\Gamma_p)$ for each position p' of length n, there is a strategy $\sigma_i^{p'} \in \Sigma_i(T_{p'})$ such that

$$\mathbf{E}_{\sigma_i^{p'}, \sigma_{-i}^{p'}} \left[f_i \mid [T_{p'}] \right] \geq \overline{v}_i(\Gamma_{p'}) - \varepsilon = \overline{v}_i(\Gamma_p) - \varepsilon.$$

Let $\sigma_i^* \in \Sigma_i(T)$ be a strategy that coincides with $\sigma_i^{p'}$ in the subgame $T_{p'}$, for every position p' of length n. Then,

$$\mathbf{E}_{\sigma_i^*, \sigma_{-i}}[f_i] = \mathbf{E}_{\sigma_i^*, \sigma_{-i}} \left[\mathbf{E}_{\sigma_i^{x_{n-1}}, \sigma_{-i}^{x_{n-1}}} \left[f_i \mid [T_{x_{n-1}}] \right] \right] \geq \overline{v}_i(\Gamma_p) - \varepsilon.$$

Since σ_{-i} is arbitrary,

$$\overline{v}_i(\Gamma) = \inf_{\sigma_{-i} \in \Sigma_{-i}(T)} \sup_{\sigma_i \in \Sigma_i(T)} \mathbf{E}_{\sigma_i, \sigma_{-i}}[f_i] \geq \overline{v}_i(\Gamma_p) - \varepsilon.$$

$\square$

As we next show, when the payoff functions of the players are tail measurable, an ε-equilibrium exists for every $\varepsilon > 0$.

Theorem 6.62 (Ashkenazi-Golan, Flesch, Predtechinski, Solan (2022)). *Let $\Gamma = (N, (X_i)_{i \in N}, (f_i)_{i \in N})$ be a multiplayer repeated game. If the functions $(f_i)_{i \in N}$ are tail measurable, then the game admits an ε-equilibrium, for every $\varepsilon > 0$.*

Proof. **Step 1:** Identifying a good play x^*.

By Theorem 6.26, there exists a vector of behavior strategies $\sigma \in \Sigma(T)$ such that for every player $i \in N$ and every position $p \in T$,

$$\mathbf{E}_\sigma \left[f_i \mid [T_p] \right] \geq \overline{v}_i(\Gamma_p) - \varepsilon = \overline{v}_i(\Gamma) - \varepsilon,$$

where the equality follows from Theorem 6.60. Lévy's zero-one law (Theorem 2.48) implies that

$$\mathbf{P}_\sigma \left(f_i(x) \geq \overline{v}_i(\Gamma) - \varepsilon \right) = 1, \quad \forall i \in N.$$

Since the set N of players is finite, it follows that there exists a play $x^* = \langle \vec{a}_0^*, \vec{a}_1^*, \dots \rangle \in [T]$ such that

$$f_i(x^*) \geq \overline{v}_i(\Gamma) - \varepsilon, \quad \forall i \in N. \tag{6.32}$$

We will show that the vector of behavior strategies σ^* in which the players follow x^*, and the first player who deviates from x^* is punished at her minmax value, is a 2ε-equilibrium. We start by formally defining σ^*.

Step 2: Definition of σ^*.

For each $j \in N$ and every position $p \in T$, let $\sigma_{-j}^{j,p} \in \Sigma_{-j}(T_p)$ be an ε-minmax strategy of the players $N \setminus \{j\}$ against player j in the subgame Γ_p, so that

$$\sup_{\sigma_j^p \in \Sigma_j(T_p)} \mathbf{E}_{\sigma_j^p, \sigma_{-j}^{j,p}}[f_j \mid [T_p]] \leq \overline{v}_j(\Gamma_p) + \varepsilon = \overline{v}_j(\Gamma) + \varepsilon.$$

For each position $p \in T$, denote by $n(p, x^*)$ the first stage in which p differs from x^*:

$$n(p, x^*) := \min\{k \leq n - 1 \colon \vec{a}_k \neq \vec{a}_k^*\}, \quad \forall p = \langle \vec{a}_0, \vec{a}_1, \ldots, \vec{a}_{n-1} \rangle \in T.$$

The minimum of an empty set is ∞, so that $n(p, x^*) = \infty$ when p is a prefix of x^*. If $n(p, x^*) < \infty$, denote by $j(p, x^*)$ the player who caused the play under p to deviate from x^*:

$$j(p, x^*) := \min\{i \in N \colon a_{n(p,x^*),i} \neq a_{n(p,x^*),i}^*\}.$$

Since there may be several players who deviate from x^* at stage $n(p, x^*)$, yet only one player can be punished, to properly define $j(p, x^*)$ we order the elements in N in some arbitrary way and take the minimum among the deviating players. Because the definition of an ε-equilibrium takes into account only deviations by a single player, the definition of the strategy if more than one player deviates from x^* is not important.

For each player $i \in N$, let σ_i^* be the behavior strategy defined as follows:

$$\sigma_i^*(p) := \begin{cases} a_{n,i}^*, & \text{if } p = \langle \vec{a}_0^*, \vec{a}_1^*, \ldots \vec{a}_{n-1}^* \rangle, \\ \sigma_i^{j(p,x^*), p_{n(p,x^*)}}(p), & \text{if } p \not\prec x. \end{cases}$$

That is, the players follow the play x^* as long as no player deviates, and once at least one player deviates, they punish player $j(p, x^*)$.

Step 3: σ^* is a 2ε-equilibrium.

By Equation (6.32),

$$\mathbf{E}_{\sigma^*}[f_i] = f_i(x^*) \geq \overline{v}_i(\Gamma) - \varepsilon.$$

To show that no player can profit more than 2ε by deviating, fix a player $i \in N$ and a strategy $\sigma_i \in \Sigma_i(T)$. Denote by μ the probability that the play under $(\sigma_i, \sigma_{-i}^*)$ is x^*:

$$\mu := \mathbf{P}_{\sigma_i, \sigma_{-i}^*}(x = x^*). \tag{6.33}$$

Here, x is a random variable that is equal to the play generated by the players. By Theorem 6.60, the payoff to player i once she deviates and the other players start punishing her is at most $\overline{v}_i(\Gamma) + \varepsilon$, hence

$$\mathbf{E}_{\sigma_i, \sigma_{-i}^*}[f_i] \leq \mu f_i(x^*) + (1 - \mu)(\overline{v}_i(\Gamma) + \varepsilon) \leq \mathbf{E}_{\sigma^*}[f_i] + 2\varepsilon.$$

$\square$

Remark 6.63 (Repeated games with tail-measurable payoffs and countably many players). *The model of repeated games can be generalized to a situation when the number of players is countably infinite. As Ashkenazi-Golan, Flesch, Predtetchinski, and Solan (2022) show, Theorem 6.26 is valid in repeated games with tail-measurable payoffs and countably many players, and hence Theorem 6.62 is valid in this setup as well.*

Remark 6.64 (On the set of equilibrium payoffs in repeated games with tail-measurable payoffs). *Theorem 6.60 states that when the payoff functions are tail measurable, the minmax value (resp., the maxmin value) in all subgames is the same. Ashkenazi-Golan, Flesch, Predtetchinski, and Solan (2025) showed that in this case, the set of equilibrium payoffs in all subgames is the same as well.*

6.8   THE BIG MATCH WITH TAIL-MEASURABLE PAYOFFS

We already encountered the Big Match, which is a two-player zero-sum repeated game, in Example 5.44. In this section we study a nonzero-sum version of the game, and prove that it admits an ε-equilibrium, for every $\varepsilon > 0$.

Definition 6.65 (Big Match game). *Let $u : \{L, R\}^{\mathbb{N}} \to \mathbb{R}^2$ be bounded and Borel-measurable function. The* Big Match game $\Gamma(u) = (\{I, II\}, (\{L, R\}, \{L, R\}), (f_I, f_{II}))$ *is a repeated game where*

- *The set of players is $N = \{I, II\}$.*

- *The set of moves of each player at each position is $\{L, R\}$.*

- *For every play $x = \langle \vec{a}_0, \vec{a}_1, \dots \rangle \in (\{L, R\}, \{L, R\})^{\mathbb{N}}$, let $n_*(x) := \min\{n \in \mathbb{N} : a_{n,I} = L\}$ be the first stage in which Player I selects the move L. The payoffs $f_I(x)$ and $f_{II}(x)$ are defined as follows:*

  - *If $n_*(x) < \infty$, then*

$$f_I(x) = \begin{cases} 1, & \text{if } a_{n_*,II} = L, \\ 0, & \text{if } a_{n_*,II} = R, \end{cases}$$

$$f_{II}(x) = \begin{cases} 0, & \text{if } a_{n_*,II} = L, \\ 1, & \text{if } a_{n_*,II} = R. \end{cases}$$

  - *If $n_* = \infty$,*
$$f_i(x) = u_i(\langle a_{0,II}, a_{1,II}, \dots \rangle), \quad \forall i \in \{I, II\}.$$

In words, if Player I ever selects the move L, the payoff is as in the Big Match in Example 5.44; if, on the other hand, Player I always selects the move R, the payoff is given by u, see Figure 22.

		Player II	
		L	R
Player I	R		
	L	$1, 0$ *	$0, 1$ *

Figure 22: The payoffs in the Big Match game when Player I ever selects L.

Note that even when u is tail measurable, the payoff function of the Big Match game $\Gamma(u)$ is not tail measurable, because, e.g.,

$$f(\langle(L, L), (R, R), (R, R), \dots\rangle) = (1, 0),$$

while

$$f(\langle(L, R), (R, R), (R, R), \dots\rangle) = (0, 1)$$

and

$$f(\langle(R, R), (R, R), (R, R), \dots\rangle) = u(\langle R, R, R, \dots\rangle).$$

The main result we prove in this section is the following.

Theorem 6.66 (Ashkenazi-Golan, Flesch, and Solan, 2022). *Every Big Match game $\Gamma(u)$ where the function u is tail measurable admits an ε-equilibrium, for every $\varepsilon > 0$.*

The rest of this section is devoted to the proof of Theorem 6.66. In particular, we fix throughout a bounded, Borel-measurable, and tail-measurable function $u : \{L, R\}^{\mathbb{N}} \to \mathbb{R}^2$ and consider the Big Match game $\Gamma(u)$.

Recall that $\underline{v}_i(\Gamma(u))$ is player i's maxmin value in the game $\Gamma(u)$, and $\overline{v}_i(\Gamma(u))$ is her minmax value in this game. As discussed in Remark 6.19, $\underline{v}_i(\Gamma(u)) = \overline{v}_i(\Gamma(u))$ for $i = \mathrm{I}, \mathrm{II}$.

The proof will be divided into several steps, as indicated in Figure 23.

1. $\overline{v}_{\mathrm{I}}(\Gamma(u)) \geq 0$ (Lemma 6.68).

2. $\overline{v}_{\mathrm{II}}(\Gamma(u)) \leq 1$ (Lemma 6.69).

3. If $\overline{v}_{\mathrm{I}}(\Gamma(u)) > 1$, then an ε-equilibrium exists for every $\varepsilon > 0$ (Lemma 6.70).

4. If $\overline{v}_{\mathrm{I}}(\Gamma(u)) \in [0, 1]$ and $\overline{v}_{\mathrm{I}}(\Gamma(u)) + \overline{v}_{\mathrm{II}}(\Gamma(u)) \leq 1$, then an ε-equilibrium exists for every $\varepsilon > 0$ (Lemma 6.71).

5. If $\overline{v}_{\mathrm{I}}(\Gamma(u)) + \overline{v}_{\mathrm{II}}(\Gamma(u)) > 1$, then an ε-equilibrium exists for every $\varepsilon > 0$ (Lemma 6.72).

In each of the three cases in the above list where we have an ε-equilibrium, the ε-equilibrium will have a different structure:

1. When $\overline{v}_{\mathrm{I}}(\Gamma(u)) > 1$, Player I will always select the move R, and Player II will select a sequence of moves that maximizes her payoff under u_{II}.

2. When $\overline{v}_{\mathrm{I}}(\Gamma(u)) + \overline{v}_{\mathrm{II}}(\Gamma(u)) \leq 1$, we will devise an ε-equilibrium where the sum of the payoffs of the two players is 1: Player I will select the move L at a random time.

3. When $\overline{v}_{\mathrm{I}}(\Gamma(u)) + \overline{v}_{\mathrm{II}}(\Gamma(u)) > 1$, the sum of the payoffs of the two players in an ε-equilibrium must be larger than 1. We will construct an ε-equilibrium where Player I always plays R, and Player II follows a subgame ε-optimal strategy. To ensure that Player II does not deviate from that strategy, we will add a test that identifies such deviations and triggers a punishment.

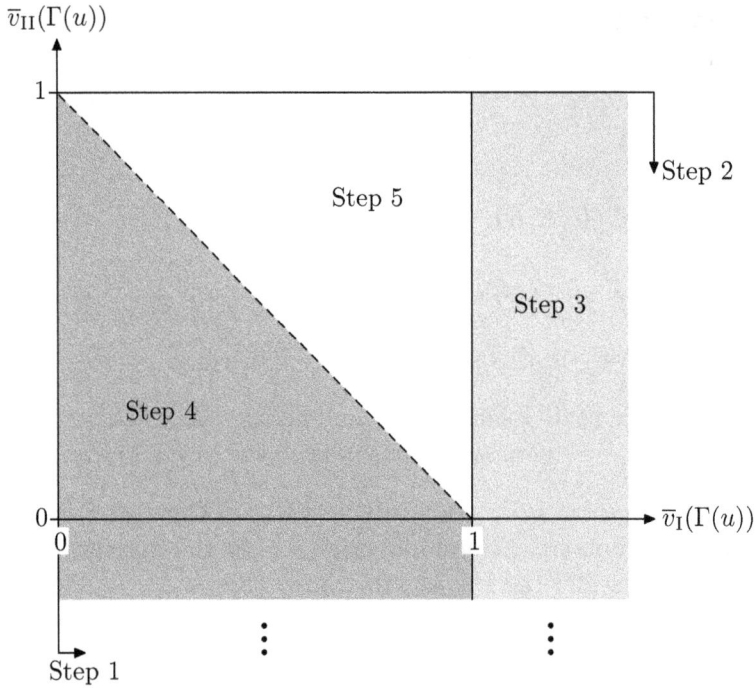

Figure 23: The steps of the proof of Theorem 6.66.

The fact that the ε-equilibria have different nature allows us to exhibit a variety of ideas to construct ε-equilibria in Blackwell games. We turn now to the formal proof.

Since u is tail measurable, the maxmin value and the minmax value of a player are independent of the position, as long as Player I selects R. This property, which is a variant of Theorem 6.60, is expressed by the following result, whose proof is left to the reader (Exercise 6.21).

For every position $p \in T = \left(\{L, R\} \times \{L, R\}\right)^{<\mathbb{N}}$, we denote the subgame that starts at position p by $\Gamma_p(u)$. Denote by $T^R \subset T$ the set of all the positions in T where Player I always selects the move R:

$$T^R := \{(R, L), (R, R)\}^{<\mathbb{N}}.$$

Lemma 6.67. *For each player* $i \in \{\mathrm{I}, \mathrm{II}\}$ *and every position* $p \in T^R$,

$$\overline{v}_i(\Gamma_p(u)) = \underline{v}_i(\Gamma_p(u)) = \overline{v}_i(\Gamma(u)) = \underline{v}_i(\Gamma(u)).$$

By selecting the move L in stage 0, Player I guarantees a payoff of at least 0. Therefore, we have:

Lemma 6.68. *The minmax value of Player* I *is non-negative:*

$$\overline{v}_{\mathrm{I}}(\Gamma(u)) \geq 0.$$

By selecting the move L in stage 0, Player I guarantees that the payoff of Player II is at most 1. Therefore, we have:

Lemma 6.69. *The minmax value of Player* II *is at most* 1:

$$\overline{v}_{\mathrm{II}}(\Gamma(u)) \leq 1.$$

Lemma 6.70. *If* $\overline{v}_{\mathrm{I}}(\Gamma(u)) > 1$, *then the game* $\Gamma(u)$ *admits an* ε-*equilibrium, for every* $\varepsilon > 0$.

Proof. Since $\overline{v}_{\mathrm{I}}(\Gamma(u)) > 1$, the function u_{I} must be strictly larger than 1. Indeed, if $u_{\mathrm{I}}(\langle a_{0,\mathrm{II}}, a_{1,\mathrm{II}}, \ldots \rangle) \leq 1$ for some sequence $\langle a_{0,\mathrm{II}}, a_{1,\mathrm{II}}, \ldots \rangle \in \{L, R\}^{\mathbb{N}}$, then by selecting this sequence of moves Player II guarantees that the payoff of Player I is at most 1, which contradicts the assumption that $\overline{v}_{\mathrm{I}}(\Gamma(u)) > 1$.

Fix $\varepsilon > 0$. Let $\langle a_{0,\mathrm{II}}^*, a_{1,\mathrm{II}}^*, \ldots \rangle$ be a sequence of moves of Player II that maximizes u_{II} up to ε:

$$u_{\mathrm{II}}(\langle a_{0,\mathrm{II}}^*, a_{1,\mathrm{II}}^*, \ldots \rangle) > \sup_{\langle a_{0,\mathrm{II}}, a_{1,\mathrm{II}}, \ldots \rangle \in \{L,R\}^{\mathbb{N}}} u_{\mathrm{II}}(\langle a_{0,\mathrm{II}}, a_{1,\mathrm{II}}, \ldots \rangle) - \varepsilon. \tag{6.34}$$

An ε-equilibrium $\sigma^* = (\sigma_{\mathrm{I}}^*, \sigma_{\mathrm{II}}^*)$ is as follows:

- Player I always selects the move R, that is, $\sigma_{\mathrm{I}}^*(p) = R$ for every $p \in T$.

- Player II select the sequence of moves $\langle a_{0,\mathrm{II}}^*, a_{1,\mathrm{II}}^*, \ldots \rangle$, that is, $\sigma_{\mathrm{II}}^*(p) = a_{\mathrm{len}(p),\mathrm{II}}^*$ for every position $p \in T$.

Indeed, under σ^*, the play is $\langle a_{0,\mathrm{II}}^*, a_{1,\mathrm{II}}^*, \ldots \rangle$, and therefore the players' payoffs are $u_{\mathrm{I}}(\langle a_{0,\mathrm{II}}^*, a_{1,\mathrm{II}}^*, \ldots \rangle) > 1$ and $u_{\mathrm{II}}(\langle a_{0,\mathrm{II}}^*, a_{1,\mathrm{II}}^*, \ldots \rangle)$, respectively. Player I cannot profit by deviating, since by deviating her payoff is either 1 or 0, depending on whether Player II selected L or R at the stage in which Player I deviated and selected L. By Equation (6.34), Player II cannot profit more than ε by deviating. It follows that σ^* is an ε-equilibrium. $\quad\square$

Lemma 6.71. *If* $\overline{v}_{\mathrm{I}}(\Gamma(u)) \leq 1$ *and* $\overline{v}_{\mathrm{I}}(\Gamma(u)) + \overline{v}_{\mathrm{II}}(\Gamma(u)) \leq 1$, *then the game admits an* ε-*equilibrium, for every* $\varepsilon > 0$.

Proof. **General idea:**
We start by explaining the main ideas behind the proof. Let $N \in \mathbb{N}$ be sufficiently large, and define a strategy pair $\sigma^N = (\sigma_{\mathrm{I}}^N, \sigma_{\mathrm{II}}^N)$ as follows.

- At the outset of the game, σ_{I}^N selects an integer $k \in \{0, 1, \ldots, N-1\}$. It then selects the move L in stage k, and the move R in all other stages.

- At every stage $n \in \{0, 1, \ldots, N-1\}$, σ_{II}^N selects the move L with probability $\overline{v}_{\mathrm{I}}(\Gamma(u))$ and the move R with probability $1 - \overline{v}_{\mathrm{I}}(\Gamma(u))$. Note that by Lemma 6.68 and by assumption, $\overline{v}_{\mathrm{I}}(\Gamma(u)) \in [0, 1]$.

Under this pair of strategies, the payoffs of the two players are $\overline{v}_{\mathrm{I}}(\Gamma(u))$ and $1 - \overline{v}_{\mathrm{I}}(\Gamma(u))$, respectively. Since $\overline{v}_{\mathrm{I}}(\Gamma(u)) + \overline{v}_{\mathrm{II}}(\Gamma(u)) \leq 1$, Player II's payoff is not less than $\overline{v}_{\mathrm{II}}(\Gamma(u))$.

Note that we did not define σ^N at and after stage N. This strategy pair is not necessarily an ε-equilibrium. Player II can profit by always selecting R, in which case her payoff will be 1, because Player I will select the move L before stage N. If we are not careful, and define σ^N in such a way that the payoff to Player I if she selects R in the first N stages is higher than $\bar{v}_I(\Gamma(u))$, then Player I may profit by selecting R in the first N stages.

How can we deter the players from deviating? Detering Player I is easy: If Player I deviates yet still selects the move L in the first N stages, her payoff is still $\bar{v}_I(\Gamma(u))$, and she does not gain. If Player I deviates and selects only the move R in the first N stages, then at stage N Player II can start punishing Player I, by switching to an ε-minmax strategy against Player I in $\Gamma_p(u)$, where p is the position at stage N. This ensures that Player I's payoff does not increase by more than ε.

Detering Player II is more intricate. Player II must select the move R with frequency $\bar{v}_{II}(\Gamma(u))$, and she gains by increasing this frequency. To ensure that Player II does not increase the frequency of R, at every stage Player I will check whether the percentage of past stages in which Player II selected R is not much higher than $\bar{v}_{II}(\Gamma(u))$. If this percentage is much higher than $\bar{v}_{II}(\Gamma(u))$, Player I will declare Player II as a deviator, and will start punishing Player II, by following an ε-minmax strategy against Player II in $\Gamma_p(u)$, where p is the position at the stage in which Player II is declared a deviator.

Step 1: Defining a strategy pair σ^*.

Fix $\varepsilon > 0$. Let $(X_k)_{k \in \mathbb{N}}$ be a sequence of Bernoulli i.i.d. random variables with parameter $\bar{v}_I(\Gamma(u))$. By the strong law of large numbers, there is $N_0 \in \mathbb{N}$ sufficiently large such that

$$\mathbf{P}\left(\left|\frac{1}{N}\sum_{k=0}^{N-1} X_k - \bar{v}_I(\Gamma(u))\right| < \varepsilon, \quad \forall N \geq N_0\right) \geq 1 - \varepsilon. \qquad (6.35)$$

Let $N_1 := \lceil \frac{N_0}{\varepsilon} \rceil$ be the smallest integer larger than $\frac{N_0}{\varepsilon}$.

For every $n \in \mathbb{N}$, denote the number of stages before stage n in which Player II selected the move L by ℓ_n:

$$\ell_n := \#\{k \in \{0, 1, \ldots, n-1\}: a_{k,II} = L\}.$$

For every position $p \in T$, let $n_*(p) \geq N_0$ be the first stage along p after stage N_0 in which ℓ_n/n is far from $\bar{v}_I(\Gamma(u))$:

$$n_*(p) := \min\{n \in \{N_0, N_0 + 1, \ldots, \text{len}(p)\}: |\ell_n/n - \bar{v}_I(\Gamma(u))| \geq \varepsilon\}.$$

As usual, the minimum of an empty set is ∞. At stage $n_*(p)$, Player I will start punishing Player II.

For every $p \in T$, let $\sigma_I^p \in \Sigma_I(T_p)$ be an ε-minmax strategy of Player I against Player II in $\Gamma_p(u)$, and let $\sigma_{II}^p \in \Sigma_{II}(T_p)$ be an ε-minmax strategy of Player II against Player I in $\Gamma_p(u)$.

Let $\sigma_I^* \in \Sigma_I(T)$ be the strategy of Player I which

- Uniformly selects an integer $k \in \{0, 1, \ldots, N_1\}$.

- At stage n, when the position is p:

  - If $n_*(p) = \infty$ (no deviation of Player II was announced so far) and $n = k$, select the move L.
  - If $n_*(p) = \infty$ (no deviation of Player II was announced so far) and $n \neq k$, select the move R.
  - If $n_*(p) = n$ (a deviation of Player II is presently announced), start following the strategy σ_{I}^p, that is,

$$\sigma_{\mathrm{I}}^*(p') := \sigma_{\mathrm{I}}^p(p'), \quad \forall p' \succeq p.$$

Let $\sigma_{\mathrm{II}}^* \in \Sigma_{\mathrm{II}}(T)$ be the following strategy of Player II:

$$\sigma_{\mathrm{II}}^*(p) := \begin{cases} [\overline{v}_{\mathrm{I}}(\Gamma(u))(L), (1 - \overline{v}_{\mathrm{I}}(\Gamma(u)))(R)], & \text{if } \mathrm{len}(p) < N_1, \\ \sigma_{\mathrm{I}}^{p_{N_1}}(p), & \text{if } \mathrm{len}(p) \geq N_1. \end{cases}$$

Thus, σ_{II}^* selects the mixed move $[\overline{v}_{\mathrm{I}}(\Gamma(u))(L), (1 - \overline{v}_{\mathrm{I}}(\Gamma(u)))(R)]$ in the first N_1 stages, and then switches to a punishment strategy against Player I.

Step 2: The expected payoff under $\sigma^* := (\sigma_{\mathrm{I}}^*, \sigma_{\mathrm{II}}^*)$.

Suppose the players follow the strategy pair σ^*. By Equation (6.35), the probability that Player II fails the statistical test is at most ε. If Player II does not fail the test, then at the stage in which Player I selects L, Player II selects the mixed move $[\overline{v}_{\mathrm{I}}(\Gamma(u))(L), (1 - \overline{v}_{\mathrm{I}}(\Gamma(u)))(R)]$. Therefore,

$$\mathbf{E}_{\sigma^*}[f_{\mathrm{I}}] \geq \overline{v}_{\mathrm{I}}(\Gamma(u)) - \varepsilon, \tag{6.36}$$

and

$$\mathbf{E}_{\sigma^*}[f_{\mathrm{II}}] \geq 1 - \overline{v}_{\mathrm{I}}(\Gamma(u)) - \varepsilon \geq \overline{v}_{\mathrm{II}}(\Gamma(u)) - \varepsilon.$$

Step 3: Player I cannot profit more than 3ε by deviating.

Let $\sigma_{\mathrm{I}} \in \Sigma_{\mathrm{I}}(T)$ be any strategy of Player I. Denote by μ the probability that Player I selects the move L at least once before stage N_1:

$$\mu := \mathbf{P}_{\sigma_{\mathrm{I}}, \sigma_{\mathrm{II}}^*}(a_{\mathrm{I},n} = L \text{ for some } n < N_1).$$

Since at stage N_1 Player II switches to an ε-minmax strategy in the remaining subgame,

$$\begin{aligned} \mathbf{E}_{\sigma_{\mathrm{I}}, \sigma_{\mathrm{II}}^*}[f_{\mathrm{I}}] &\leq \mu \cdot \overline{v}_{\mathrm{I}}(\Gamma(u)) + (1 - \mu)(\overline{v}_{\mathrm{I}}(\Gamma(u)) + \varepsilon) \tag{6.37} \\ &\leq \overline{v}_{\mathrm{I}}(\Gamma(u)) + \varepsilon \\ &\leq \mathbf{E}_{\sigma^*}[f_{\mathrm{I}}] + 2\varepsilon, \tag{6.38} \end{aligned}$$

where Equation (6.37) holds by the definition of μ and σ_{II}^*, and Equation (6.38) follows from Equation (6.36).

Step 4: Player II cannot profit more than 4ε by deviating.

By Remark 6.48, it suffices to show that Player II cannot profit more than ε by deviating to any pure strategy $s_{\mathrm{II}} \in S_{\mathrm{II}}(T)$. Let $\langle a_{0,\mathrm{II}}, a_{1,\mathrm{II}}, \ldots, a_{N_1-1,\mathrm{II}} \rangle$ be the sequence of moves selected by s_{II} in the first N_1 stages, when Player I always selects the move R. Recall that k is the random stage in which Player I selects L.

Denote
$$p = \langle (R, a_{0,\mathrm{II}}), (R, a_{1,\mathrm{II}}), \ldots, (R, a_{N_1-1,\mathrm{II}}) \rangle.$$

Note that the play after stage k is irrelevant for the payoffs of the players under $(\sigma_{\mathrm{I}}^*, s_{\mathrm{II}})$, since under σ_{I}^* Player I selects the move L at stage k.

By the definition of $n_*(p)$, in stages $0, 1, \ldots, n_*(p) - 1$, the number of times Player II selects L is at least $n_*(p) \cdot (\overline{v}_{\mathrm{I}}(\Gamma(u)) - \varepsilon)$. Therefore, the number of times Player II selects R in stages $0, 1, \ldots, n_*(p) - 1$ is at most $n_*(p) \cdot (1 - \overline{v}_{\mathrm{I}}(\Gamma(u)) + \varepsilon)$. Hence, the expected payoff given $k < n_*(p)$ is at most $1 - \overline{v}_{\mathrm{I}}(\Gamma(u)) + \varepsilon$.

Since k is independent of the choices of Player II, the probability that $k < n_*(p)$ is $\frac{n_*(p)}{N_1}$. Therefore,

$$\mathbf{E}_{\sigma_{\mathrm{I}}^*, \sigma_{\mathrm{II}}}[f_{\mathrm{II}}] \leq \mathbf{E}_{\sigma_{\mathrm{I}}^*, \sigma_{\mathrm{II}}} \left[\frac{N_0}{N_1} + \mathbf{1}_{\{n_*(p) < N_1\}} \frac{n_*(p) - N_0}{N_1} (1 - \overline{v}_{\mathrm{I}}(\Gamma(u)) + \varepsilon) \right.$$
$$\left. + \frac{1}{N_1} + \mathbf{1}_{\{n_*(p) = \infty\}} (\overline{v}_{\mathrm{II}}(\Gamma(u)) + \varepsilon) \right] \tag{6.39}$$
$$\leq \mathbf{E}_{\sigma_{\mathrm{I}}^*, \sigma_{\mathrm{II}}^*}[f_{\mathrm{II}}] + 3\varepsilon, \tag{6.40}$$

where the first term in Equation (6.39) bounds the payoff when $k < N_0$ and there were not enough stages to check whether Player II selects her moves randomly, the second term bounds the payoff when Player I selects L before Player II is announced as a deviator, the third term bounds the payoff when Player I selects L at the same stage Player II is announced a deviator, and the fourth term in Equation (6.39) bounds the payoff when Player I did not select L before or at the stage in which Player II is announced a deviator. It follows that Player II cannot profit more than 4ε by deviating from σ_{II}^*. □

It remains to handle the case where the sum of minmax values of the two players is high.

Lemma 6.72. *If $v_{\mathrm{I}}(\Gamma(u)) + v_{\mathrm{II}}(\Gamma(u)) > 1$, then the game admits an ε-equilibrium, for every $\varepsilon > 0$.*

Proof. Theorem 6.49 implies that in all ε-equilibria, the sum of payoffs of the players is at least $\overline{v}_{\mathrm{I}}(\Gamma(u)) + \overline{v}_{\mathrm{II}}(\Gamma(u)) - 2\varepsilon$. Since $v_{\mathrm{I}}(\Gamma(u)) + v_{\mathrm{II}}(\Gamma(u)) > 1$, when ε is sufficiently small the sum of payoffs in all ε-equilibria is more than 1, yet 1 is the sum of payoffs when Player I selects the move L. We will construct an ε-equilibrium in which Player I always selects the move R with high probability.

The idea of the construction is as follows. Suppose that Player II follows a subgame ε-maxmin strategy $\widehat{\sigma}_{\mathrm{II}}^\varepsilon$. Then in all subgames Γ_p for $p \in T^R$, Player II's expected payoff is at least $\overline{v}_{\mathrm{II}}(\Gamma(u)) - \varepsilon$. If Player I selects the move L for the first time at position p, then Player II's payoff in the subgame $\Gamma_p(u)$ is $(\widehat{\sigma}_{\mathrm{II}}^\varepsilon(p))(R)$, the probability with which she selects the move R at position p. This implies that this probability is at least $\overline{v}_{\mathrm{II}}(\Gamma(u)) - \varepsilon$. Since $v_{\mathrm{I}}(\Gamma(u)) + v_{\mathrm{II}}(\Gamma(u)) > 1 + 2\varepsilon$, provided ε is sufficiently small, Player I's payoff in this case is below $v_{\mathrm{I}}(\Gamma(u)) - \varepsilon$, and hence Player I's best response to $\widehat{\sigma}_{\mathrm{II}}^{\varepsilon}$ is the strategy σ_{I}^{R} that always selects the move R. The pair $(\sigma_{\mathrm{I}}^{R}, \widehat{\sigma}_{\mathrm{II}}^{\varepsilon})$ is not an ε-equilibrium, because $\widehat{\sigma}_{\mathrm{II}}^{\varepsilon}$ might not be Player II's best response to σ_{I}^{R}: Player II will prefer to select a sequence of moves $\langle a_{0,\mathrm{II}}, a_{1,\mathrm{II}}, \ldots \rangle$ that maximizes u_{II}. Using the Borel measurability of u_{II} we will find a position p^* such that in the subgame $[T_{p^*}]$, under $\widehat{\sigma}_{\mathrm{II}}^{\varepsilon}$ the value of u_{II} is almost constant. The tail-measurability of u_{II} implies that the moves selected in the first $\mathrm{len}(p^*)$ stages do not affect the payoff. We will therefore have Player II select the move R in the first $\mathrm{len}(p^*)$ stages (so that Player I will not have incentive to select L in these stages), and afterwards follow the continuation of $\widehat{\sigma}_{\mathrm{II}}^{\varepsilon}$, assuming the position p^* was reached in stage $\mathrm{len}(p^*)$. So that Player II does not deviate from this strategy, we will add to Player I's strategy instructions that punish Player II if a deviation is detected.

Step 1: Definition of a behavior strategy $\widehat{\sigma}_{\mathrm{II}}^{\varepsilon} \in \Sigma_{\mathrm{II}}(T)$.

Fix $\varepsilon > 0$ such that

$$\varepsilon < \frac{1}{2}\big(\overline{v}_{\mathrm{I}}(\Gamma(u)) + \overline{v}_{\mathrm{II}}(\Gamma(u)) - 1\big). \tag{6.41}$$

Let $\widehat{\sigma}_{\mathrm{II}}^{\varepsilon}$ be a subgame ε-maxmin strategy of Player II in $\Gamma(u)$. By Lemma 6.67,

$$\mathbf{E}_{\sigma_{\mathrm{I}}, \widehat{\sigma}_{\mathrm{II}}^{\varepsilon}}[f_{\mathrm{II}} \mid [T_p]] \geq \overline{v}_{\mathrm{II}}(\Gamma_p(u)) - \varepsilon = \overline{v}_{\mathrm{II}}(\Gamma(u)) - \varepsilon, \quad \forall \sigma_{\mathrm{I}} \in \Sigma_{\mathrm{I}}(T), \forall p \in T^R. \tag{6.42}$$

For every position $p \in T^R$, $\widehat{\sigma}_{\mathrm{II}}^{\varepsilon}(p)$ is a mixed move, namely, a probability distribution on $\{L, R\}$. Denote by $(\widehat{\sigma}_{\mathrm{II}}^{\varepsilon}(p))(a_{\mathrm{II}})$ the probability that Player II selects the move $a_{\mathrm{II}} \in \{L, R\}$ at the position p under $\widehat{\sigma}_{\mathrm{II}}^{\varepsilon}$.

Step 2: $(\widehat{\sigma}_{\mathrm{II}}^{\varepsilon}(p))(R) \geq \overline{v}_{\mathrm{II}}(\Gamma(u)) - \varepsilon$, for every position $p \in T^R$.

Fix $p \in T^R$, and denote by $\widetilde{\sigma}_{\mathrm{I}} \in \Sigma_{\mathrm{I}}(T)$ the strategy that selects R in the first $\mathrm{len}(p) - 1$ stages, and selects L afterwards. In the subgame $\Gamma_p(u)$ Player I selects L at position p, so that Player II's payoff under $(\widetilde{\sigma}_{\mathrm{I}}, \widehat{\sigma}_{\mathrm{II}}^{\varepsilon})$ is $(\widehat{\sigma}_{\mathrm{II}}^{\varepsilon}(p))(R)$. Since $\widehat{\sigma}_{\mathrm{II}}^{\varepsilon}$ is a subgame ε-maxmin strategy, we have

$$\overline{v}_{\mathrm{II}}(\Gamma(u)) - \varepsilon \leq \mathbf{E}_{\widetilde{\sigma}_{\mathrm{I}}, \widehat{\sigma}_{\mathrm{II}}^{\varepsilon}}[f_{\mathrm{II}} \mid [T_p]] = (\widehat{\sigma}_{\mathrm{II}}^{\varepsilon}(p))(R),$$

and the claim follows.

Recall that n_* is the first stage in which Player I selects the move L. We now prove that in any subgame, the expected payoff of Player I on the event that Player I ever selects L is low.

Step 3: $\mathbf{E}_{\sigma_{\mathrm{I}}, \widehat{\sigma}_{\mathrm{II}}^{\varepsilon}}[f_{\mathrm{I}} \mid n_* < \infty, [T_p]] < \overline{v}_{\mathrm{I}}(\Gamma(u)) - \varepsilon$, for every $p \in T^R$ and every $\sigma_{\mathrm{I}} \in \Sigma_{\mathrm{I}}(T)$.

The claim follows from the following chain of equaltiy and inequalities:

$$\mathbf{E}_{\sigma_\mathrm{I},\widehat{\sigma}_\mathrm{II}^\varepsilon}[f_\mathrm{I} \mid n_* < \infty, [T_p]] = 1 - \mathbf{E}_{\sigma_\mathrm{I},\widehat{\sigma}_\mathrm{II}^\varepsilon}[f_\mathrm{II} \mid n_* < \infty, [T_p]] \tag{6.43}$$

$$\leq 1 - \overline{v}_\mathrm{II}(\Gamma(u)) + \varepsilon \tag{6.44}$$

$$< \overline{v}_\mathrm{I}(\Gamma(u)) - \varepsilon, \quad \forall p \in T^R, \forall \sigma_\mathrm{I} \in \Sigma_\mathrm{I}(T), \tag{6.45}$$

where Equation (6.43) holds since the sum of payoffs of the two players when Player I selects the move L is 1, Equation (6.44) follows from Step 2, and Equation (6.45) holds by Equation (6.41).

Let σ_I^R be the behavior strategy of Player I that always selects the move R:

$$\sigma_\mathrm{I}^R(p) = R, \quad \forall p \in T.$$

We next show that σ_I^R is a best response of Player I to $\widehat{\sigma}_\mathrm{II}^\varepsilon$ in all subgames. The intuition for this result is that by Step 3, Player I's payoff by selecting L is low, hence the best response of Player I to $\widehat{\sigma}_\mathrm{II}^\varepsilon$ will never select L.

Step 4: $\mathbf{E}_{\sigma_\mathrm{I}^R,\widehat{\sigma}_\mathrm{II}^\varepsilon}[f_\mathrm{I} \mid [T_p]] \geq \mathbf{E}_{\sigma_\mathrm{I},\widehat{\sigma}_\mathrm{II}^\varepsilon}[f_\mathrm{I} \mid [T_p]]$, for every $p \in T^R$ and every $\sigma_\mathrm{I} \in \Sigma_\mathrm{I}(T)$. Fix $\delta > 0$. For every $p \in T^R$ set

$$\gamma_p := \sup_{\sigma_\mathrm{I} \in \Sigma_\mathrm{I}(T)} \mathbf{E}_{\sigma_\mathrm{I},\widehat{\sigma}_\mathrm{II}^\varepsilon}[f_\mathrm{I} \mid [T_p]].$$

This is the maximal payoff that Player I can obtain in the subgame $\Gamma_p(u)$ when Player II adopts the behavior strategy $\widehat{\sigma}_\mathrm{II}^\varepsilon$. The definition of the minmax value implies that

$$\gamma_p \geq \overline{v}_\mathrm{I}(\Gamma(u)). \tag{6.46}$$

Let $\sigma_\mathrm{I}^{\delta,p} \in \Sigma_\mathrm{I}(T)$ be a strategy that attains γ_p up to δ:

$$\mathbf{E}_{\sigma_\mathrm{I}^{\delta,p},\widehat{\sigma}_\mathrm{II}^\varepsilon}[f_\mathrm{I} \mid [T_p]] \geq \gamma_p - \delta.$$

To prove the desired inequality, fix a position $p \in T^R$, and imagine three persons, Alice, Alaya, and Almorava, who all take the part of Player I in $\Gamma_p(u)$, but use different strategies.

Alice as Player I adopts the behavior strategy $\sigma_\mathrm{I}^{\delta,p}$. Alaya as Player I adopts the following behavior strategy σ_I':

- As long as Alice selects R, Alaya also selects R.

- The first time Alice selects L, say, at position $p' \succeq p$, Alaya switches to $\sigma_\mathrm{I}^{\delta,p'}$.

Whenever Alice selects L, her payoff is at most $\overline{v}_\mathrm{I}(\Gamma(u)) - \varepsilon$, while Alaya's payoff is at least $\gamma_p - \delta \geq \overline{v}_\mathrm{I}(\Gamma(u)) - \delta$. It follows that

$$\gamma_p - \delta \leq \mathbf{E}_{\sigma_\mathrm{I}^{\delta,p},\widehat{\sigma}_\mathrm{II}^\varepsilon}[f_\mathrm{I} \mid [T_p]] \tag{6.47}$$

$$\leq \mathbf{E}_{\sigma_\mathrm{I}',\widehat{\sigma}_\mathrm{II}^\varepsilon}[f_\mathrm{I} \mid [T_p]] - \mathbf{P}_{\sigma_\mathrm{I}^{\delta,p},\widehat{\sigma}_\mathrm{II}^\varepsilon}(n_* < \infty \mid [T_p]) \cdot (\varepsilon - \delta) \tag{6.48}$$

$$\leq \gamma_p - \mathbf{P}_{\sigma_\mathrm{I}^{\delta,p},\widehat{\sigma}_\mathrm{II}^\varepsilon}(n_* < \infty \mid [T_p]) \cdot (\varepsilon - \delta), \tag{6.49}$$

where Equation (6.47) follows from the choice of $\sigma_{\mathrm{I}}^{\delta,p}$, Equation (6.48) holds since before stage n_* the behavior strategies $\sigma_{\mathrm{I}}^{\delta,p}$ and σ_{I}' coincide and since after stage n_* Alaya follows a δ-maxmin strategy, and Equation (6.49) holds by the definition of γ_p. As a result,

$$\mathbf{P}_{\sigma_{\mathrm{I}}^{\delta,p},\widehat{\sigma}_{\mathrm{II}}^{\varepsilon}}\left(n_* < \infty \mid [T_p]\right) \leq \frac{\delta}{\varepsilon - \delta}. \tag{6.50}$$

Almorava as Player I adopts the behavior strategy σ_{I}^R that always selects the move R. The play under $(\sigma_{\mathrm{I}}^{\delta,p}, \widehat{\sigma}_{\mathrm{II}}^{\varepsilon})$ and that under $(\sigma_{\mathrm{I}}^R, \widehat{\sigma}_{\mathrm{II}}^{\varepsilon})$ differ only on the event $\{n_* < \infty\}$. Since $\sigma_{\mathrm{I}}^{\delta,p}$ and σ_{I}^R coincide until stage n_*,

$$\begin{aligned}
\mathbf{E}_{\sigma_{\mathrm{I}}^R,\widehat{\sigma}_{\mathrm{II}}^{\varepsilon}}\left[f_{\mathrm{I}} \mid [T_p]\right] &\geq \mathbf{E}_{\sigma_{\mathrm{I}}^{\delta,p},\widehat{\sigma}_{\mathrm{II}}^{\varepsilon}}\left[f_{\mathrm{I}} \mid [T_p]\right] - 2\|f_{\mathrm{I}}\|_{\infty} \cdot \mathbf{P}_{\sigma_{\mathrm{I}}^{\delta,p},\widehat{\sigma}_{\mathrm{II}}^{\varepsilon}}\left(n_* < \infty \mid [T_p]\right) \\
&\geq \gamma_p - \delta - 2\|f_{\mathrm{I}}\|_{\infty} \cdot \frac{\delta}{\varepsilon - \delta}.
\end{aligned}$$

Since this inequality holds for every $\delta \in (0, \varepsilon)$, it follows that

$$\mathbf{E}_{\sigma_{\mathrm{I}}^R,\widehat{\sigma}_{\mathrm{II}}^{\varepsilon}}\left[f_{\mathrm{I}} \mid [T_p]\right] \geq \gamma_p,$$

and the claim follows.

Step 5: There exist $g \in \mathbb{R}^2$ and a position $p^* \in T^R$ such that $g_i \geq \overline{v}_i(\Gamma(u)) - 3\varepsilon\|f_i\|_{\infty}$ for $i \in \{\mathrm{I}, \mathrm{II}\}$ and $\mathbf{P}_{\sigma_{\mathrm{I}}^R,\widehat{\sigma}_{\mathrm{II}}^{\varepsilon}}\left(\|f(x) - g\|_{\infty} < \varepsilon \mid [T_{p^*}]\right) > 1 - \varepsilon$.

Since the set $[-\|f_{\mathrm{I}}\|_{\infty}, \|f_{\mathrm{I}}\|_{\infty}] \times [-\|f_{\mathrm{II}}\|_{\infty}, \|f_{\mathrm{II}}\|_{\infty}]$ is compact, there is a finite collection of vectors $\{g^0, g^1, \ldots, g^m\} \subseteq \mathbb{R}^2$ that is ε-dense in this set; that is, for every $z \in [-\|f_{\mathrm{I}}\|_{\infty}, \|f_{\mathrm{I}}\|_{\infty}] \times [-\|f_{\mathrm{II}}\|_{\infty}, \|f_{\mathrm{II}}\|_{\infty}]$ there is $j \in \{0, 1, \ldots, m\}$ such that $\|z - g^j\|_{\infty} < \varepsilon$.

It follows that for every $x \in [T]$,

$$\sum_{j=0}^{m} \mathbf{P}_{\sigma_{\mathrm{I}}^R,\widehat{\sigma}_{\mathrm{II}}^{\varepsilon}}\left(\|f(x) - g^j\|_{\infty} < \varepsilon\right) \geq 1.$$

Hence, for some $j \in \{0, 1, \ldots, m\}$ we have

$$\mathbf{P}_{\sigma_{\mathrm{I}}^R,\widehat{\sigma}_{\mathrm{II}}^{\varepsilon}}\left(\|f(x) - g^j\|_{\infty} < \varepsilon\right) > 0.$$

When Player I adopts σ_{I}^R, only positions in T^R can be reached, and hence by Corollary 2.49 there is a position $p^* \in T^R$ such that

$$\mathbf{P}_{\sigma_{\mathrm{I}}^R,\widehat{\sigma}_{\mathrm{II}}^{\varepsilon}}\left(\|f(x) - g^j\|_{\infty} < \varepsilon \mid [T_{p^*}]\right) > 1 - \varepsilon. \tag{6.51}$$

That is, the payoff under $(\sigma_{\mathrm{I}}^R, \widehat{\sigma}_{\mathrm{II}}^{\varepsilon})$ is close to g^j with high probability.

It remains to prove that $g_i^j \geq \overline{v}_i(\Gamma(u)) - 3\varepsilon\|f_i\|_{\infty}$ for $i \in \{\mathrm{I}, \mathrm{II}\}$. Equation (6.51) implies that for each $i \in \{\mathrm{I}, \mathrm{II}\}$ we have

$$\mathbf{E}_{\sigma_{\mathrm{I}}^R,\widehat{\sigma}_{\mathrm{II}}^{\varepsilon}}\left[f_i \mid [T_{p^*}]\right] \leq (1 - \varepsilon)(g_i^j + \varepsilon) + \varepsilon\|f_i\|_{\infty} \leq g_i^j + 2\varepsilon\|f_i\|_{\infty}. \tag{6.52}$$

Moreover, for Player I we have

$$
\begin{aligned}
\gamma_{p^*} &= \mathbf{E}_{\sigma_{\mathrm{I}}^R, \widehat{\sigma}_{\mathrm{II}}^\varepsilon}\left[f_{\mathrm{I}} \mid [T_{p^*}]\right] \\
&\geq (1-\varepsilon)(g_{\mathrm{I}}^j - \varepsilon) - \varepsilon \|f_{\mathrm{I}}\|_\infty \qquad (6.53) \\
&\geq g_{\mathrm{I}}^j - 2\varepsilon \|f_{\mathrm{I}}\|_\infty.
\end{aligned}
$$

Since $\widehat{\sigma}_{\mathrm{II}}^\varepsilon$ is a subgame ε-maxmin strategy, we have $\mathbf{E}_{\sigma_{\mathrm{I}}^R, \widehat{\sigma}_{\mathrm{II}}^\varepsilon}\left[f_{\mathrm{II}} \mid [T_{p^*}]\right] \geq \overline{v}_{\mathrm{II}}(\Gamma(u)) - \varepsilon$, and hence Equation (6.52) implies that

$$
g_{\mathrm{II}}^j \geq \overline{v}_{\mathrm{II}}(\Gamma(u)) - 3\varepsilon \|f_{\mathrm{I}}\|_\infty.
$$

Step 4 implies that

$$
\mathbf{E}_{\sigma_{\mathrm{I}}^R, \widehat{\sigma}_{\mathrm{II}}^\varepsilon}\left[f_{\mathrm{I}} \mid [T_{p^*}]\right] \geq \overline{v}_{\mathrm{I}}(\Gamma(u)),
$$

which, together with Equation (6.52), implies that

$$
g_{\mathrm{I}}^j \geq \overline{v}_{\mathrm{I}}(\Gamma(u)) - 2\varepsilon \|f_{\mathrm{I}}\|_\infty.
$$

The claim follows.

Step 6: Definition of a strategy pair $(\sigma_{\mathrm{I}}^*, \sigma_{\mathrm{II}}^*)$.

Let $\sigma_{\mathrm{II}}^* \in \Sigma_{\mathrm{II}}(T)$ be the behavior strategy of Player II that selects the mixed move R in the first $\mathrm{len}(p^*)$ stages, and afterwards follows $\widehat{\sigma}_{\mathrm{II}}^\varepsilon$ as if p^* is the position that was reached at stage $\mathrm{len}(p^*)$. Denoting $p^* = \langle \vec{a}_0^*, \vec{a}_1^*, \ldots, \vec{a}_{n-1}^* \rangle$, the behavior strategy σ_{II}^* is defined as follows:

$$
\sigma_{\mathrm{II}}^*(p) := \begin{cases} R, & \text{if } \mathrm{len}(p) \leq n, \\ \widehat{\sigma}_{\mathrm{II}}^\varepsilon(\langle \vec{a}_0^*, \vec{a}_1^*, \ldots, \vec{a}_{n-1}^*, \vec{a}_n, \vec{a}_{n+1}, \ldots, \vec{a}_{k-1} \rangle), & \text{if } p = \langle \vec{a}_0, \vec{a}_1, \ldots, \vec{a}_{k-1} \rangle \\ & \text{for } k > n. \end{cases}
$$

By Theorem 2.20, the probability measure $\mathbf{P}_{\sigma_{\mathrm{I}}^R, \sigma_{\mathrm{II}}^*}$ is regular. Hence, by Step 5, there exists a closed set $A \subseteq \{x \in [T] : \|f(x) - g\|_\infty < \varepsilon\}$ such that

$$
\mathbf{P}_{\sigma_{\mathrm{I}}^R, \sigma_{\mathrm{II}}^*}(A \mid [T_{p^*}]) > \mathbf{P}_{\sigma_{\mathrm{I}}^R, \sigma_{\mathrm{II}}^*}\left(\|f_{\mathrm{I}} - g\|_\infty < \varepsilon \mid [T_{p^*}]\right) - \varepsilon > 1 - 2\varepsilon. \qquad (6.54)
$$

As long as the play is in A, the payoff of Player II is close to g_{II}. We will devise σ_{I}^* in such a way that if the play is outside A, Player II will be punished at her minmax level.

The complement of A is an open set and, in particular, a union of basic open sets: there is a set $Z \subseteq T$ such that no element of Z is a prefix of any other element of Z, and

$$
A^c = \bigcup_{p \in Z} [T_p].
$$

Let $\sigma_{\mathrm{I}}^* \in \Sigma_{\mathrm{I}}(T)$ be the behavior strategy of Player I that selects R until a position in Z is reached. Once such a position is reached (which is necessarily in T^R), σ_{I}^* lowers Player II's payoff to $\overline{v}_{\mathrm{II}}(\Gamma(u)) + \varepsilon$. To formally define σ_{I}^*, we need a few notations.

For every position $p = \langle \vec{a}_0, \vec{a}_1, \ldots, \vec{a}_{k-1} \rangle \in T$ longer than p^*, namely, with $k > n$, denote by $p' = \langle \vec{a}_0^*, \vec{a}_1^*, \ldots, \vec{a}_{n-1}^*, \vec{a}_n, \vec{a}_{n+1}, \ldots, \vec{a}_{k-1} \rangle$ the position generated from p by changing the first n pairs of moves to p^*. For every such position p, denote by $n_Z(p)$ the smallest $m \in \{n, n+1, \ldots, k-1\}$ such that $p'_m \in Z$. If there is no such prefix, set $n_Z(p) = \infty$. Thus if $n_Z(p) = \text{len}(p) - 1$, then at stage $\text{len}(p)$ it is known that the generated play will not be in A.

For every position $p \in T^R$, let $\hat{\sigma}_I^p$ be an ε-minmax strategy of Player I against Player II in the subgame $\Gamma_p(u)$. The behavior strategy σ_I^* is defined as follows:

$$\sigma_I^*(p) := \begin{cases} R, & \text{if } n_Z(p) = \infty, \\ \hat{\sigma}_I^{p'_{n_Z(p)}}(p'), & \text{if } n_Z(p) < \infty. \end{cases}$$

Step 7: $(\sigma_I^*, \sigma_{II}^*)$ is a $7\varepsilon\|f\|_\infty$-equilibrium.

We start by calculating the expected payoff under $(\sigma_I^*, \sigma_{II}^*)$. For every play $x = \langle \vec{a}_0, \vec{a}_1, \ldots \rangle \in [T]$, denote by $n_Z(x)$ the smallest $m \in \{n, n+1, \ldots, k\}$ such that

$$\langle \vec{a}_0^*, \vec{a}_1^*, \ldots, \vec{a}_{n-1}^*, \vec{a}_n, \vec{a}_{n+1}, \ldots, \vec{a}_m \rangle \in Z.$$

If there is no such prefix (that is, if $x \in A$), then $n_Z(x) = \infty$. Equation (6.54) implies that

$$\mathbf{P}_{\sigma_I^*, \sigma_{II}^*}(n_Z(x) < \infty) < 2\varepsilon.$$

It follows that for each $i \in \{\text{I}, \text{II}\}$ the expected payoff under $(\sigma_I^*, \sigma_{II}^*)$ satisfies

$$\begin{aligned} \mathbf{E}_{\sigma_I^*, \sigma_{II}^*}[f_i] &\geq \mathbf{P}_{\sigma_I^*, \sigma_{II}^*}(n_Z(x) = \infty) \cdot (g_i - \varepsilon) - \mathbf{P}_{\sigma_I^*, \sigma_{II}^*}(n_Z(x) < \infty) \cdot \|f_i\|_\infty \\ &\geq (1 - 2\varepsilon)(g_i - \varepsilon) - 2\varepsilon\|f_i\|_\infty \\ &\geq g_i - 5\varepsilon\|f_i\|_\infty. \end{aligned} \tag{6.55}$$

We next show that Player I cannot profit much by deviating from σ_I^*. Fix then a behavior strategy $\sigma_I \in \Sigma_I(T)$. Denote the probability that under $(\sigma_I, \sigma_{II}^*)$ Player I selects L in the first n_* stages by

$$\mu := \mathbf{P}_{\sigma_I, \sigma_{II}^*}(n_* < n).$$

Then,

$$\begin{aligned} \mathbf{E}_{\sigma_I, \sigma_{II}^*}[f_I] &\leq \mu \cdot 0 + (1 - \mu) \cdot \gamma_{p^*} & (6.56) \\ &\leq \mu \cdot \gamma_{p^*} + (1 - \mu) \cdot \gamma_{p^*} & (6.57) \\ &= \gamma_{p^*} & (6.58) \\ &\leq g_I + 2\varepsilon\|f_I\|_\infty & (6.59) \\ &\leq \mathbf{E}_{\sigma_I^*, \sigma_{II}^*}[f_I] + 7\varepsilon\|f_I\|_\infty, & (6.60) \end{aligned}$$

where Equation (6.56) holds because on the event $\{n_* < n\}$ Player I's payoff is 0, while on the event $\{n_* \geq n\}$ Player I cannot obtain more than her maximal payoff against σ_{II}^*, which is γ_{p^*}, Equation (6.57) follows from Equation (6.46) and

Lemma 6.68, Equation (6.59) follows from Equation (6.53), and Equation (6.60) follows from Equation (6.55).

We finally show that Player II cannot profit much by deviating from σ_{II}^*. Fix then a behavior strategy $\sigma_{II} \in \Sigma_{II}(T)$. By the definition of σ_I^*, on the set $\{n_Z(x) < \infty\}$ we have

$$\mathbf{E}_{\sigma_I^*,\sigma_{II}}[f_{II} \mid \{n_Z(x) < \infty\}] \leq \bar{v}_{II}(\Gamma(u)) + \varepsilon \leq g_{II} + 2\varepsilon,$$

while on the set $\{n_Z(x) = \infty\}$ we have

$$\mathbf{E}_{\sigma_I^*,\sigma_{II}}[f_{II} \mid \{n_Z(x) = \infty\}] \leq g_{II} + \varepsilon.$$

Together with Equation (6.55), this implies that

$$\mathbf{E}_{\sigma_I^*,\sigma_{II}}[f_{II}] \leq g_{II} + 2\varepsilon \leq \mathbf{E}_{\sigma_I^*,\sigma_{II}^*}[f_I] + 7\varepsilon \|f_{II}\|_\infty.$$

It follows that Player II cannot profit more than $7\varepsilon \|f_{II}\|_\infty$ by deviating. □

Remark 6.73. *The technique we used in Step 4 of the proof of Lemma 6.72 is called coupling.*

6.9 SOLVABLE POSITIONS

As mentioned before, whether every multiplayer (or even two-player) Blackwell game admits an ε-equilibrium for every $\varepsilon > 0$ is still an open problem. We will here prove that for every $\varepsilon > 0$, at least one subgame of Γ admits an ε-equilibrium. This subgame may depend on ε.

Theorem 6.74 (Flesch and Solan, 2024). *For every multiplayer Blackwell game* $\Gamma = (N, T, (f_i)_{i \in N})$ *and every* $\varepsilon > 0$ *there is a position* $p^* \in T$ *such that the subgame* Γ_{p^*} *admits an* ε-*equilibrium.*

Example 6.75 (Big Match games). *In a Big Match game* $\Gamma(u)$, *for any position* p *along which Player I selected the move* L *at least once, the payoffs in the subgame* $\Gamma_p(u)$ *are constant, and, in particular, any pair of strategies in* $\Gamma_p(u)$ *is a* 0-*equilibrium. As Theorem 6.66 states, for all other positions* p, *the subgame* $\Gamma_p(u)$ *has an* ε-*equilibrium for any* $\varepsilon > 0$, *provided the function* u *is tail measurable.*

Proof of Theorem 6.74. Assume without loss of generality that payoffs are between 0 and 1, and fix $\varepsilon \in (0, 1]$. For each $i \in N$, let H_i be the function given by Theorem 6.24 for player i. Denote by $\hat{\sigma}$ the vector of behavior strategies constructed in the proof of Theorem 6.26 (in that theorem we denoted it σ^*): for each position $p \in T$, $\hat{\sigma}(p)$ is a 0-equilibrium of the one-shot game $(N, (X_i^p)_{i \in N}, (H_i(\langle p, * \rangle))_{i \in N})$.

Recall that $(\mathcal{F}^n)_{n \in \mathbb{N}}$ is the standard filtration.

Notation 6.76 ($\varphi(p)$). *Throughout the proof we will define various functions* φ *whose domain is* $[T]$ *and which satisfy the following condition: there is a set* $Z \subseteq T$ *such that for every* $p \in Z$, *the function* φ *is constant on* $[T_p]$. *In such a case, we will denote by* $\varphi(p)$ *the value of* φ *on* $[T_p]$.

Two cases that satisfy the condition presented in Notation 6.76 are:

1. When $(Y_n)_{n\in\mathbb{N}}$ is a stochastic process defined on $[T]$ and adapted to the standard filtration $(\mathcal{F}^n)_{n\in\mathbb{N}}$, for each $n, k \in \mathbb{N}$ such that $k \geq n$, the function Y_n satisfies the condition in Notation 6.76 with respect to the set Z that consists of all positions of length k.

2. When $\theta : [T] \to \mathbb{N} \cup \{\infty\}$ is a stopping time adapted to the standard filtration $(\mathcal{F}^n)_{n\in\mathbb{N}}$, θ satisfies the condition in Notation 6.76 with respect to the set Z that consists of all positions $x_{\theta(x)}$ for which $\theta(x) < \infty$.

General idea:

The vector of behavior strategies $\hat{\sigma}$ is not an ε-equilibrium, because there are some ways in which a player, say player i, may be able to profit by deviating from $\hat{\sigma}$:

- Player i's payoff may not be constant on the support of $\hat{\sigma}$, hence player i may prefer some moves to others. Lévy's zero-one law will imply that for some $n \in \mathbb{N}$, with $\mathbf{P}_{\hat{\sigma}}$-probability higher than $1 - \varepsilon$, $f(x)$ is close to $\mathbf{E}_{\hat{\sigma}}[f \mid \mathcal{F}^n](x)$; that is, for most positions p of length n, with high probability the payoff in the subgame Γ_p is almost constant.

  This will imply that there is a position p^* of length n and a vector $c \in \mathbb{R}^N$ (c will be equal to $\mathbf{E}_{\hat{\sigma}}[f \mid \mathcal{F}^n](p)$) such that with $\mathbf{P}_{\hat{\sigma}}(\cdot \mid [T_{p^*}])$-probability higher than $1 - \varepsilon$, $f(x)$ is close to c by at most ε. Since all probability measures over $[T]$ are regular, there is a closed set D such that $\mathbf{P}_{\hat{\sigma}}(D \mid [T_{p^*}]) > 1 - 2\varepsilon$ and on D the payoff function is ε-close to c. To ensure that no player deviates to a play that yields her a high payoff, we will instruct the players to punish a deviator if a play outside D is reached. Since D is closed, the fact that the realized play is outside D is known in finite time. To identify the deviator, we will use the method described in Section 6.4.

- Punishment is effective only if player i's minmax value at the position where the punishment starts is at most $c_i + \varepsilon$. Yet, player i may select moves in a way that leads the play to a position where her minmax value is higher than $c_i + \varepsilon$. We will add a test that verifies that no player i plays in a way that increases the probability of reaching a position where her minmax value is higher than $c_i + \varepsilon$. A player who fails this test will be punished in an effective way.

Step 1: Representing the probability to reach certain positions.

In this step we prove a certain formula that involves the probability to reach a given set of positions. Let $Z \subseteq T$ be a set of positions such that no element of Z is a prefix of another element in Z, let σ be a vector of behavior strategies, and let θ be a stopping time. We will provide a formula for the calculation of $\mathbf{P}_\sigma(x_k \in Z$ for some $k \leq \theta(x))$, i.e., the probability under σ that the play reaches a position in Z before being stopped by θ.

Recall that for every position $p = \langle \vec{a}_0, \vec{a}_1, \ldots, \vec{a}_{n-1} \rangle \in T$ we denoted by $n_Z(p)$ the smallest $m \leq n$ such that

$$\langle \vec{a}_0, \vec{a}_1, \ldots, \vec{a}_m \rangle \in Z,$$

so that $p_{n_z(p)}$ is the shortest prefix of p that lies in Z, provided $n_Z(p) < \infty$.

Fix a player $i \in N$. For every position $p \in T$, and every move $a_i \in X_i^p$, define

$$\Lambda_i(p, a_i) := \begin{cases} 0, & \text{if } n_Z(p) < \infty, \\ \sum_{\{a_{-i} \in \prod_{j \neq i} X_j^p : \langle p, (a_i, a_{-i}) \rangle \in Z\}} \left(\prod_{j \neq i} (\sigma_j(p))(a_j) \right), & \text{if } n_Z(p) = \infty. \end{cases}$$
(6.61)

The meaning of the quantity $\Lambda_i(p, a_i)$ is as follows. If one of the prefixes of p is in Z, then $\Lambda_i(p, a_i) = 0$. Otherwise, $\Lambda_i(p, a_i)$ measures the probability that the position in the next stage is in Z, when player i selects the move a_i and the other players follow σ_{-i}. Though the quantity $\Lambda_i(p, a_i)$ depends on the vector of behavior strategies σ_{-i}, to lighten the notation we omit this dependence.

For every play $x = \langle \vec{a}_0, \vec{a}_1, \dots \rangle \in [T]$, every $n \geq 1$, and every $\ell \in \mathbb{N} \cup \{\infty\}$, define

$$\zeta_i(x, n, \ell) := \sum_{k=n}^{\ell} \Lambda_i(x_{k-1}, a_{k,i}).$$
(6.62)

When $n > \ell$, the sum in Equation (6.62) is empty, and $\zeta_i(x, n, \ell) = 0$.

Example 6.77. *Consider a two-player Blackwell game over the alphabet $\mathcal{A} = \{a, a'\}$. Let Z be the set of all positions $p = \langle \vec{a}_0, \vec{a}_1, \dots, \vec{a}_{n-1} \rangle$ where $\vec{a}_k = (a, a)$ if and only if $k = n - 1$; that is, the set of all positions where only in the last pair of moves both players simultaneously select a. Let Q be the set of all positions where the players never play a simultaneously:*

$$Q := \{p = \langle \vec{a}_0, \vec{a}_1, \dots, \vec{a}_{n-1} \rangle : \vec{a}_k \neq (a, a), \quad \forall k = 0, 1, \dots, n-1\}.$$

Let $\sigma_1 \in \Sigma_1(T)$ and $\sigma_2 \in \Sigma_2(T)$ be the behavior strategies in which at every position, the player selects each move with probability $\frac{1}{2}$:

$$(\sigma_1(p))(a) = (\sigma_1(p))(a') = \frac{1}{2}, \quad (\sigma_2(p))(a) = (\sigma_2(p))(a') = \frac{1}{2}, \quad \forall p \in T.$$

For every $p \in Q$ and $i \in \{I, II\}$ we have $\Lambda_i(p, a) = \frac{1}{2}$ and $\Lambda_i(p, a') = 0$, and for every $p \notin Q$ we have $\Lambda_i(p, a) = \Lambda_i(p, a') = 0$. Therefore, for every play $x = \langle \vec{a}_0, \vec{a}_1, \dots \rangle \in (\mathcal{A}^2)^{\mathbb{N}}$, we have $\zeta_i(x, n, \ell) = k(x)/2$, where $k(x)$ is the number of stages $k \geq n$ that satisfy the following properties:

- *$k \leq \ell$, and until stage k it never happened that both players selected a at the same stage, and*

- *at stage k, player i selects the move a.*

As Example 6.77 shows, the quantity $\zeta_i(x, n, \ell)$ may be larger than 1. Nevertheless, it can be thought of as a fictitious probability that the play could have reached a position in Z at any of the stages $n, \dots, \ell$, given the actual play x and assuming that player i's opponents follow σ_{-i}. As we now show, the *expectation* of ζ_i is indeed the probability to reach Z. Specifically, let $p \in T$ be a position such that none of its prefixes lies in Z, and denote $n = \text{len}(p)$. We claim that for every stopping time $\theta \geq n$,

$$\mathbf{E}_\sigma[\zeta_i(x, n, \theta(x)) \mid [T_p]] = \mathbf{P}_\sigma(x_k \in Z \text{ for some } k \leq \theta(x) \mid [T_p]).$$
(6.63)

Indeed,

$$\mathbf{E}_\sigma\left[\zeta_i(x, n, \theta(x)) \mid [T_p]\right] = \mathbf{E}_{\sigma_i}\left[\sum_{k=n}^{\theta(x)-1} \Lambda_i(x_k, a_{k,i}) \mid [T_p]\right] \tag{6.64}$$

$$= \sum_{k=n}^{\infty} \sum_{\substack{p \preceq p' \in T \\ \mathrm{len}(p')=k<\theta(p')}} \sum_{a_i \in X_i^{p'}} \mathbf{P}_\sigma([T_{p'}] \mid [T_p]) \cdot (\sigma_i(p'))(a_i) \cdot \Lambda_i(x_k, a_i) \tag{6.65}$$

$$= \mathbf{P}_\sigma(x_k \in Z \text{ for some } n \le k \le \theta(x) \mid [T_p]), \tag{6.66}$$

where Equation (6.64) follows from the definition of ζ_i, Equation (6.65) follows from changing the order of summation, and Equation (6.66) holds by the definition of Λ_i.

Step 2: Identifying the subgame Γ_{p^*} and a target equilibrium payoff c.

Fix for the moment a player $i \in N$. For each $n \in \mathbb{N}$ define a random variable $Y_i^n : [T] \to [0, 1]$ by

$$Y_i^n(x) := H_i(x_n).$$

By property (M.2) in Theorem 6.24, the process $(Y_i^n)_{n \in \mathbb{N}}$ is a sub-martingale under $\mathbf{P}_{\widehat\sigma}$, and hence it converges $\mathbf{P}_{\widehat\sigma}$-a.s. to a limit Y_i^∞. Denote

$$Y^n(x) := (Y_i^n(x))_{i \in N} \in \mathbb{R}^{|N|}, \quad \forall x \in [T], \forall n \in \mathbb{N}.$$

By Corollary 2.49, there is $n_1 \in \mathbb{N}$ such that[2]

$$\mathbf{P}_{\widehat\sigma}\left(\|f(x) - \mathbf{E}_{\widehat\sigma}[f \mid [T_{x_n}]]\|_\infty \le \varepsilon, \ \forall n \ge n_1\right) > \frac{2}{3}. \tag{6.67}$$

Since the sequence $(Y^n(x))_{n \in \mathbb{N}}$ converges $\mathbf{P}_{\widehat\sigma}$-a.s., there is $n_2 \in \mathbb{N}$ such that

$$\mathbf{P}_{\widehat\sigma}\left(\|Y^k(x) - Y^n(x)\|_\infty \le \varepsilon, \ \forall k \ge n \ge n_2\right) > \frac{2}{3}. \tag{6.68}$$

For each player $i \in N$ denote

$$W_i^n := \mathbf{E}_{\widehat\sigma}[Y_i^{n+1} \mid \mathcal{F}^n], \quad \forall n \in \mathbb{N}.$$

The sequence $(W_i^n)_{k \in \mathbb{N}}$ is a sub-martingale under $\mathbf{P}_{\widehat\sigma}$. Indeed,

$$\begin{aligned} E_{\widehat\sigma}[W_i^{n+1} \mid \mathcal{F}^n] &= E_{\widehat\sigma}[E_{\widehat\sigma}[Y_i^{n+2} \mid \mathcal{F}^{n+1}] \mid \mathcal{F}^n] \\ &\ge E_{\widehat\sigma}[Y_i^{n+1} \mid \mathcal{F}^n] = W_i^n, \end{aligned}$$

where the inequality holds since $(Y_i^n)_{n \in \mathbb{N}}$ is a sub-martingale under $\widehat\sigma$. Since $(Y_i^n)_{k \in \mathbb{N}}$ is a sub-martingale under $\widehat\sigma$, we also have $Y_i^n \le W_i^n$ for every $n \in \mathbb{N}$. Since $(Y_i^n)_{n \in \mathbb{N}}$

---

[2]As will be clear in the derivation of Equation (6.71), the constant $\frac{2}{3}$ in Equations (6.67), (6.68), and (6.69) can be replaced by any three constants whose sum is at least 2.

converges $\mathbf{P}_{\widehat{\sigma}}$-a.s. to Y_i^∞, the sequence $(W_i^n)_{n \in \mathbb{N}}$ converges $\mathbf{P}_{\widehat{\sigma}}$-a.s. to Y_i^∞ as well. As in the previous paragraph, there is $n_3 \in \mathbb{N}$ such that

$$\mathbf{P}_{\widehat{\sigma}}\left(\|W^k(x) - W^n(x)\|_\infty \leq \varepsilon, \ \forall k \geq n \geq n_3\right) > \frac{2}{3}. \tag{6.69}$$

Set $n_0 := \max\{n_1, n_2, n_3\}$, and define

$$\widehat{D}_1 := \left\{ x \in [T] : \begin{array}{l} \left\| f(x) - \mathbf{E}_{\widehat{\sigma}}[f \mid [T_{x_n}]] \right\|_\infty \leq \varepsilon, \quad \forall n \geq n_0, \\ \|Y^k(x) - Y^n(x)\|_\infty \leq \varepsilon, \quad \forall k \geq n_0 \\ \|W^k(x) - W^n(x)\|_\infty \leq \varepsilon, \quad \forall k \geq n_0 \end{array} \right\}. \tag{6.70}$$

By Equations (6.67), (6.68), and (6.69),

$$\mathbf{P}_{\widehat{\sigma}}(\widehat{D}_1) > 0. \tag{6.71}$$

By Corollary 2.49, there is $n \geq n_0$ and a position p^* of length $\operatorname{len}(p^*) = n$ such that

$$\mathbf{P}_{\widehat{\sigma}}(\widehat{D}_1 \mid [T_{p^*}]) > 1 - \varepsilon. \tag{6.72}$$

Denote

$$c := \mathbf{E}_{\widehat{\sigma}}[f(x) \mid [T_{p^*}]] \in \mathbb{R}^{|N|}. \tag{6.73}$$

By property (M.3) of Theorem 6.24,

$$c_i = \mathbf{E}_{\widehat{\sigma}}[f_i(x) \mid [T_{p^*}]] \geq H_i(p^*), \quad \forall i \in N. \tag{6.74}$$

We will construct a $(7\varepsilon + 2^{3/2}(|N| + 1)\varepsilon^{1/4})$-equilibrium in the subgame Γ_{p^*} with payoff close to c.

Step 3: The probability to reach a position with high minmax value.

For each $i \in N$ denote

$$Q_i := \{p \in T : \overline{v}_i(\Gamma_p) > c_i + 3\varepsilon\}.$$

Since player i's minmax value in subgames Γ_p for $p \in Q_i$ is high, in the subgame Γ_{p^*} player i would like to drive the game towards a position in Q_i, where she can obtain a payoff at least $c_i + 3\varepsilon$.

For each play $x \in [T]$, define the entry time to Q_i by

$$n_{Q_i}(x) := \min\{k \in \mathbb{N} : x_k \in Q_i\}.$$

As before, the minimum of an empty set is ∞. For each $i \in N$, define the quantities $\Lambda_i(p, a_i)$ and $\zeta_i(x, n, l)$ as in Equations (6.61) and (6.62) with respect to the set Q_i.

Taking $n = \operatorname{len}(p^*)$ and $\theta = \infty$ in Equation (6.63), we obtain that for every $i \in N$ and every behavior strategy $\sigma_i \in \Sigma_i(T)$,

$$\begin{aligned} \mathbf{E}_{\sigma_i, \sigma_{-i}}[\zeta_i(x, n, \infty) \mid [T_{p^*}]] &= \mathbf{P}_{\sigma_i, \widehat{\sigma}_{-i}}(x_k \in Q_i \text{ for some } n \leq k < \infty \mid [T_{p^*}]) \\ &= \mathbf{P}_{\sigma_i, \widehat{\sigma}_{-i}}(n_{Q_i}(x) < \infty \mid [T_{p^*}]). \end{aligned} \tag{6.75}$$

For each player $i \in N$, let $\nu_i : [T] \to \mathbb{N} \cup \{\infty\}$ be the stopping time that indicates the first stage after stage n at which ζ_i is higher than $\sqrt{\varepsilon}$:

$$\nu_i(x) := \min\{k \geq n : \zeta_i(x, n, k) \geq \sqrt{\varepsilon}\}.$$

Step 4: Identifying a good set of plays D.

Fix for the moment $i \in N$. A useful property of the set $\widehat{D}_1$ is that player i's minmax value in subgames Γ_{x_k}, where $x \in \widehat{D}_1$ and $k \geq n$, are not high:

$$\overline{v}_i(\Gamma_{x_k}) \leq Y_i^n(x) + 2\varepsilon \leq c_i + 2\varepsilon, \quad \forall x \in \widehat{D}_1, \forall k \geq n. \tag{6.76}$$

Indeed,

$$\begin{aligned}
\overline{v}_i(\Gamma_{x_k}) &\leq H_i(x_k) + \varepsilon &(6.77)\\
&= Y_i^k(x) + \varepsilon &(6.78)\\
&\leq Y_i^n(x) + 2\varepsilon &(6.79)\\
&= H_i(p^*) + 2\varepsilon &(6.80)\\
&\leq c_i + 2\varepsilon, &(6.81)
\end{aligned}$$

where Equation (6.77) holds by properties (M.1) and (M.2) of Theorem 6.24, Equation (6.78) follows from the definition of Y_i^k, Equation (6.79) holds since $x \in \widehat{D}_1$, Equation (6.80) holds by the definition of Y_i^n, and Equation (6.81) holds by Equation (6.74).

Equation (6.76) implies in particular that $[T_p] \cap \widehat{D}_1 = \emptyset$ for every $p \in Q_i$. Hence, Equation (6.72) implies that

$$\mathbf{P}_{\widehat{\sigma}}(n_{Q_i}(x) < \infty \mid [T_{p^*}]) \leq \mathbf{P}_{\widehat{\sigma}}((\widehat{D}_1)^c \mid [T_{p^*}]) < \varepsilon. \tag{6.82}$$

Substituting $\sigma = \widehat{\sigma}$ in Equation (6.75) and using Equation (6.82), we obtain

$$\mathbf{E}_{\widehat{\sigma}}[\zeta_i(x, n, \infty) \mid [T_{p^*}]] < \varepsilon. \tag{6.83}$$

The random variable $\zeta_i(\cdot, n, \infty)$ is non-negative, and hence, by Markov's inequality and Equation (6.83),

$$\mathbf{P}_{\widehat{\sigma}}(\zeta_i(x, n, \infty) \geq \sqrt{\varepsilon} \mid [T_{p^*}]) \leq \frac{\mathbf{E}_{\widehat{\sigma}}[\zeta_i(x, n, \infty) \mid [T_{p^*}]]}{\sqrt{\varepsilon}} < \sqrt{\varepsilon}. \tag{6.84}$$

That is, under $\widehat{\sigma}$, with high probability ζ_i is small.

Define

$$\widehat{D} := \widehat{D}_1 \cap \{x \in [T_{p^*}] : \zeta_i(x, n, \infty) < \sqrt{\varepsilon}, \quad \forall i \in N\} \subseteq [T_{p^*}].$$

By Equations (6.72) and (6.84),

$$\mathbf{P}_{\widehat{\sigma}}(\widehat{D} \mid [T_{p^*}]) > 1 - \varepsilon - |N|\sqrt{\varepsilon}.$$

Since $\mathbf{P}_{\widehat{\sigma}}$ is regular, there is a closed set $D \subseteq \widehat{D} \cap [T_{p^*}]$ such that

$$\mathbf{P}_{\widehat{\sigma}}(D \mid [T_{p^*}]) > 1 - (|N| + 1)\sqrt{\varepsilon}. \tag{6.85}$$

For each $x \in D$ for which $\lim_{k\to\infty} Y^k(x)$ and $\lim_{k\to\infty} W^k(x)$ exist and coincide, we have $\|Y^n(x) - Y^\infty(x)\| < \varepsilon$ and $\|W^n(x) - Y^\infty(x)\| < \varepsilon$. Since $\lim_{k\to\infty} Y^k(x)$ and $\lim_{k\to\infty} W^k(x)$ exist and coincide $\mathbf{P}_{\widehat{\sigma}}$-a.s. on $[T_{p^*}]$,

$$\|Y^n(p^*) - W^n(p^*)\|_\infty \leq 2\varepsilon. \tag{6.86}$$

Denote

$$\widehat{\varepsilon} := 2^{3/2}(|N| + 1)\varepsilon^{1/4},$$

and note that when $\varepsilon \leq 1$ we have $\varepsilon < \widehat{\varepsilon}$. By Theorem 6.37, there is a function $g : D^c \to N$ such that

$$\begin{aligned}
\mathbf{P}_{\sigma_i, \widehat{\sigma}_{-i}}(D^c \text{ and } \{g(x) \neq i\} \mid [T_{p^*}]) &\leq 2^{3/2}(|N| + 1)\varepsilon^{1/4} \\
&= \widehat{\varepsilon}, \quad \forall i \in N, \ \forall \sigma_i \in \Sigma_i(T).
\end{aligned} \tag{6.87}$$

Since the set D is closed, its complement is open. Hence $D^c = \bigcup_{p \in Z}[T_p]$, for some set $Z \subseteq T$. We assume without loss of generality that no position in Z is a subset of any other position in Z. As before, define the stopping time $n_Z : [T] \to \mathbb{N} \cup \{\infty\}$ as the stage in which the play reaches a position in Z:

$$n_Z(x) := \min\{k \in \mathbb{N} : x_k \in Z\},$$

where the minimum of an empty set is ∞. When $n_Z(x) < \infty$, at stage $n_Z(x)$ it is clear that the play is going to be in D^c. Since D is closed,

$$D^c \cap [T_{p^*}] = \{x \in [T_{p^*}] : n_Z < \infty\}.$$

Recall from the proof of Theorem 6.37 that the function g which satisfies Equation (6.87) is determined by the prefix of the play that lies in Z (see Remark 6.45).

Step 5: Definition of a vector of behavior strategies σ^*.

We are now ready to define a vector of behavior strategies σ^* in the subgame Γ_{p^*}. We will then prove that σ^* is a $(7\varepsilon + \widehat{\varepsilon})$-equilibrium in Γ_{p^*}.

The vector of behavior strategies σ^* coincides with $\widehat{\sigma}$, with one modification that ensures that no player can profit too much by deviating. Suppose the current position is $p \succeq p^*$.

- If no prefix of p is in Z, then each player $i \in N$ selects the mixed move $\widehat{\sigma}_i(p)$. Denote by $\vec{a} = (\vec{a}_i)_{i \in \mathbb{N}}$ the vector of moves that was actually selected.

- If $p \in Z$ (so that a deviation is detected at stage $\text{len}(p) - 1$), player $g(p)$ is declared the deviator. From stage $\text{len}(p)$ and on, the players punish player $g(p)$ at her minmax level, that is, all players in $N \setminus \{g(p)\}$ switch to a vector of behavior strategies $\widehat{\sigma}_{-g(p)}^{g(p),p}$ for which

$$\sup_{\sigma_{g(p)} \in \Sigma_{g(p)}(T_p)} \mathbf{E}_{\sigma_{g(p)}, \widehat{\sigma}_{-g(p)}^{g(p),p}}[f_{g(p)} \mid [T_p]] \leq \overline{v}_{g(p)}(\Gamma_p) + \varepsilon.$$

Step 6: σ^* is a $(5\varepsilon + 3\widehat{\varepsilon})$-equilibrium in Γ_{p^*}.

We start by calculating the expected payoff under σ^*. By Equation (6.73) and since payoffs are between 0 and 1,

$$\mathbf{E}_{\sigma^*}\big[f_i \mid [T_{p^*}]\big] \geq c_i - (|N| + 1)\sqrt{\varepsilon} \cdot \|f_i\|_1 \geq c_i - 2(|N| + 1)\sqrt{\varepsilon}. \tag{6.88}$$

We next show that no player can profit more than $7\varepsilon + \widehat{\varepsilon}$ by deviating from σ^*. Fix then a player $i \in N$ and a behavior strategy $\sigma_i \in \Sigma_i(T_{p^*})$. To calculate $\mathbf{E}_{\sigma_i,\sigma_{-i}^*}\big[f_i \mid [T_{p^*}]\big]$, we will divide the set $[T_{p^*}]$ into four subsets, and bound from above the expectation of player i's payoff on each of these sets.

- **First subset:** The set D.

  Since $D \subseteq \widehat{D} \subseteq \widehat{D}_1$, on the set D we have $f_i(x) \leq c_i + \varepsilon$, see Equations (6.70) and (6.73). Hence,

$$\mathbf{E}_{\sigma_i,\sigma_{-i}^*}\big[f_i \cdot \mathbf{1}_D \mid [T_{p^*}]\big] \leq (c_i + \varepsilon)\mathbf{P}_{\sigma_i,\sigma_{-i}^*}(D \mid [T_{p^*}]). \tag{6.89}$$

  On the complement of D, player $g(x)$ is punished at her minmax level.

  The next subset handles the case that another player is wrongly punished.

- **Second subset:** The set $E_0 := [T_{p^*}] \cap D^c \cap \{g(x) \neq i\}$.

  On E_0, a player different from player i is punished, and the only upper bound we have on player i's payoff is the maximal payoff, 1. Fortunately, the event E_0 occurs with small probability. Indeed, by Equation (6.87),

$$\mathbf{P}_{\sigma_i,\sigma_{-i}^*}(E_0 \mid [T_{p^*}]) \leq \widehat{\varepsilon}. \tag{6.90}$$

  The third subset handles the case that player i is punished because ζ_i became large.

- **Third subset:** The set $E_1 := [T_{p^*}] \cap D^c \cap \{g(x) = i\} \cap \{\nu_i(x) = n_Z(x)\}$.

  Let us explain the intuition in this case. On E_1, the play leaves the set D, and the function g blames player i. Hence, on E_1, according to the definition of σ^*, player i is punished. Also, on E_1, one has $\nu_i(x) = n_Z(x)$, which means that the quantity ζ_i becomes large (it exceeds $\sqrt{\varepsilon}$), and this happens at the stage $n_Z(x)$ when the play x leaves the set D. In other words, the past moves draw the suspicion to player i for having the play leave D.[3] Since in stage $n_Z(x) - 1$ the stopping time n_Z is not yet triggered, the properties of $\widehat{D}_1$ imply that the expected value of H_i in stage $n_Z(x)$ cannot be much larger than in the beginning of the subgame Γ_{p^*}. This implies that player i can be punished effectively.

---

[3]We do not rule out the case that $\nu_j(x) = n_Z(x)$ for some additional player $j \neq i$, but the probability that g blames an innocent player for leaving the set D is low.

We now turn this intuition into a formal argument. On the set E_1, leaving D is caused by player i: $\zeta_i(x, n, n_Z(x))$ is at least $\sqrt{\varepsilon}$.

Let $p \in T_{p^*}$ be a position such that no prefix of p is in Z. That is, there is an extension x of p that lies in D. Let $k < \mathrm{len}(p)$, and denote by $a_{k,i}$ player i's move at stage k along p. Suppose that from stage $k + 1$ and on, the players $N \setminus \{i\}$ will punish player i at her minmax level. Then player i's payoff will be at most

$$\sum_{a_{-i} \in \prod_{j \neq i} X_j^{p_{k-1}}} (\hat{\sigma}_{-i}(p_{k-1}))(a_{-i}) \cdot (\overline{v}_i(\langle p_{k-1}, a_{k,i}, a_{-i} \rangle) + \varepsilon) \tag{6.91}$$

$$\leq \sum_{a_{-i} \in \prod_{j \neq i} X_j^{p_{k-1}}} (\hat{\sigma}_{-i}(p_{k-1}))(a_{-i}) \cdot (H_i(\langle p_{k-1}, a_{k,i}, a_{-i} \rangle) + 2\varepsilon) \tag{6.92}$$

$$\leq W_i^{k-1}(p_{k-1}) + 2\varepsilon \tag{6.93}$$

$$\leq W_i^n(p^*) + 3\varepsilon \tag{6.94}$$

$$\leq Y_i^n(p^*) + 5\varepsilon \tag{6.95}$$

$$\leq c_i + 5\varepsilon, \tag{6.96}$$

where Equation (6.92) holds by properties (M.1) and (M.2) in Theorem 6.24, Equation (6.93) holds since $\hat{\sigma}(p_{k-1})$ is an equilibrium in the one-shot game with payoffs $(h_j^{p_{k-1}})_{j \in \mathbb{N}}$, Equation (6.94) holds since on D we have $|W_i^{k-1}(x) - W_i^n(x)| < \varepsilon$, Equation (6.95) follows from Equation (6.86), and Equation (6.96) follows from Equation (6.76).

Let now $x \in E_1$ and set $k := n_Z(x)$. Denote by $p := x_k$ the prefix of x at which the play leaves D, and write $p = \langle p^*, \vec{a}_n, \vec{a}_{n+1}, \ldots, \vec{a}_k \rangle$. Since $\nu_i(x) = n_Z(x)$, we have $\zeta_i(p, n, k) \geq \sqrt{\varepsilon}$.

The position p_{k-1} is a strict prefix of p. The minimality of p implies that $\zeta_i(p_{k-1}, n, k - 1) < \sqrt{\varepsilon}$. The definition of ζ_i implies that $\zeta_i(p, n, k)$ does not depend on the moves selected by the *other* players at the last stage of p, see Equation (6.62). That is, for each vector of moves $a_{-i} \in \prod_{j \neq i} X_j^{p_{k-1}}$ we have

$$\zeta_i(\langle p_{k-1}, a_{k,i}, a_{-i} \rangle, n, k) = \zeta_i(p, n, k) \geq \sqrt{\varepsilon}.$$

Therefore, $\langle p_{k-1}, a_{k,i}, a_{-i} \rangle \notin D$ for each $a_{-i} \in \prod_{j \neq i} X_j^{p_{k-1}}$. In particular, if at position p_{k-1} player i selects the move $a_{k,i}$, then whatever moves the other players select, the play will leave D, and some player (not necessarily player i) will be punished. Yet, by Equation (6.87), the probability that a player in $N \setminus \{i\}$ will be punished is at most $\hat{\varepsilon}$.

This implies that

$$\mathbf{E}_{\sigma_i, \sigma_{-i}^*}[f_i \cdot \mathbf{1}_{E_1}] \leq (c_i + 5\varepsilon)\mathbf{P}_{\sigma_i, \sigma_{-i}^*}(E_1) + \hat{\varepsilon}. \tag{6.97}$$

The last subset handles the case that player i is punished yet ζ_i is small.

- **Fourth subset:** The set $E_2 := [T_{p^*}] \cap D^c \cap \{g(x) = i\} \cap \{\nu_i(x) > n_Z(x)\}$.

  Let us explain the intuition in this case. On E_2, the play leaves the set D, and the function g blames player i; hence, player i is punished. Also, on E_2 we have $\nu_i(x) > n_Z(x)$, which means that at stage $n_Z(x)$, when the play x leaves the set D, the quantity ζ_i is lower than $\sqrt{\varepsilon}$. Due to this property, with high probability, player i's minmax value is not much higher than the target payoff c_i. This implies that, with high probability, player i can be punished effectively.

  We now formalize this intuition. The set E_2 is the disjoint union of the sets $[T_p]$ over all $p \in Z$ where the blame function g declares player i as the deviator (this does not depend on the continuation of the play after p) and ζ_i at p is still below $\sqrt{\varepsilon}$. Let $Z_2 \subseteq Z$ denote the set of these positions. At a position $p \in Z_2$, punishment against player i is effective if player i's minmax value is not high, i.e., $p \notin Q_i$, and thus the set of positions in Z_2 where punishment is not effective is $Q_i^2 := Z_2 \cap Q_i$. We have

  $$\mathbf{P}_{\sigma_i,\sigma_{-i}^*}\big(x_k \in Q_i^2 \text{ for some } k \in \{n, n+1, \ldots, n_Z(x)\} \mid [T_{p^*}]\big)$$
  $$\leq \mathbf{E}_{\sigma_i,\sigma_{-i}^*}[\zeta_i(x, n, n_Z(x)) \mid [T_{p^*}]] \tag{6.98}$$
  $$\leq \sqrt{\varepsilon}, \tag{6.99}$$

  where Equation (6.98) follows from Equation (6.63), and Equation (6.99) holds since on E_2 we have $\zeta_i < \sqrt{\varepsilon}$.

Equations (6.89), (6.90), (6.97), and (6.99) imply that

$$\mathbf{E}_{\sigma_i,\sigma_{-i}^*}[f_i \mid [T_{p^*}]] \leq c_i + 5\varepsilon + \widehat{\varepsilon}.$$

Together with Equation (6.88), this shows that σ^* is a $(5\varepsilon + 3\widehat{\varepsilon})$-equilibrium, as claimed. $\qquad\qquad\square$

6.10 DISCUSSION

Classical papers on alternating-move games and Blackwell games studied zero-sum games. The study of multiplayer Blackwell games was initiated only recently, and the results presented in this chapter are from the current decade. In this section we will summarize the results we proved in this chapter, review a few additional related results, and list a few open problems. Several recent books and surveys can help readers who would like to expand their knowledge about classes of multiplayer sequential games that are related to Blackwell games. These include, among others, Grädel and Ummels (2008), Chatterjee and Henzinger (2012), Solan and Vieille (2015), Jaśkiewicz and Nowak (2018), Levy and Solan (2020), Bruyère (2021), and Solan (2022).

We studied three ways to play well in multiplayer Blackwell games:

- Good vectors of behavior strategies, which ensure that the players obtain a high payoff. This concept, which was introduced in Solan (2018), was studied in the context of multiplayer Blackwell games by Flesch and Solan (2024), who proved the following.

Theorem 6.26 (Flesch and Solan (2024)). *Let $\Gamma = (N, T, (f_i)_{i \in N})$ be a multiplayer Blackwell game. For every $\varepsilon > 0$ there is an ε-acceptable vector of strategies.*

- Goals, which are pairs (σ^*, D) of a vector of behavior strategies and a subset of plays, such that under σ^* with high probability the play lies in D. In this set up, we would like to identify a deviator who causes the play to leave D. This concept was presented and studied by Alon, He, Gunby, Shmaya, and Solan (2024) who proved the following.

Theorem 6.36 (Alon, Gunby, He, Shmaya, and Solan (2024)). *Let (σ^*, D) be a goal such that $\mathbf{P}_{\sigma^*}(D) = 1$. Then (σ^*, D) is 0-testable.*

Theorem 6.37 (Alon, Gunby, He, Shmaya, and Solan (2024)). *Let $\varepsilon > 0$, and let (σ^*, D) be a goal such that $\mathbf{P}_{\sigma^*}(D) > 1 - \varepsilon$. Then (σ^*, D) is $2\sqrt{(|N| - 1)\varepsilon}$-testable.*

- ε-equilibria, which are vectors of behavior strategies σ^* such that no player can gain more than ε by unilaterally deviating from σ^*. We proved the existence of ε-equilibria in two classes of Blackwell games.

Theorem 6.62 (Ashkenazi-Golan, Flesch, Predtechinski, Solan (2022)). *Let $\Gamma = (N, (X_i)_{i \in N}, (f_i)_{i \in N})$ be a multiplayer repeated game. If the functions $(f_i)_{i \in N}$ are tail measurable, then the game admits an ε-equilibrium, for every $\varepsilon > 0$.*

Theorem 6.66 (Ashkenazi-Golan, Flesch, and Solan, 2022). *Every Big Match game $\Gamma(u)$ where the function u is tail measurable admits an ε-equilibrium, for every $\varepsilon > 0$.*

Alternative proofs for Theorem 6.62 can be found in Flesch and Solan (2025). Whether every multiplayer Blackwell game admits an ε-equilibrium for any $\varepsilon > 0$ is one of the most challenging open problems in the area of dynamic games to date. Nevertheless, Flesch and Solan (2024) proved the following.

Theorem 6.74 (Flesch and Solan, 2024). *For every multiplayer Blackwell game $\Gamma = (N, T, (f_i)_{i \in N})$ and every $\varepsilon > 0$ there is a position $p^* \in T$ such that the subgame Γ_{p^*} admits an ε-equilibrium.*

Theorem 6.74 was proved by Thuijsman and Vrieze (1991) in the setup of two-player nonzero-sum stochastic games with the long-run average payoff, and by Vieille (2000c) in the setup of multiplayer stochastic games with the long-run average payoff. The technique we developed in this chapter shows that this result extends to a far more general setup.

To prove the existence of the value in two-player zero-sum Blackwell games, we defined for every $u \in \mathbb{R}$ an auxiliary alternating-move game with a winning set G_u, and proved that there is $u^* \in \mathbb{R}$ such that Player 1 wins G_u for every $u < u^*$, Player 2 wins G_u for every $u > u^*$, and that u^* is the value of the original Blackwell game. One could wonder whether the same approach can be used to prove the existence of an ε-equilibrium in nonzero-sum Blackwell games. That is, one could attempt to define a parameterized family of appropriate auxiliary alternating-move games with a winning set that satisfies the following property: If there is some parameter u^* such that, in each of its open neighborhoods, there are two parameter values u and u' such that Player 1 wins the game with parameter u and Player 2 wins the game with parameter u', then u^* is an ε-equilibrium payoff. At present we do not know how to define such a family of auxiliary games.

Theorem 6.62 states that every repeated game with tail-measurable payoffs admits an ε-equilibrium for every $\varepsilon > 0$. Denote by $Q_\varepsilon(\Gamma) \subseteq \mathbb{R}^{|N|}$ the set of all ε-equilibrium payoffs in the game Γ. The sets $Q_\varepsilon(\Gamma)$ are increasing: $Q_\varepsilon(\Gamma) \subseteq Q_{\varepsilon'}(\Gamma)$ whenever $\varepsilon < \varepsilon' < 0$. Ashkenazi-Golan, Flesch, Predtetchinski, and Solan (2025) characterized the limit set $\bigcap_{\varepsilon>0} Q_\varepsilon(\Gamma)$ as follows.

Theorem 6.78 (Ashkenazi-Golan, Flesch, Predtetchinski, and Solan (2025)). *Let $\Gamma = (N, (X_i)_{i\in N}, (f_i)_{i\in N})$ be a multiplayer repeated game, where the functions $(f_i)_{i\in N}$ are tail measurable. For every $\varepsilon > 0$ denote*

$$R_\varepsilon(\Gamma) := \{f(x) \colon x \in [T], f_i(x) \geq \bar{v}_i - \varepsilon, \quad \forall i \in N\}.$$

Then

$$\bigcap_{\varepsilon>0} Q_\varepsilon(\Gamma) = \bigcap_{\varepsilon>0} \mathrm{conv}(\mathrm{closure}\,(R_\varepsilon(\Gamma)).$$

Four properties shared by both Big Match games with tail-measurable payoffs and stopping games are

(R.1) Both players have two available moves until the play effectively terminates.

(R.2) The play is effectively terminated once a pair of moves in a certain set of pairs of moves is selected.

(R.3) The payoff of the players when the play effectively terminates is determined by the vector of moves selected at the termination stage, and is independent of the time in which this vector of moves was selected for the first time.

(R.4) And, if the play never effectively terminates, the payoff is some tail-measurable function of the realized play.

Ashkenazi-Golan, Flesch, and Solan (2022) extended Theorem 6.62 to games that share properties (R.2)–(R.4), but in which players may have more than two possible moves until termination occurs, and the set of pairs of moves that lead to termination may be an arbitrary subset of $X_{\mathrm{I}} \times X_{\mathrm{II}}$.

Theorem 6.79 (Ashkenazi-Golan, Flesch, and Solan (2022)). *Let $\Gamma = (N, (X_i)_{i\in N}, (f_i)_{i\in N})$ be a two-player repeated game that satisfies the following property: There is a subset $A \subseteq X_{\mathrm{I}} \times X_{\mathrm{II}}$ and a function $u \colon A \to \mathbb{R}^2$, such that for every play $x = \langle \vec{a}_0, \vec{a}_1, \dots \rangle \in [T]$,*

- *If $n_0(x) := \min\{n \in \mathbb{N} : \vec{a}_n \in A\} < \infty$, then $f_i(x) = u_i(\vec{a}_{n_0(x)})$ for each $i \in N$.*

- *On the set $\{x \in [T] : n_0(x) = \infty\}$, the functions $(f_i)_{i \in N}$ are tail measurable.*

Then the game admits an ε-equilibrium, for every $\varepsilon > 0$.

Shmaya, Solan, and Vieille (2003) studied two-player stopping games, where the payoff upon termination depends on the set of players who stop at the termination stage, as well as on the stage in which termination occurred, and proved the existence of an ε-equilibrium, for every $\varepsilon > 0$.

Theorem 6.80 (Shmaya, Solan, and Vieille (2003)). *Let $\Gamma = \big(N, (X_i)_{i \in N}, (f_i)_{i \in N}\big)$ be a two-player repeated game, that satisfies the following properties:*

- *$X_I = X_{II} = \{C, S\}$.*

- *There is a bounded function $u : \mathbb{N} \times (2^N \setminus \emptyset) \to \mathbb{R}^2$ such that $f_i(x) = u_i(n_0(x), S_0(x))$ for every play $x = \langle \vec{a}_0, \vec{a}_1, \dots \rangle \in [T]$ such that $n_0(x) < \infty$ and every player $i \in N$, where $n_0(x) := \min\{n \in \mathbb{N} : a_{n,I} = S \text{ or } a_{n,II} = S\}$ and $S_0(x) := \{i \in N : a_{n_0(x),i} = S\}$.*

Then the game admits an ε-equilibrium, for every $\varepsilon > 0$.

To date it is not known whether Theorem 6.80 can be extended to games that include more than three players.

Let $\Gamma = \big(N, (X_i)_{i \in N}, (f_i)_{i \in N}\big)$ be a multiplayer repeated game. We say that two positions $p, p' \in T$ are *equivalent* if the subgames Γ_p and $\Gamma_{p'}$ are similar: For every $x \in (\prod_{i \in I} X_i)^{\mathbb{N}}$, we have $f_i(p \circ x) = f_i(p' \circ x)$ for each $i \in N$. A Blackwell game where this equivalence relation has finitely many equivanece classes, is called a *stochastic game with general payoff function*.[4] For a detailed treatment of stochastic games the reader is referred to the textbooks Filar and Vrieze (1997) or Solan (2022).

The payoff function f_i is called *shift invariant* if $f_i(x) = f_i(\vec{a} \circ x)$ for every $\vec{a} \in \mathcal{A}^N$ and every $x \in [T]$. Flesch and Solan (2024) proved[5] that if the payoff functions of the two players are shift-invariant, then an ε-equilibrium exists for every $\varepsilon > 0$.

Theorem 6.81 (Flesch and Solan (2024)). *Every two-player stochastic game with shift-invariant payoff functions admits an ε-equilibrium, for every $\varepsilon > 0$.*

Whether every multiplayer stochastic game (whose payoff functions are not necessarily shift invariant) admits an ε-equilibrium for every $\varepsilon > 0$ is also an open problem.

---

[4]This class of games was already mentioned in Section 5.6.

[5]In fact, Flesch and Solan (2024) allow for the introduction of Nature, namely, a player who is not strategic and follows a given behavior strategy, provided that this behavior strategy respects the equivalence relation: it selects the same mixed move at positions p and p' whenever p and p' are equivalent.

6.11 EXERCISES

Exercise 6.3 is used in Exercise 6.16. Exercise 6.4 is used in Exercise 6.6. Exercise 6.14 is used in Exercise 6.16. Exercise 6.15 is an extension of Theorem 5.46, and is used in the proof of Lemma 7.43. Exercise 6.25 is an extension of Exercise 5.21.

Exercise 6.14 is adapted from Solan and Vieille (2002). The technique used in the solution to this exercise provides an alternative proof to a result by Mashiah-Yaakovi (2015). Exercise 6.20 is based on correspondence with Alexander Kechris. Exercise 6.27 is based on Ashkenazi-Golan, Flesch, Predtetchinski, and Solan (2025).

Exercises marked with a star are more challenging than unmarked ones, while those marked with two stars are even more difficult.

1. Prove Theorem 6.13: Let $(\{I, II\}, (X_I, X_{II}), (h_I, h_{II}))$ be a one-shot game where $h_I + h_{II} = 0$. Prove that the pair of mixed moves $x^* = (x_I^*, x_{II}^*)$ is an equilibrium if and only if x_I^* and x_{II}^* are 0-optimal in the two-player zero-sum game (X_I, X_{II}, h_I) according to Definition 5.13.

2. (*) In this exercise we prove Nash's Theorem (Theorem 6.6). Let $(N, (X_i)_{i \in N}, (h_i)_{i \in N})$ be a one-shot game where the sets N and $(X_i)_{i \in N}$ are finite.

   For every $i \in N$, every $a_i \in X_i$, and every $z \in \prod_{j \in N} \Delta(X_j)$, define

   $$g_{i,a_i}(z) := \max\{0, h_i(a_i, z_{-i}) - h_i(z)\}.$$

   This is the gain (if such a gain exists) to player i if she selects the move a_i instead of the mixed move z_i, when all other players select the mixed moves z_{-i}. Define a function $f_i : \prod_{j \in N} \Delta(X_j) \to \Delta(X_i)$ by

   $$(f_i(z))(a_i) := \frac{z_i(a_i) + g_{i,a_i}(z)}{1 + \sum_{a_i' \in X_i} g_{i,a_i'}(z)}, \quad \forall a_i \in X_i.$$

   (a) Prove that the image of f_i is included in $\Delta(X_i)$, for each $i \in N$.
   (b) Prove that f_i is continuous, for each $i \in N$.
   (c) Prove that if z^* is a 0-equilibrium of $(N, (X_i)_{i \in N}, (h_i)_{i \in N})$, then $f_i(z^*) = z_i^*$ for each $i \in N$.
   (d) Prove that there is $z^* \in \prod_{i \in N} \Delta(X_i)$ such that $f_i(z^*) = z_i^*$ for each $i \in N$.
   (e) Let $z^* \in \prod_{i \in N} \Delta(X_i)$ satisfy $f_i(z^*) = z_i^*$ for each $i \in N$. Prove that z^* is a 0-equilibrium of $(N, (X_i)_{i \in N}, (h_i)_{i \in N})$.

3. (*) In this exercise, we extend Theorem 6.6 to the situation in which one player has a finite set of moves and the other player has a countably infinite set of moves.

   Let $(\{I, II\}, (X_I, X_{II}), (h_I, h_{II}))$ be a two-player one-shot game where X_I is finite, X_{II} is countably infinite, and the payoff functions h_I and h_{II} are bounded. Prove that for every $\varepsilon > 0$ the game has an ε-equilibrium.

   Hint: Reduce the problem to the setup of Theorem 6.6.

4. In this exercise we present a three-player one-shot game where the maxmin value of a player differs from her minmax value.

Consider a three-player one-shot game where each player has two moves: Player 1 selects T or B, Player 2 selects L or R, and Player 3 selects W or E. The payoff function of Player 1 is (see Figure 24 for a graphical depiction, where Player 1 selects a row, Player 2 selects a column, and Player 3 selects the left-hand side matrix or the right-hand side matrix):

$$h(T, L, W) = h(B, R, E) = 0,$$
$$h(T, R, W) = h(T, L, E) = h(T, R, E) = 1,$$
$$h(B, L, W) = h(B, R, W) = h(B, L, E) = 1.$$

The payoff functions of Players 2 and 3 will play no role in this exercise.

Prove that Player 1's minmax value is $\frac{3}{4}$, while her maxmin value is $\frac{1}{2}$.

	L	R
T	0	1
B	1	1

W

	L	R
T	1	1
B	1	0

E

Figure 24: Player 1's payoff function in the game in Exercise 6.4.

5. Consider the three-player alternating-move game that is described in Exercise 4.13. For each $i = \text{I}, \text{II}, \text{III}$, let σ_i^* be the following behavior strategy: In all positions of the form $\langle c, c, \ldots, c \rangle$ where player i selects a move, she selects s with probability $\frac{1}{2}$ and c with probability $\frac{1}{2}$.

   (a) What is the payoff to the three players under $(\sigma_\text{I}^*, \sigma_\text{II}^*, \sigma_\text{III}^*)$?

   (b) Prove that $(\sigma_\text{I}^*, \sigma_\text{II}^*, \sigma_\text{III}^*)$ is a 0-equilibrium.

6. (*) Let $\Gamma = (N, T, f)$ be a Blackwell game, where $N = \{\text{I}, \text{II}, \text{III}\}$ and $f_\text{I} = \mathbf{1}_W$ for some Borel set W.

For every position $p \in T$, Player I's *maxmin value* at p is

$$\underline{v}_\text{I}(\Gamma_p) = \sup_{\sigma_\text{I}^p \in \Sigma_\text{I}(T_p)} \quad \inf_{(\sigma_\text{II}^p, \sigma_\text{III}^p) \in \Sigma_\text{II}(T_p) \times \Sigma_\text{III}(T_p)} \mathbf{P}_{\sigma_\text{I}^p, \sigma_\text{II}^p, \sigma_\text{III}^p}(x \in W),$$

and Player I's *minmax value* at p is

$$\overline{v}_\text{I}(\Gamma_p) := \inf_{(\sigma_\text{II}^p, \sigma_\text{III}^p) \in \Sigma_\text{II}(T_p) \times \Sigma_\text{III}(T_p)} \quad \sup_{\sigma_\text{I}^p \in \Sigma_\text{I}(T_p)} \mathbf{P}_{\sigma_\text{I}^p, \sigma_\text{II}^p, \sigma_\text{III}^p}(x \in W),$$

where here x is the random variable that indicates the realized play. Suppose that $\overline{v}_\text{I}(\langle \, \rangle) = 1$.

   (a) Prove that Player I has a move a_I^0 such that $\overline{v}_\text{I}(\Gamma_{\langle a_\text{I}^0, a_\text{II}^0, a_\text{III}^0 \rangle}) = 1$, for every pair of moves $(a_\text{II}^0, a_\text{III}^0) \in X_\text{II}^{\langle \, \rangle} \times X_\text{III}^{\langle \, \rangle}$.

(b) Prove that there is $\sigma_I \in \Sigma_I(T)$ such that $\overline{v}_I(\Gamma_p) = 1$ for every pair of strategies $(\sigma_{II}, \sigma_{III}) \in \Sigma_{II}(T) \times \Sigma_{III}(T)$ and every position p that satisfies $\mathbf{P}_{\sigma_I,\sigma_{II},\sigma_{III}}([T_p]) > 0$.

(c) Provide an example of a three-player Blackwell game over the alphabet $\{0, 1\}$ where the payoff function is the characteristic function of some Borel set, $\overline{v}_I(\langle\ \rangle) = 1$, and $\underline{v}_I(\langle\ \rangle) = 0$.

(d) Explain why Part (c) does not contradict Part (b).

Hint: For Part (c), recall Exercise 6.4.

7. (**) Let $\sigma^* \in \Sigma(T)$ be the vector of behavior strategies constructed in the proof of Theorem 6.26 for a given $\varepsilon > 0$. Fix a player $i \in N$. For each position $p \in T$ let σ_{-i}^p be a vector of ε-minmax behavior strategies of players $N \setminus \{i\}$ in the subgame Γ_p; that is, we have

$$\mathbf{E}_{\sigma_i,\sigma_{-i}^p}[f_i \mid [T_p]] \leq \overline{v}_i(\Gamma_p) + \varepsilon, \quad \forall \sigma_i \in \Sigma_i(T).$$

Fix a position $p \in T$, and a move $\widehat{a}_i \in X_i^p$. Let $\widehat{\sigma}_i$ be a strategy that coincides with σ_i^* in all positions p' that do *not* extend p, and at p it selects the move $\widehat{a}_i$. For each player $j \neq i$, let $\widehat{\sigma}_j$ be the strategy that coincides with σ_j^* at all positions that do *not* extend p, as well as at p, and, for every $\vec{a} \in X(p)$, in the subgame $\Gamma_{\langle p,\vec{a}\rangle}$, it coincides with $\sigma_{-i,j}^{\langle p,\vec{a}\rangle}$. In words, under the vector of behavior strategies $\widehat{\sigma} = (\widehat{\sigma}_k)_{k \in I}$ the players follow σ^* until the play reaches the position p. Once p is reached, player i selects the move $\widehat{a}_i$, and in the following stage all other players punish player i.

(a) Prove that
$$\mathbf{E}_{\widehat{\sigma}}[f_i] \leq \mathbf{E}_{\sigma^*}[f_i] + 3\varepsilon.$$

(b) Let $\widehat{\sigma}$ be the following vector of behavior strategies: The players follow σ^* until the first stage in which one of the players selects a move that has probability 0 to be selected. Once this happens, say, at position p, all players other than player i switch to the vector of behavior strategies σ_{-i}^p. Does Part (a) imply that $\widehat{\sigma}$ is a 3ε-equilibrium? Explain your answer.

8. Larry claims that in Example 6.39, the quantity $q = \mathbf{P}_{\sigma^*}(g = I \mid D^c)$ must be either 0 or 1, because g is a function and the play is in D^c if and only if both players select 1 at stage 0. Is Larry correct? Justify your answer.

9. (*) In this exercise, we prove a version of Theorem 6.37 for a random blame function.

Let (σ^*, D) be a goal where the target set D is Borel. A *random blame function* is a Borel-measurable function $\varphi : D^c \to \Delta(I)$, with the interpretation that if the play x is not in D, then each player i is blamed with probability $\varphi_i(x)$. The goal (σ^*, D) is δ-*randomly testable* if in the definition of δ-testability we allow for random blame functions. Prove that if $\mathbf{P}_{\sigma^*}(D) > 1 - \varepsilon$, then (σ^*, D) is $\sqrt{(|N| - 1)\varepsilon}$-randomly testable.

10. (*) Adriana claims that she can extend Theorem 6.37 to the case $\mathbf{P}_{\sigma^*}(D) = 1 - \varepsilon$. "You see," she claims, "$\mathbf{P}_{\sigma^*}(D) > 1 - \varepsilon - \frac{1}{k}$ for every $k \in \mathbb{N}$, and hence, by Theorem 6.37, (σ^*, D) is $2\sqrt{(|N| - 1)(\varepsilon + \frac{1}{k})}$-testable. Denote by g_k the corresponding blame function. As in the proof of Theorem 6.36, define $g(x) = i$ if i is the minimal index such that $g_k(x) = i$ for infinitely many k's. Then g is a blame function that correctly identifies the deviation with probability $1 - \varepsilon$." Is Adriana's argument valid? Justify your answer.

11. (**) In this exercise we prove a version of Theorem 6.37 for a setup where several players may deviate simultaneously.

    Let (σ^*, D) be a goal, and let $\delta > 0$. The goal (σ^*, D) is δ-*testable for groups* if there is a Borel-measurable blame function $g : D^c \to N$ such that for every subset $J \subset N$ of players and every vector of behavior strategies $\sigma_J = (\sigma_i)_{i \in J}$, we have $\mathbf{P}_{\sigma_J, \sigma^*_{-J}}(D^c \text{ and } \{g \notin J\}) \leq \delta$, where $\sigma^*_{-J} = (\sigma^*_i)_{i \notin J}$.

    Consider the zero-sum game between the adversary and the statistician defined in the proof of Theorem 5.32, where a pure strategy of the adversary is a pair (J, σ^J), in which $J \subset N$ is a nonempty set of players and $\sigma^J = (\sigma^J_i)_{i \in J} \in \prod_{i \in J} \Sigma_j(T)$ is a vector of behavior strategies for the players in J. Suppose that D is closed, and set $\varepsilon := 1 - \mathbf{P}_{\sigma^*}(D) > 0$. Since D is closed, $D^c = \bigcup_{p \in Z}[T_p]$ for some finite or countably infinite set $Z \subseteq T$, with the property that no position in Z is a prefix of some other position in Z.

    (a) For each pure strategy (J, σ^J) of the adversary and each $i \in J$, write down the likelihood function ℓ_i^{J, σ^J}.

    (b) For each $J \subset N$, let $\sigma^J = (\sigma^J_i)_{i \in J} \in \prod_{i \in J} \Sigma_i(T)$ be a vector of behavior strategies, and let $\alpha = (\alpha^J)_{\emptyset \neq J \subset N}$ be a probability distribution on the nonempty strict subsets of N. Suppose the adversary adopts the mixed strategy that selects each pure strategy (J, σ^J) with probability α^J. Suppose further that for every $p \in Z$, for every play that extends p the statistician blames a player i for which the likelihood function at p is maximal (think well how to mathematically phrase this condition). Using Hölder's inequality instead of the Cauchy-Schwarz inequality, prove that under (J, σ^J), the probability that an honest player is blamed is $O(\varepsilon^{1/(|N|+1)})$.

    (c) Does Part (b) imply that the goal (σ^*, D) is $O(\varepsilon^{1/(|N|+1)})$-testable for groups? Explain your answer.

12. (**) In this exercise you are asked to identify by yourself a goal and a blame function of your choosing.

    Find a *non-trivial* example of a goal (σ^*, D), $\varepsilon \geq 0$, and $\delta \geq 0$, such that $\mathbf{P}_{\sigma^*}(D) \geq 1 - \varepsilon$ and for which you can identify an explicit blame function g with the property that $\mathbf{P}_{\sigma_i, \sigma^*_{-i}}(D^c \text{ and } \{g \neq i\}) \leq \delta(\varepsilon)$, for every $i \in N$ and every $\sigma_i \in \Sigma_i(T)$.

13. (**) In Theorem 6.37 we proved that if $\mathbf{P}_{\sigma^*}(D) > 1 - \varepsilon$, then (σ^*, D) is $2\sqrt{(|N| - 1)\varepsilon}$-testable. Prove that in fact (σ^*, D) is $(2\lceil \ln_2(|N|) \rceil \sqrt{\varepsilon})$-testable.

14. (**) In this exercise, we study extensive-form correlated equilibria in Blackwell games. Let $\Gamma = (N, T, (f_i)_{i \in N})$ be a multiplayer Blackwell game. A *mediator* is an entity that can send a private message to each player at each stage, which can depend on the current position and on the past messages it sent.

Formally, the set of messages of the mediator is some finite set M. Denote by

$$\mathcal{T} := \{(\vec{m}_0, \vec{a}_0, \vec{m}_1, \vec{a}_1, \ldots, \vec{m}_{n-1}, \vec{a}_{n-1}) \in (M^N \times \mathcal{A})^n : \langle \vec{a}_0, \vec{a}_1, \ldots, \vec{a}_{n-1} \rangle \in T\}$$

the set of all possible positions in the game, together with the messages the mediator could send to the players.

A *mediator* is a function $m : \mathcal{T} \to \Delta(M^N)$. The game with the mediator is played as follows: if the current position is $p = \langle \vec{a}_0, \vec{a}_1, \ldots, \vec{a}_{n-1} \rangle$, and the messages the mediator sent in the past are $\vec{m}_0, \vec{m}_1, \ldots, \vec{m}_{n-1}$, then (i) the mediator selects a vector of messages $\vec{m}_n = (m_{n,i})_{i \in N} \in M^N$ according to $m(\vec{m}_0, \vec{a}_0, \vec{m}_1, \vec{a}_1, \ldots, \vec{m}_{n-1}, \vec{a}_{n-1})$, (ii) the mediator sends to each player $i \in N$ the private message $m_{n,i}$, and (iii) the players simultaneously select moves from their sets of available moves at p.

An ε-equilibrium in the game with mediator is called an *extensive-form correlated ε-equilibrium* of the original game Γ. Using Theorem 6.26, prove that in every multiplayer Blackwell game there exists an ε-equilibrium, for every $\varepsilon > 0$.

Hint: Use Exercise 6.7.

15. In this exercise we show that the construction in Exercise 6.14 cannot be turned into an ε-equilibrium without the mediator.

Consider the following two-player Blackwell game, which essentially lasts two stages. The set of moves is $\mathcal{A} = \{b, c\}$, and the payoff functions of the players are as follows. For every play $x = \langle \vec{a}_0, \vec{a}_1, \ldots \rangle$,

$$f_1(x) = \begin{cases} 2, & \vec{a}_0 = (b, c), \\ 2, & \vec{a}_0 = (c, b), \\ 1, & \vec{a}_0 = (c, c), \\ 1, & \vec{a}_0 = (b, c), a_{1,2} = b, \\ 0, & \vec{a}_0 = (b, c), a_{1,2} = c, \end{cases} \qquad f_2(x) = \begin{cases} 0, & \vec{a}_0 = (b, c), \\ 0, & \vec{a}_0 = (c, b), \\ 1, & \vec{a}_0 = (c, c), \\ 1, & \vec{a}_0 = (b, c), a_{1,2} = b, \\ 0, & \vec{a}_0 = (b, c), a_{1,2} = c. \end{cases}$$

In words, if at least one player selects the move c in stage 0, the payoff is determined according to the pair of moves selected in stage 0; If both players selected b in stage 0, the payoff is determined by the move of Player 2 in stage 1.

(a) Calculate the players' minmax values at the positions $\langle \, \rangle$ and $\langle (b, b) \rangle$.

(b) Suggest functions H_1 and H_2 that satisfy the conclusion of Theorem 6.24.

(c) Is the vector of strategies σ^* that is defined in the proof of Theorem 6.26 relative to the functions H_1 and H_2 of Part (b) an ε-equilibrium in the absence of a mediator?

16. (*) Let $\Gamma = (N, T, (f_i)_{i \in N})$ be a Blackwell game where $N = \{\mathrm{I}, \mathrm{II}\}$, for all positions $p \in T$ at least one of the sets X_{I}^p and X_{II}^p is finite (and the other may be countably infinite), and $(f_i)_{i \in N}$ are bounded and Borel measurable. Is it true that for every $\varepsilon > 0$ the game admits an extensive-form correlated ε-equilibrium, where the set M of messages may be general (and not necessarily finite)? Justify your answer.

For the definition of an extensive-form correlated ε-equilibrium, see Exercise 6.14.

17. In the game defined in Exercise 4.11, calculate all 0-equilibria in behavior strategies, together with the corresponding set of 0-equilibrium payoffs.

18. Let $T = \mathcal{A}^{\mathbb{N}}$ be a tree over an alphabet $\mathcal{A}$. Prove that $\mathcal{F}^{\mathrm{tail}}(T)$ coincides with the collection of all tail sets in $[T]$.

19. Let $T = \mathcal{A}^{\mathbb{N}}$ be a tree over the alphabet $\mathcal{A}$. Among the following real-valued functions from $[T]$, which ones are tail measurable? Explain your answer.

(a) Let $A \subseteq \mathcal{A}$, and let

$$Z_A := \{\langle a_0, a_1, \ldots \rangle \in [T] : \text{if } a_n = a \text{ infinitely often, then } a \in A\}$$

be the set of all plays along which only elements in A can appear infinitely often. The following function f_A is termed the *Müller winning condition*:

$$f_A(\langle a_0, a_1, \ldots \rangle) := \begin{cases} 1, & \text{if } \langle a_0, a_1, \ldots \rangle \in Z_A, \\ 0, & \text{otherwise.} \end{cases}$$

(b) Let $\lambda \in (0, 1]$ and let $u : \mathcal{A} \to [0, 1]$. The following function f_λ is termed the λ-*discounted evaluation*:

$$f_\lambda(\langle a_0, a_1, \ldots \rangle) := \lambda \sum_{n=0}^{\infty} (1 - \lambda)^n u(a_n).$$

20. (**) Let $T = \{0, 1\}^{<\mathbb{N}}$. In this exercise we will prove that for every ordinal $\alpha \geq 2$ there is a tail set that lies in $\Sigma_\alpha([T])$, but not in $\Pi_\alpha([T])$.

We start by constructing a closed set $C \subseteq [T]$ that is homeomorphic to $[T]$ and such that every two points in C are different in infinitely many coordinates.

For every $k \in \mathbb{N}$ define a set of positions S_k as in the following figure:

S_0	S_1	S_2	S_3
$\langle 0 \rangle$	$\langle 0,0,0 \rangle$	$\langle 0,0,0,0,0,0 \rangle$	$\langle 0,0,0,0,0,0,0,0,0,0 \rangle$
			$\langle 0,0,0,0,0,0,0,0,0,1 \rangle$
		$\langle 0,0,0,0,0,1 \rangle$	$\langle 0,0,0,0,0,1,0,0,1,0 \rangle$
			$\langle 0,0,0,0,0,1,0,0,1,1 \rangle$
	$\langle 0,0,1 \rangle$	$\langle 0,0,1,0,1,0 \rangle$	$\langle 0,0,1,0,1,0,0,1,0,0 \rangle$
			$\langle 0,0,1,0,1,0,0,1,0,1 \rangle$
		$\langle 0,0,1,0,1,1 \rangle$	$\langle 0,0,1,0,1,1,0,1,1,0 \rangle$
			$\langle 0,0,1,0,1,1,0,1,1,1 \rangle$
$\langle 1 \rangle$	$\langle 1,1,0 \rangle$	$\langle 1,1,0,1,0,0 \rangle$	$\langle 1,1,0,1,0,0,1,0,0,0 \rangle$
			$\langle 1,1,0,1,0,0,1,0,0,1 \rangle$
		$\langle 1,1,0,1,0,1 \rangle$	$\langle 1,1,0,1,0,1,1,0,1,0 \rangle$
			$\langle 1,1,0,1,0,1,1,0,1,1 \rangle$
	$\langle 1,1,1 \rangle$	$\langle 1,1,1,1,1,0 \rangle$	$\langle 1,1,1,1,1,0,1,1,0,0 \rangle$
			$\langle 1,1,1,1,1,0,1,1,0,1 \rangle$
		$\langle 1,1,1,1,1,1 \rangle$	$\langle 1,1,1,1,1,1,1,1,1,0 \rangle$
			$\langle 1,1,1,1,1,1,1,1,1,1 \rangle$

The set S_k contains 2^{k+1} positions, all of them of length $1 + 2 + \cdots + (k+1) = \frac{(k+1)(k+2)}{2}$. The positions in S_{k+1} are extensions of positions in S_k: for each position in S_k there are exactly two extensions in S_{k+1}. When ordering the positions in S_k lexicographically, the two sequences of bits added to the l'th position in S_k are the $(2l)$'th and $(2l+1)$'th elements of length $k+2$ (when the 2^{k+2} sequences of length $k+2$ are also ordered lexicographically ordering).

Define
$$C_k := \bigcup_{p \in S_k} [T_p], \quad \forall k \in \mathbb{N},$$
and
$$C := \bigcap_{k \in \mathbb{N}} C_k.$$

(a) Prove that C is nonempty and closed in $[T]$.

(b) Prove that C is homeomorphic to $[T]$.

(c) Prove that if both $x = \langle a_0, a_1, \dots \rangle$ and $y = \langle b_0, b_1, \dots \rangle$ belong to C, then there are infinitely many indices $n \in \mathbb{N}$ such that $a_n \neq b_n$.

(d) Prove that for every $A \subseteq [T]$, if $A \in \Sigma_\alpha([T])$, then $A \cap C \in \Sigma_\alpha(C)$.

Fix an ordinal $\alpha \geq 2$, and let $B \subseteq C$ be a set that lies in $\Sigma_\alpha(C)$ and such that $B^c \cap C$ does not lie in $\Sigma_\alpha(C)$.

(e) Prove that B lies in $\Sigma_\alpha([T])$.

For each finite subset $R \subseteq \mathbb{N}$, define B_R to be the set of all plays in $[T]$ that differ from some play in C exactly in the stages in R:

$$B_R := \left\{ y = \langle b_0, b_1, \dots \rangle \in [T] \colon \begin{array}{l} \text{there is } x = \langle a_0, a_1, \dots \rangle \in B \text{ satisfying} \\ a_n \neq b_n \text{ if and only if } n \in R \end{array} \right\}.$$

Note that $B_\emptyset = B$.

(f) Prove that B_R lies in $\Sigma_\alpha([T])$, for every finite subset $R \subset \mathbb{N}$.

Let

$$A := \left(B \cup \bigcup_{\emptyset \neq R \subset \mathbb{N}, R \text{ finite}} B_R \right)^c$$

be the set of all plays that differ from all plays in B in infinitely many coordinates.

(g) Prove that A is a tail set.

(h) Prove that $A \in \Pi_\alpha([T])$.

(i) Prove that $A \notin \Sigma_\alpha([T])$.

21. (*) Prove Theorem 6.67: Let $u : \{L, R\}^{\mathbb{N}} \to \mathbb{R}^2$ be a bounded, Borel measurable, and tail-measurable function. Prove that for each $i \in \{\mathrm{I}, \mathrm{II}\}$ and every position $p \in T^R = (\{R\} \times \{L, R\})^{<\mathbb{N}}$,

$$\overline{v}_i(\Gamma_p(u)) = \overline{v}_i(\Gamma(u)), \quad \underline{v}_i(\Gamma_p(u)) = \underline{v}_i(\Gamma(u)).$$

22. Let $\Gamma = (N, (X_i)_{i \in N}, (f_i)_{i \in N})$ be a multiplayer repeated game where the payoff function is tail measurable, let $i \in N$, and let $\varepsilon > 0$. Define

$$B_i := \{x \in [T] \colon f_i(x) \geq \underline{v}_i(\Gamma) - \varepsilon\}.$$

Prove that player i's maxmin value in the Blackwell game $(T, \mathbf{1}_{B_i})$ is 1. Where in the proof did you use the assumption that the payoff function is tail measurable?

23. (*) In this exercise we show that the conclusion of Exercise 6.22 does not hold without the condition that the payoff functions are tail measurable.

Provide an example of a two-player repeated game $G = (T, f_\mathrm{I}, f_\mathrm{II})$ such that the value of the Blackwell game $(T, \mathbf{1}_B)$ is *not* 1, where

$$B := \{x \in [T] \colon f_\mathrm{I}(x) \geq \underline{v}_\mathrm{I}(\Gamma) - \varepsilon\}.$$

24. (*) To prove that multiplayer alternating-move games admit an ε-equilibrium for every $\varepsilon > 0$, we constructed a vector of behavior strategies in which each player i follows a subgame ε-maxmin strategy (which is pure), and a deviation is punished by an indefinite punishment (see Exercise 4.12). Explain why this argument cannot be applied to prove the existence of an ε-equilibrium in repeated games with tail-measurable payoffs.

25. (*) In this exercise we provide an alternative proof to Theorem 6.62 in the case of two-player games.

    Let Γ be a two-player repeated game with tail-measurable payoffs. Fix $\varepsilon > 0$. For each player $i \in \{\mathrm{I}, \mathrm{II}\}$, set

    $$B_i := \{x \in [T]: f_i(x) \geq \underline{v}_i(\Gamma) - \varepsilon\}, \qquad (6.100)$$
    $$C_i := \{x \in [T]: f_i(x) \geq \underline{v}_i(\Gamma) + \varepsilon\}. \qquad (6.101)$$

    (a) Prove that the value of the game $(N, X_1, X_2, \mathbf{1}_{B_1}, -\mathbf{1}_{B_1})$ is 1.
    (b) Prove that the value of the game $(N, X_1, X_2, \mathbf{1}_{C_1}, -\mathbf{1}_{C_1})$ is 0.
    (c) Prove that there is a play $x^* \in [T]$ such that $f_i(x^*) \geq \overline{v}_i - \varepsilon$ for each $i \in \{1, 2\}$.
    (d) Construct a 2ε-equilibrium that is based on the play x^*.

26. In Theorem 6.62 we constructed a vector of behavior strategies σ^* that is a 2ε-equilibrium. Does σ^* induce a 2ε-equilibrium in all subgames? That is, is it true that for all positions $p \in T$, the vector of behavior strategies $\sigma_p^* := (\sigma_{i,p}^*)_{i \in N}$ is a 2ε-equilibrium in the subgame Γ_p? Justify your answer.

27. (*) Let $\Gamma = (N, T, (f_i)_{i \in N})$ be a Blackwell game where the payoff functions are characteristic functions: for each $i \in N$ there is a Borel set $W_i \subseteq [T]$ such that $f_i = \mathbf{1}_{W_i}$. Suppose that $\sum_{i \in N} \overline{v}_i(\Gamma) > n - 1$. In this exercise we will prove that in this case Γ admits a 0-equilibrium.

    (a) Prove that for every probability space $(\Omega, \mathcal{F}, \mathbf{P})$ and for every collection of measurable sets $(A_i)_{i=1}^n$ we have

    $$\mathbf{P}(A_1 \cap A_2 \cap \cdots \cap A_n) \geq \mathbf{P}(A_1) + \mathbf{P}(A_2) + \ldots \mathbf{P}(A_n) - n + 1.$$

    (b) Explain why for every $i \in N$ and every $\varepsilon > 0$ there is a closed set $C_i \subseteq W_i$ such that $\overline{v}_i(N, T, (\mathbf{1}_{C_i})_{i \in N}) \geq \overline{v}_i(\Gamma) - \varepsilon$.

    (c) For every closed set $C_i \subseteq [T]$ and every $m \in \mathbb{N}$ define

    $$C_i^m := \bigcup \{[T_p]: \mathrm{len}(p) = m, [T(p)] \cap C_i \neq \emptyset\}.$$

    In the game $(N, T, (\mathbf{1}_{C_i^m})_{i \in N})$, player i's payoff is 1 as soon as the position at stage m is a prefix of some play in C_i.
    Prove that $C_i \subseteq C_i^m$ for every $m \in \mathbb{N}$, and, moreover, that the game $(N, T, (\mathbf{1}_{C_i^m})_{i \in N})$ admits a 0-equilibrium σ^m.

    (d) Prove that by properly choosing the sets $(C_i)_{i \in N}$ in Part (b), we can ensure that the sum of the players' equilibrium payoffs in the game $(N, T, (\mathbf{1}_{C_i^m})_{i \in N})$ is more than $n - 1$ for every $m \in \mathbb{N}$:

    $$\sum_{i \in N} \mathbf{E}_{\sigma^m}[\mathbf{1}_{C_i^m}] > n - 1.$$

    (e) Conclude that $\bigcap_{i \in N} C_i \neq \emptyset$.

    (f) Prove that the game Γ admits a 0-equilibrium.

28. (*) For every $\varepsilon > 0$, find an ε-equilibrium in the Big Match game $\Gamma(u)$, where

$$u_{\mathrm{I}}(z) = 1 - \mu(z), \quad u_{\mathrm{II}}(z) = 2\mu(z), \quad \forall z \in \{L, R\}^{\mathbb{N}},$$

where

$$\mu(\langle a_{0,\mathrm{II}}, a_{1,\mathrm{II}}, \dots \rangle) := \limsup_{n \to \infty} \frac{1}{n} \cdot \#\{k < n \colon a_{k,\mathrm{II}} = L\}$$

is the long-run frequency of L in the sequence of moves of Player II.

29. (**) Repeat Exercise 6.28 for the function u defined as follows:

$$u_{\mathrm{I}}(z) = \frac{1}{2}(1 - \mu_0 + \mu_1), \quad u_{\mathrm{II}}(Z) = 1 + \mu_0 - \mu_1, \quad \forall z \in \{L, R\}^{\mathbb{N}},$$

where

$$\mu_0(\langle a_{0,\mathrm{II}}, a_{1,\mathrm{II}}, \dots \rangle) := \limsup_{n \to \infty} \frac{1}{n} \cdot \#\{k < n \colon a_{2k,\mathrm{II}} = L\}$$

is the long-run frequency of L selected by Player II in even stages and

$$\mu_1(\langle a_{0,\mathrm{II}}, a_{1,\mathrm{II}}, \dots \rangle) := \limsup_{n \to \infty} \frac{1}{n} \cdot \#\{k < n \colon a_{2k+1,\mathrm{II}} = L\}$$

is the long-run frequency of L selected by Player II in odd stages.

30. Let $\Gamma = (N, T, f)$ be a multiplayer Blackwell game. Prove that for every $\varepsilon > 0$ and every position $p \in T$ there is a position $p' \succeq p$ such that in the subgame $\Gamma_{p'}$ there is an ε-equilibrium.

31. (*) Theorem 6.74 states that in any multiplayer Blackwell game, for every $\varepsilon > 0$, there exists a position p such that the subgame Γ_p has an ε-equilibrium. Constantine proposes the following exercise: construct a game where there exist $\varepsilon_1 > \varepsilon_2 > 0$ and a position p such that Γ_p has an ε_1-equilibrium but not an ε_2-equilibrium. As Constantine explains, "This exercise will illustrate that the position p corresponding to a given ε may change with ε."

Explain why this exercise is overly challenging.

CHAPTER 7

Games with Eventual Perfect Monitoring

All games we studied so far were games of *perfect information*, that is, whenever a player selects a move, she knows the moves made by all players in all previous stages. The mathematical way to capture this assumption is to define a strategy of a player as a function from the set of positions which she controls to her set of available moves. The fact that the strategy depends on the current position implies that the player knows the current position, so in particular she knows the past moves that were selected by all players.

In this chapter we relax this assumption, and study alternating-move games in which players learn the moves made by other players with some delay. We will prove that in two-player alternating-move win/lose games, under certain conditions on the learning process, the game still has a value.

Since at a given stage, a player does not necessarily know all moves selected by the other player in previous stages, when a player selects a move, she does not necessarily know the current position, and in particular, she may not know the set of moves available to her. Yet, a natural assumption in games is that a player always knows the set of moves available to her, because it is not clear what it means to select a move that is not available at the current position. To overcome this difficulty, we will assume in this chapter that the set of moves of a player is the same in all positions (as we did for repeated games, see Definition 6.57). For simplicity of exposition, we will assume that the two players have the same finite set of moves, denoted $\mathcal{A}$, in all positions. In particular, we will study games over the tree $T = \mathcal{A}^{<\mathbb{N}}$.

The analysis in this chapter will use Exercises 5.15 and 5.25. Material that could prepare the reader for this chapter includes Exercise 3.10.

7.1 INFORMATION PARTITIONS

In this section, we present and study the concepts of information partition and information structure, which represent the knowledge that the players have along the game. Recall that at stage n, the game is in some position $p = \langle a_0, a_1, \ldots, a_{n-1} \rangle$ of length $\text{len}(p) = n$.

DOI: 10.1201/9781003582106-7

Definition 7.1 (Information partition). *Let $T = \mathcal{A}^{<\mathbb{N}}$ and let $n \in \mathbb{N}$. An* information *partition for stage n is a partition P_n of the set $\{p \in T \colon \mathrm{len}(p) = n\}$.*

The interpretation of an information partition P_n is that the player whose turn it is to select a move at stage n does not know the exact position; all she knows is the element of P_n that contains the current position. The assumption that P_n contains only positions of length n implies that players know the stage of the game.

Notation 7.2 ($\pi(p, P_n)$). *Let P_n be an information partition. For each position p of length n, denote by $\pi(p, P_n)$ the element of P_n that contains p.*

Remark 7.3 (Weakening the assumption of a constant set of moves). *As mentioned in the introduction, it is natural to assume that at any position, the player who controls the position knows the set of moves available to her. Hence, if we remove the requirement that the set of moves is the same in all positions, we would still need to require that whenever a player cannot distinguish between two positions, she has the same set of moves. That is, for every $n \in \mathbb{N}$, every element of the information partition P_n, and every two positions p, p' in this element, the set of available moves at p coincides with the set of available moves at p'.*

Definition 7.4 (Information structure). *Let $T = \mathcal{A}^{<\mathbb{N}}$. An* information structure *is a sequence $P = (P_n)_{n \in \mathbb{N}}$, where P_n is an information partition for stage n, for each $n \in \mathbb{N}$.*

An information structure represents the players' information throughout the game.

Example 7.5 (Perfect information). *Let $T = \mathcal{A}^{<\mathbb{N}}$. Suppose that for each $n \in \mathbb{N}$ the partition P_n contains only singletons: for every $p \in T$ of length n, P_n contains the set $\{p\}$. This definition corresponds to the situation where the players know the current position, so in effect we are in the situation studied in Chapter 3.*

Example 7.6 (Complete forgetfulness). *Let $T = \mathcal{A}^{<\mathbb{N}}$. Suppose that for each n the partition P_n consists of a single element – the set $\{p \in T \colon \mathrm{len}(p) = n\}$. This definition corresponds to the situation where a player only knows the calendarial time – the stage of the game, but she does not remember the moves she selected in the past and she does not know the moves selected by the other player in previous stages.*

Notation 7.7 ($C_n(a_{k_0}^*, a_{k_1}^*, \ldots, a_{k_m}^*)$). *For given $n \in \mathbb{N}$, $0 \le k_1 < k_2 < \cdots < k_m < n$, and a finite vector of moves $(a_{k_0}^*, a_{k_1}^*, \ldots, a_{k_m}^*) \in \mathcal{A}^m$ it will be convenient to denote the set of all positions of length n, in which the move at stage k_i is $a_{k_i}^*$, for each $0 \le i \le m$, by $C_n(a_{k_0}^*, a_{k_1}^*, \ldots, a_{k_m}^*)$:*

$$C_n(a_{k_0}^*, a_{k_1}^*, \ldots, a_{k_m}^*) := \{\langle a_0,\ a_1, \ldots, a_{n-1} \rangle \in T \colon a_{k_0} = a_{k_0}^*, a_{k_1} = a_{k_1}^*, \ldots, a_{k_m} = a_{k_m}^*\}.$$

Example 7.8 (Ignorance of the other player's moves). *Let $T = \mathcal{A}^{<\mathbb{N}}$. Suppose that for each $n \in \mathbb{N}$, the partition P_{2n} consists of all sets $C_{2n}(a_0^*, a_2^*, \ldots, a_{2n-2}^*)$, for every $(a_0^*, a_2^*, \ldots, a_{2n-2}^*) \in \mathcal{A}^n$. This is the set of all positions where the moves of Player I* *are some fixed moves. Suppose that for each $n \in \mathbb{N}$, the partition P_{2n+1} consists of all sets $C_{2n+1}(a_1^*, a_3^*, \ldots, a_{2n-1}^*)$, for every $(a_1^*, a_3^*, \ldots, a_{2n-1}^*) \in \mathcal{A}^n$. The information structure $P = (P_n)_{n \in \mathbb{N}}$ corresponds to the situation where a player only knows her own past moves, and she does not know what were the moves of the other players.*

Example 7.9 (Fixed delay). *Let $T = \mathcal{A}^{<\mathbb{N}}$ and let $k > 0$. Suppose that for each $n \leq k$, the partition P_{2n} contains only singletons, as in Example 7.5, and for each $n > k$, the partition P_{2n} contains all sets*

$$C_{2n}(a_0^*, a_1^*, a_2^*, a_3^*, \ldots, a_{2n-2k-1}^*, a_{2n-2k}^*, a_{2n-2k+2}^*, \ldots, a_{2n-2}^*),$$

for every $(a_0^*, a_1^*, a_2^*, a_3^*, \ldots, a_{2n-2k-1}^*, a_{2n-2k}^*, a_{2n-2k+2}^*, \ldots, a_{2n-2}^*) \in \mathcal{A}^{2n-k}$. These information partitions correspond to delayed information: Player I knows all past moves, except of the last k moves of Player II.

Example 7.10 (Games with simultaneous moves). *Let $T = \mathcal{A}^{<\mathbb{N}}$. Suppose that for each $n \in \mathbb{N}$, the partition P_{2n} contains only singletons, as in Example 7.5. Suppose also that for each $n \in \mathbb{N}$, the partition P_{2n+1} consists of all sets $C_{2n+1}(a_0^*, a_1^*, \ldots, a_{2n-1}^*)$ for every $(a_0^*, a_1^*, \ldots, a_{2n-1}^*) \in \mathcal{A}^{2n}$. The information structure $P = (P_n)_{n \in \mathbb{N}}$ corresponds to the situation where Player I knows all past moves, while Player II knows all past moves except the last move made by Player I. This information structure corresponds to simultaneous move games, as studied in Chapter 5.*

Notation 7.11 ($x \approx_n x'$). *Let $T = \mathcal{A}^{<\mathbb{N}}$, let $P = (P_n)_{n \in \mathbb{N}}$ be an information structure, let $x, x' \in [T]$, and let $n \in \mathbb{N}$. If x_n and x_n' lie in the same element of P_n, then we write $x \approx_n x'$.*

In words, $x \approx_n x'$ if the two plays are equivalent at stage n: the information available to the player who selects the move at stage n does not distinguish between x and x'. Note that $\approx_n$ is an equivalence relation: it is reflexive, symmetric, and transitive.

We now define two properties of information structures that will be useful below. The first property is called *perfect recall*, and it requires that a player does not forget her own past moves, nor does she forget anything she learned in the past.

Definition 7.12 (Perfect recall). *Let $T = \mathcal{A}^{<\mathbb{N}}$, and let $P = (P_n)_{n \in \mathbb{N}}$ be an information structure. The information structure P satisfies* perfect recall *if the following conditions hold:*

1. *Players know their own past moves: If x and x' differ in the move made in stage k, then $x \not\approx_n x'$ for every $n > k$ that has the same parity as k.*

2. *Players do not forget information: for every $x, x' \in [T]$, if $x \not\approx_n x'$ then $x \not\approx_{n+2} x'$.*

Remark 7.13 (Perfect recall of only one player). *Definition 7.12 requires that* both *players do not forget their own past moves, nor information they learned in the past.*

We could have conceived of situations where this property applies to only one player, while the other may forget her own past moves or information she learned in the past. In this case, we would define the concept of a player who has perfect recall, which would apply to $(P_{2n})_{n\in\mathbb{N}}$ (if the player who has perfect recall is Player I) or to $(P_{2n+1})_{n\in\mathbb{N}}$ (if the player who has perfect recall is Player II).

The second property of information structure we define is called *eventual perfect monitoring*, and it requires that both players can eventually distinguish between any two plays.

Definition 7.14 (Eventual perfect monitoring). *Let $T = \mathcal{A}^{<\mathbb{N}}$, and let $P = (P_n)_{n\in\mathbb{N}}$ be an information structure. The information structure P satisfies* eventual perfect monitoring *if for every two distinct plays $x, x' \in [T]$ there is $n \in \mathbb{N}$ such that $x \not\approx_k x'$ for every $k \geq n$.*

The following result is a consequence of eventual perfect monitoring. It states that for every $m \in \mathbb{N}$ there is $n > m$ which has a different parity than m, such that the move selected at stage m is known by the player who moves at stage n.

Theorem 7.15. *Let $\mathcal{A}$ be a finite alphabet, let $T = \mathcal{A}^{<\mathbb{N}}$, and let $P = (P_n)_{n\in\mathbb{N}}$ be an information structure that satisfies perfect recall and eventual perfect monitoring. Then, for every $n \in \mathbb{N}$ there is $k(n) > n$ such that (a) n and $k(n)$ have a different parity, and (b) for every two distinct plays $x = \langle a_0, a_1, \ldots \rangle$ and $\widehat{x} = \langle \widehat{a}_0, \widehat{a}_1, \ldots \rangle$ in $[T]$, if $x \approx_{k(n)} \widehat{x}$, then $a_n = \widehat{a}_n$.*

In the proof of Theorem 7.15 we will consider converging sequences in $\mathcal{A}^{\mathbb{N}}$. As we mentioned at the end of Section 2.1, when $\mathcal{A}$ is finite, the space $\mathcal{A}^{\mathbb{N}}$ is compact in the product topology, and a sequence $(x^n)_{n\in\mathbb{N}}$ of points in $\mathcal{A}^{\mathbb{N}}$ converges to $x \in \mathcal{A}^{\mathbb{N}}$ if every prefix of x is a prefix of all but finitely many elements of $(x^n)_{n\in\mathbb{N}}$.

Proof. Suppose that the claim is not true. Then there is $n \in \mathbb{N}$ such that for every $m > n$ with a different parity than n, there are two distinct plays, $x^m = \langle a_0^m, a_1^m, \ldots \rangle$ and $\widehat{x}^m = \langle \widehat{a}_0^m, \widehat{a}_1^m, \ldots \rangle$, in $[T]$ for which $x^m \approx_m \widehat{x}^m$, but $a_n^m \neq \widehat{a}_n^m$. Recall that $x_m^m = \langle a_0^m, a_1^m, \ldots, a_m^m \rangle$ and $\widehat{x}_m^m = \langle \widehat{a}_0^m, \widehat{a}_1^m, \ldots, \widehat{a}_m^m \rangle$.

Consider the following four sequences:

- $(a_n^m)_{m\in N}$ and $(\widehat{a}_n^m)_{m\in N}$, which are sequences of moves in $\mathcal{A}$.

- $(x^m)_{m\in\mathbb{N}}$ and $(\widehat{x}^m)_{m\in\mathbb{N}}$, which are sequences of plays in $[T] = \mathcal{A}^{\mathbb{N}}$.

The set $\mathcal{A}$ is finite, and the set $\mathcal{A}^{\mathbb{N}}$ is compact in the product topology. It follows that there is an increasing subsequence $(m_l)_{l\in\mathbb{N}}$ with $m_0 > n$ along which all four sequences $(a_n^{m_l})_{l\in\mathbb{N}}$, $(\widehat{a}_n^{m_l})_{l\in\mathbb{N}}$, $(x^{m_l})_{l\in\mathbb{N}}$, and $(\widehat{x}^{m_l})_{l\in\mathbb{N}}$ converge. Denote the limits of these sequences by a^*, $\widehat{a}^*$, x^*, and $\widehat{x}^*$, respectively. Then $a_n^{m_l} = a^*$ for every $l \in \mathbb{N}$ sufficiently large, $\widehat{a}_n^{m_l} = \widehat{a}^*$ for every $l \in \mathbb{N}$ sufficiently large, every prefix of x^* is a prefix of all x^{m_l} for every $l \in \mathbb{N}$ sufficiently large, and every prefix of $\widehat{x}^*$ is a prefix of all $\widehat{x}^{m_l}$ for every $l \in \mathbb{N}$ sufficiently large.

Since for every $l \in \mathbb{N}$ the parity of m_l is different than the parity of n, all numbers in the sequence $(m_l)_{l \in \mathbb{N}}$ have the same parity. By dropping the first elements of the subsequence $(m_l)_{l \in \mathbb{N}}$, we can assume that for every $l \in \mathbb{N}$,

1. $a_n^{m_l} = a^*$.

2. $\widehat{a}_n^{m_l} = \widehat{a}^*$.

3. The prefix of length l of x^* is a prefix of x^{m_k}, for all $k \geq l$.

4. The prefix of length l of $\widehat{x}^*$ is a prefix of $\widehat{x}^{m_k}$, for all $k \geq l$.

Note that $a_n^* = a^*$ and $\widehat{a}_n^* = \widehat{a}^*$. Since $a^* \neq \widehat{a}^*$, it follows that $x^* \neq \widehat{x}^*$. As the information structure P satisfies eventual perfect monitoring, this implies that $x^* \not\approx_m \widehat{x}^*$ for every $m \in \mathbb{N}$ sufficiently large. To derive a contradiction, we will prove that for every $m > n$ that has a different parity from n, and every l sufficiently large,

$$x^* \approx_m x^{m_l} \approx_m \widehat{x}^{m_l} \approx_m \widehat{x}^*. \tag{7.1}$$

Fix then $m > n$ that has a different parity from n. In particular, m has the same parity as all elements in the sequence $(m_l)_{l \in \mathbb{N}}$.

We start with the second equivalence in Equation (7.1). For every $l \in \mathbb{N}$ we have $x^{m_l} \approx_{m_l} \widehat{x}^{m_l}$. Since m and m_l have the same parity, and since P satisfies perfect recall, $x^{m_l} \approx_m \widehat{x}^{m_l}$, provided $m_l \geq m$.

We turn to the first and third equivalences in Equation (7.1). By (iii) and (iv), $x_m^{m_l} = x_m^*$ and $\widehat{x}_m^{m_l} = \widehat{x}_m^*$, provided l is sufficiently large. In particular, $x^{m_l} \approx_m x^*$ and $\widehat{x}^{m_l} \approx_m \widehat{x}^*$ for every l sufficiently large. □

Remark 7.16 (Monotonicity of $n \mapsto k(n)$). *While eventual perfect monitoring ensures that for each $n \in \mathbb{N}$ there exists $k(n) \in \mathbb{N}$ such that at stage $k(n)$ the players know the move selected at stage n, it does not guarantee that the function $n \mapsto k(n)$ is monotone. Indeed, it can be that, e.g., at stage 3 Player II learns Player I's move at stage 2, and at stage 5 Player II learns Player I's move at stages 0 and 4, and afterwards, in each stage $2n + 1$ Player II learns Player I's move at stage $2n$. Then,*

$$k(0) = 5, \quad k(2) = 3, \quad k(2n) = 2n + 1, \quad \forall n \geq 2.$$

Theorem 7.15 states that if the information structure P satisfies perfect recall and eventual perfect monitoring, then there is a stage $k(n)$ at which both players know the move that was selected in stage n. Since the number of stages up to stage n is finite, by setting $\widehat{k}(n) := \max_{m \leq n} k(m)$, we obtain that if the information structure P satisfies perfect recall and eventual perfect monitoring, then at stage $\widehat{k}(n)$ both players know all moves that were selected up to stage n. This is the content of the following result.

Corollary 7.17. *Let $\mathcal{A}$ be a finite alphabet, let $T = \mathcal{A}^{<\mathbb{N}}$, and let $P = (P_n)_{n \in \mathbb{N}}$ be an information structure that satisfies perfect recall and eventual perfect monitoring. Then, for every $n \in \mathbb{N}$ there is $\widehat{k}(n) > n$ such that (a) n and $\widehat{k}(n)$ have a different parity, and (b) for every two distinct plays $x, \widehat{x} \in [T]$, if $x \approx_{\widehat{k}(n)} \widehat{x}$, then $x_n = \widehat{x}_n$.*

7.2 GAMES WITH IMPERFECT MONITORING

We now define the games that we will study in this chapter, which are win/lose games as we treated in Chapter 3, equipped with an information structure.

Definition 7.18 (Games with imperfect monitoring). *A game with imperfect monitoring (over an alphabet $\mathcal{A}$) is a triplet $G = (T, W, P)$ where (T, W) is an alternating-move win/lose game (over the alphabet $\mathcal{A}$) and $P = (P_n)_{n \in \mathbb{N}}$ is an information structure.*

As the information a player has is the element of the current partition that contains the actual position, we have to adapt the definition of strategies.

Definition 7.19 (Pure strategy). *Let $G = (T, W, P)$ be a game with imperfect monitoring. A pure strategy of Player I is a function $s_I : \bigcup_{n \in \mathbb{N}} P_{2n} \to \mathcal{A}$. A pure strategy of Player II is a function $s_{II} : \bigcup_{n \in \mathbb{N}} P_{2n+1} \to \mathcal{A}$.*

Definition 7.20 (Behavior strategy). *Let $G = (T, W, P)$ be a game with imperfect monitoring. A behavior strategy of Player I is a function $\sigma_I : \bigcup_{n \in \mathbb{N}} P_{2n} \to \Delta(\mathcal{A})$. A behavior strategy of Player II is a function $\sigma_{II} : \bigcup_{n \in \mathbb{N}} P_{2n+1} \to \Delta(\mathcal{A})$.*

Remark 7.21 (The domain of strategies). *The fact that the domain of a strategy is $\bigcup_{n \in \mathbb{N}} P_{2n}$ (for Player I) or $\bigcup_{n \in \mathbb{N}} P_{2n+1}$ (for Player II) reflects the property that the information available to a player when selecting her move at stage n is the atom of P_n that includes the current position. Since all positions in this atom have the same length – n, players know calendarial time.*

Notation 7.22 ($S_i(T, P)$, $\Sigma_i(T, P)$). *Let $G = (T, W, P)$ be a game with imperfect monitoring, and let $i \in \{I, II\}$. The set of pure strategies of player i is denoted by $S_i(T, P)$, and the set of her behavior strategies is denoted by $\Sigma_i(T, P)$.*

Remark 7.23 (Kuhn's Theorem for games with imperfect monitoring). *Recall that a probability distribution on pure strategies is called a mixed strategy. In Remark 5.32 we mentioned that when monitoring is perfect, the set of probability distributions on pure strategies is equivalent to the set of behavior strategies. As proven by Kuhn (1957), see also Chapter 6 in Maschler, Solan, and Zamir (2020), this equivalence holds also in games with imperfect monitoring, provided the information structure satisfies perfect recall.*

Notation 7.24 ($\mathbf{P}_{\sigma_I, \sigma_{II}}$). *Every pair of behavior strategies (σ_I, σ_{II}) induces a probability measure on $[T]$, which is denoted by $\mathbf{P}_{\sigma_I, \sigma_{II}}$.*

The concepts of minmax value, maxmin value, and value, are analogous to these concepts in win/lose games and in two-player zero-sum simultaneous-move games.

Definition 7.25 (Minmax value, maxmin value). *Let $G = (T, W, P)$ be a game with imperfect monitoring. The minmax value of G is*

$$\overline{v}(G) := \inf_{\sigma_{II} \in \Sigma_{II}(T,P)} \sup_{\sigma_I \in \Sigma_I(T,P)} \mathbf{P}_{\sigma_I, \sigma_{II}}(W).$$

The maxmin value *of G is*

$$\underline{v}(G) := \sup_{\sigma_{\mathrm{I}} \in \Sigma_{\mathrm{I}}(T,P)} \inf_{\sigma_{\mathrm{II}} \in \Sigma_{\mathrm{II}}(T,P)} \mathbf{P}_{\sigma_{\mathrm{I}},\sigma_{\mathrm{II}}}(W).$$

Definition 7.26 (Value). *Let $G = (T, W, P)$ be a game with imperfect monitoring. We say that the real number $v = v(G)$ is the* value *of G if $v = \underline{v}(G) = \overline{v}(G)$.*

The main result we prove in this chapter is:

Theorem 7.27 (Shmaya, 2011). *Let $G = (T, W, P)$ be a game with imperfect monitoring over a finite alphabet $\mathcal{A}$. If W is Borel, and if P satisfies perfect recall and eventual perfect monitoring, then G has a value.*

Remark 7.28 (Games with finite horizon). *Suppose that the game with imperfect monitoring $G = (T, W, P)$ lasts only finitely many stages. That is, there is $n \in \mathbb{N}$ such that the identity of the winner is determined by the prefix of length n of the play. By identifying pure strategies that coincide in the moves they select in the first n stages, we obtain that the set of pure strategies of each player is finite. It follows that the game G is equivalent to a one-shot game, in which the value exists when players are restricted to mixed strategies (see Theorem 5.17). By Kuhn's Theorem (see Remark 7.23), the value exists also when players are restricted to behavior strategies, provided both players have perfect recall.*

When the information structure does not satisfy perfect recall, the value need not exist when the players are restricted to behavior strategies, see Exercise 7.7. Theorem 7.27 states that when the game does not have a finite horizon, assuming eventual perfect monitoring in addition to perfect recall is sufficient to guarantee that the value exists.

7.3   EXAMPLE: STOPPING GAME

In this section, we study a certain stopping game with different information structures, and see that the minmax value, the maxmin values, and the existence of the value depend on the information structure.

The set of moves of the two players consists of two moves, continue and stop:

$$\mathcal{A} = \{c, s\}.$$

For every play $x = \langle a_0, a_1, \dots \rangle \in [T]$, denote the first stage in which Player I stops by

$$n_{\mathrm{I}}(x) := \min\{k \in \mathbb{N} : k \text{ even}, a_k = s\},$$

and the first stage in which Player II stops by

$$n_{\mathrm{II}}(x) := \min\{k \in \mathbb{N} : k \text{ odd}, a_k = s\}.$$

The winning set is

$$W := \{x \in [T] : n_{\mathrm{II}}(x) < n_{\mathrm{I}}(x) < \infty \text{ or } n_{\mathrm{I}}(x) < \infty = n_{\mathrm{II}}(x)\}. \tag{7.2}$$

Thus, Player I wins if she stops after Player II, or if she stops but Player II never stops.

7.3.1 Fixed Delay

We here study the game in which players observe their opponent's moves with a fixed delay.

Let $k \in \mathbb{N}$ be odd. Consider the game with the following information structure: at each stage n, the player whose turn it is to move knows her past moves, as well as the moves of her opponent in all stages m for $m \leq n - k$ (see Example 7.9).

The value of this game is 0. Indeed, the pure strategy s_{II} that stops whenever Player II observes that Player I stopped ensures that Player II wins. Formally, s_{II} is defined as follows:

- For each stage $2n + 1$ such that $2n + 1 < k$, continue.

- For each stage $2n + 1$ such that $2n + 1 \geq k$, copy Player I's move in stage $2n + 1 - k$.

Indeed, if Player I always select the move c, Player II also always selects c. The realized play is then $x = \langle c, c, c, \dots \rangle$, so that $n_I(x) = n_{II}(x) = \infty$ and Player I loses. If Player I selects the move s for the first time in stage $2n$, then Player II will select the move c until stage $2n + k - 1$ and will select the move s in stage $2n + k$. We will then have $n_I(x) = 2n$ and $n_{II}(x) = 2n + k$, so that once again Player I loses.

7.3.2 Ignorance of the Other Player's Moves

Suppose now that the information structure is the one in Example 7.8, where each player observes only her own moves, but is ignorant of the moves of the other player.

We argue that
$$\overline{v} = 1, \quad \underline{v} = 0,$$
so that the game does not have a value. We will prove only that $\overline{v} = 1$. The proof that $\underline{v} = 0$ is analogous. Fix then a strategy $\sigma_{II} \in \Sigma_{II}(T, P)$. Since Player II does not observe Player I's moves, the probability that Player II stops in a finite time after stage n, namely, $\mathbf{P}_{\sigma_I, \sigma_{II}}(n < n_{II}(x) < \infty)$, is independent of σ_I, and so can be written as $\mathbf{P}_{\sigma_{II}}(n < n_{II}(x) < \infty)$. Note that

$$\lim_{n \to \infty} \mathbf{P}_{\sigma_{II}}(n < n_{II}(x) < \infty) = \mathbf{P}_{\sigma_{II}}\left(\bigcap_{n \in \mathbb{N}} \{n < n_{II}(x) < \infty\}\right) = \mathbf{P}_{\sigma_{II}}(\emptyset) = 0,$$

hence for every $\varepsilon > 0$ there is $n = n(\sigma_{II}, \varepsilon) \in \mathbb{N}$ such that

$$\mathbf{P}_{\sigma_{II}}(n < n_{II}(x) < \infty) < \varepsilon.$$

Let $\widehat{\sigma}_I \in \Sigma_I(T, P)$ be the strategy that continues until stage n and stops at the first even stage after stage n. Under $(\widehat{\sigma}_I, \sigma_{II})$ Player I wins with probability at least $1 - \varepsilon$, hence

$$\sup_{\sigma_I \in \Sigma_I(T, P)} \mathbf{P}_{\sigma_I, \sigma_{II}}(W) \geq 1 - \varepsilon.$$

Since this inequality holds for every $\sigma_{\mathrm{II}} \in \Sigma_{\mathrm{II}}(T, P)$, it follows that

$$\overline{v} = \inf_{\sigma_{\mathrm{II}} \in \Sigma_{\mathrm{II}}(T,P)} \sup_{\sigma_{\mathrm{I}} \in \Sigma_{\mathrm{I}}(T,P)} \mathbf{P}_{\sigma_{\mathrm{I}},\sigma_{\mathrm{II}}}(W) \geq 1 - \varepsilon.$$

Since ε is arbitrary, $\overline{v} = 1$.

7.3.3 Player II is Ignorant about Player I's Moves

Suppose now that the information structure is such that Player I observes the moves of Player II (as in Example 7.5), while Player II is ignorant of Player I's moves (as in Example 7.8).

As in Subsection 7.3.2,

$$\overline{v} = 1.$$

In Exercise 7.5 we ask the reader to prove that $\underline{v} = \frac{1}{2}$, so that the game does not have a value. In fact, the following behavior strategy is a 0-maxmin strategy of Player I:

- In stage 0, select each of the moves c and s with probability $\frac{1}{2}$.

- For each $n \geq 1$, in stage $2n$ select the move that Player II selected in stage $2n - 1$.

7.4 PROOF OF SHMAYA'S THEOREM – MAIN IDEAS

We start by providing the main ideas of the proof. A behavior strategy of a player, say, Player I, in a game with imperfect monitoring G, is a function $\sigma_{\mathrm{I}} : \bigcup_{n \in \mathbb{N}} P_{2n} \to \Delta(\mathcal{A})$. When Player I adopts a strategy σ_{I}, at each stage n she knows the atom $Q \in P_n$ that contains the current position, and selects a move according to $\sigma_{\mathrm{I}}(Q)$. Instead of doing the randomization herself, Player I could use a mediator who possesses the same information she has: at each stage n she could transmit to the mediator the restriction of σ_{I} to P_n, and let the mediator select Player I's move according to $\sigma_{\mathrm{I}}(Q)$.

Suppose both players use the mediator. In that case, we obtain an alternating-move game G', where at each even stage n Player I selects a function $\sigma_{\mathrm{I},n} : P_n \to \Delta(\mathcal{A})$, at each odd stage n Player II selects a function $\sigma_{\mathrm{II},n} : P_n \to \Delta(\mathcal{A})$, and at every stage the mediator selects the actual moves of the players.

So that the game G' properly reflects the original game G, in G' the players must not know the actual moves that the mediator selected in the past. Otherwise, they could condition their choices on past choices of the mediator, which might not be known to them in the original game.

Given the functions $\sigma_{\mathrm{I}} = (\sigma_{\mathrm{I},2n})_{n \in \mathbb{N}}$ and $\sigma_{\mathrm{II}} = (\sigma_{\mathrm{II},2n+1})_{n \in \mathbb{N}}$ that the players transmit to the mediator along G', we can calculate the probability that the play selected by the mediator lies in W, in which case Player I wins G'. In other words, to every pair of strategies σ_{I} and σ_{II} in G' we can assign a payoff $-\mathbf{P}_{\sigma_{\mathrm{I}},\sigma_{\mathrm{II}}}(W)$. This way we transform the game with imperfect monitoring into an alternating-move game as we studied in Chapter 3.

An important property of G' is that it satisfies the conditions of Theorem 4.15, and hence it has a value. We argue that

$$\underline{v}(G) \leq v(G') \leq \overline{v}(G). \tag{7.3}$$

Let us see why the left-hand side inequality holds; the right-hand side inequality holds for analogous reasons. Fix then $\varepsilon > 0$, and consider a strategy σ_{I} in G that guarantees $\underline{v}(G) - \varepsilon$, that is, $\mathbf{P}_{\sigma_{\mathrm{I}},\sigma_{\mathrm{II}}}(W) \geq \underline{v}(G) - \varepsilon$ for every $\sigma_{\mathrm{II}} \in S_{\mathrm{II}}(T, P)$. Using the very same strategy in G', namely, transmitting to the mediator $(\sigma_{\mathrm{I}}(Q))_{Q \in P_n}$ in each even stage n, Player I guarantees $\underline{v}(G) - \varepsilon$ in G'.

Equation (7.3) implies that *if* the game G has a value, *then* its value coincides with the value of G'. However, we do not know that the value of G exists. In fact, this is what we set out to prove.

To complete the proof, we need to show that $v(G') \leq \underline{v}(G)$. We will not be able to show that for the game G' defined above, but we will be able to show it for a variation of this game.

Consider again the auxiliary game G', and assume that instead of selecting the move for stage n at stage n, the mediator selects that move at stage $k(n)$, and immediately reveals it to the players. This additional information that the players obtain has no effect on the value of G', because for $m \geq k(n)$ the information partition P_m allows the player who controls stage m to condition her move on the move selected at stage n. Call the new auxiliary game G''. The significance of this alteration of the game is that the play in G'' determines the play in G, and hence the identity of the winner (because the moves selected by the mediator are part of the play in G'', and they form the play in G).

By Exercise 5.15, the game G'' has a value, and as in Exercise 5.25, we can approximate the winning set by a compact subset. As we will see, games with imperfect monitoring and a compact winning set have a value. We will show that the limit of the values of these approximating games, as the approximation improves, is the value of G.

The rest of this section is devoted to the formal proof, which is organized as follows. In Section 7.5 we present the concept of upper semi-continuous functions, and extend Von Neumann's Minmax Theorem (Theorem 5.17) to two-player one-shot games with compact sets of moves and upper-semi-continuous payoff function. As we will see, the characteristic function of a compact set is upper semi-continuous. In Section 7.6 we prove that games with imperfect monitoring and compact winning set have a value. In Section 7.7 we show that we can approximate any strategy by a strategy that uses only finitely many mixed moves. In Section 7.8 we define the auxiliary game and use the tools developed earlier to prove Theorem 7.27.

7.5 ONE-SHOT GAMES WITH UPPER SEMI-CONTINUOUS PAYOFFS

Von Neumann's Minmax Theorem (Theorem 5.17) states that every two-player zero-sum one-shot game where the players have finite sets of moves has a value, and both players have 0-optimal mixed moves. As Example 5.18 shows, when the sets of moves of the two players are countably infinite, the value need not exist. In this section, we present (without a proof) an extension of von Neumann's Theorem, to the case where the sets of moves of the players are convex subsets of $[0,1]^{\mathbb{N}}$, provided the payoff function satisfies a property called upper semi-continuity.

Definition 7.29 (Upper semi-continuous function). *Let X be a topological space. For $x_0 \in X$, a function $f : X \to \mathbb{R}$ is upper semi-continuous at x_0 if for every $y > f(x_0)$ there is an open neighborhood U of x_0 such that $f(x) < y$ for every $x \in U$. The function f is upper semi-continuous if it is upper semi-continuous at all points $x_0 \in X$.*

Remark 7.30 (Upper semi-continuity in metric spaces). *When X is also a metric space, an equivalent definition to upper semi-continuity is that for every sequence $(x_n)_{n \in \mathbb{N}}$ that converges to x_0 we have $\limsup_{n \to \infty} f(x_n) \leq f(x_0)$.*

Example 7.31 (Continuous functions). *Every continuous function $f : X \to \mathbb{R}$ is upper semi-continuous.*

Example 7.32 (The characteristic function of a closed set). *Let X be a topological space and let $W \subseteq X$ be closed. The function $f := \mathbf{1}_W$ is upper semi-continuous.*

Indeed, for $x_0 \in W$, the function f is upper semi-continuous at x_0 since it attains at x_0 the value 1, which is its maximal value. Since W is closed, W^c is open, and hence For every $x_0 \notin W$ there is an open neighborhood U of x_0 that is contained in W^c. For every x in this neighborhood we have $f(x) = 0 = f(x_0)$.

Von Neumann's Minmax Theorem has been extended by Fan (1952) and Kneser (1952) to upper semi-continuous functions. The following theorem is a special case of these results. A simple proof of this theorem can be found in Borwein and Zhuang (1986).

Theorem 7.33 (Fan, 1952; Kneser, 1952). *Let X and Y be two convex subsets of $[0,1]^{\mathbb{N}}$ endowed with the product topology, and let $f : X \times Y \to [0,1]$ be a function such that the section $x \mapsto f(x,y)$ is upper semi-continuous for every fixed $y \in Y$. If X is compact, then*

$$\sup_{\mu \in \Delta(X)} \inf_{y \in Y} \mathbf{E}_{\mu}[f(x,y)] = \inf_{\nu \in \Delta(Y)} \sup_{x \in X} \mathbf{E}_{\nu}[f(x,y)].$$

7.6 PROOF OF SHMAYA'S THEOREM FOR COMPACT WINNING SETS

In this section we prove Theorem 7.27 for compact winning sets. As we will see, for this result we do not need the information structure to satisfy eventual perfect monitoring. We start by showing that when the winning set is closed, all games with imperfect monitoring and perfect recall have a value.

Theorem 7.34. *All games with imperfect monitoring, perfect recall, and closed winning sets have a value.*

Proof. Let $G = (T, W, P)$ be a game with imperfect monitoring, where the information structure satisfies perfect recall and where the winning set W is closed.

For each $i \in \{I, II\}$, the set of player i's pure strategies $S_i(T, P)$ is compact in the product topology. The function $x : S_I(T, P) \times S_{II}(T, P) \to [T]$ that assigns to each pair of pure strategies $(s_I, s_{II}) \in S_I(T, P) \times S_{II}(T, P)$ the play $x(s_I, s_{II}) \in [T]$ that they induce is continuous in the product topology (Exercise 7.8). Since W is closed, it follows that the set $x^{-1}(W) \subseteq S_I(T, P) \times S_{II}(T, P)$ is closed as well. As we saw in Example 7.32, since $x^{-1}(W)$ is closed, the function $\mathbf{1}_{x^{-1}(W)}$ is upper semi-continuous.

The function $\mathbf{1}_{x^{-1}(W)}$ is the characteristic function of the set $x^{-1}(W)$: it attains the value 1 for every pair of strategies for which the corresponding play is in W, and the value 0 otherwise. Therefore, for every pair of mixed strategies $(\mu_I, \mu_{II}) \in \Delta(S_I(T, P)) \times \Delta(S_{II}(T, P))$,

$$\mathbf{E}_{\mu_I, \mu_{II}} \left[\mathbf{1}_{x^{-1}(W)}(s_I, s_{II}) \right] = \mathbf{P}_{\mu_I, \mu_{II}}(x(s_I, s_{II}) \in W). \tag{7.4}$$

Theorem 7.33 and Equation (7.4) imply that

$$\sup_{\mu_I \in \Delta(S_I(T,P))} \inf_{s_{II} \in S_{II}(T,P)} \mathbf{P}_{\mu_I}(x(s_I, s_{II}) \in W)$$
$$= \inf_{\mu_{II} \in \Delta(S_{II}(T,P))} \sup_{s_I \in S_I(T,P)} \mathbf{P}_{\mu_{II}}(x(s_I, s_{II}) \in W).$$

As discussed in Remark 7.28, since the information structure P satisfies perfect recall, the set $\Delta(S_i(T, P))$ of mixed strategies of each player i, is equivalent to her set of behavior strategies, $\Sigma_i(T, P)$. It follows that the game G has a value. □

Remark 7.35 (Games without perfect recall and mixed strategies). *When the game does not have perfect recall, the set of mixed strategies is not equivalent to the set of behavior strategies. Nevertheless, the proof of Theorem 7.34 shows that all games with imperfect monitoring and a closed winning set have a value when the players are restricted to mixed strategies.*

Remark 7.36 (On approximating the winning set by a closed subset). *When the winning set W is not closed, one could hope to approximate it from inside by a closed subset W_0 (which is compact since the space of plays $[T]$ is compact in the product topology), and argue that the value of the game is close to the value of the game with winning set W_0. The difficulty with this approach is that approximating a set by a closed subset is done relative to a fixed probability distribution, see Definition 2.19, whereas to assert the existence of the value we need the approximation to be valid for every strategy of the opponent.*

7.7 APPROXIMATING BEHAVIOR STRATEGIES

In this section we will approximate a behavior strategy of a player with a behavior strategy that satisfies a special property: the mixed move in positions of length n is taken from a finite set, which depends on n. The approximation will be done by selecting for every $n \in \mathbb{N}$ a finite subset of mixed moves in $\Delta(\mathcal{A})$ that is $\frac{\varepsilon}{2^n}$-dense, for some small ε. This is done so that the set of moves in the auxiliary game we construct will be finite, allowing us to apply to it results previously proved for games with finite sets of moves.

Recall that every probability distribution $\mu \in \Delta(\mathcal{A})$ is a vector in $\mathbb{R}^{\mathcal{A}}$. For every two probability distributions $\mu_1, \mu_2 \in \Delta(\mathcal{A})$, define

$$\|\mu_1 - \mu_2\|_1 := \sum_{a \in \mathcal{A}} |\mu_1(a) - \mu_2(a)|.$$

Definition 7.37 (Distance between behavior strategies). *Let $G = (T, W, P)$ be a game with imperfect monitoring, and let $\sigma_{\mathrm{I}}, \sigma_{\mathrm{I}}' \in \Sigma_{\mathrm{I}}(T, P)$ be two behavior strategies of Player I. The* distance *between σ_{I} and σ_{I}' is*

$$d(\sigma_{\mathrm{I}}, \sigma_{\mathrm{I}}') := \sum_{n \in \mathbb{N}} \max_{\{p \in T: \, \mathrm{len}(p)=2n\}} \|\sigma_{\mathrm{I}}(p) - \sigma_{\mathrm{I}}'(p)\|_1.$$

The function d thus introduced can take the value ∞, and it satisfies the properties that characterize a distance: it is non-negative, symmetric, satisfies the triangle inequality, and the distance between two behavior strategies is 0 if and only if these strategies coincide.

The following result, whose proof is left to the reader (Exercise 7.9), relates the distance between strategies to the difference in the probabilities to reach the winning set.

Theorem 7.38. *Let $G = (T, W, P)$ be a game with imperfect monitoring, and let $\sigma_{\mathrm{I}}, \sigma_{\mathrm{I}}' \in \Sigma_{\mathrm{I}}(T, P)$ be two behavior strategies. Then*

$$\left| \mathbf{P}_{\sigma_{\mathrm{I}}, \sigma_{\mathrm{II}}}(W) - \mathbf{P}_{\sigma_{\mathrm{I}}', \sigma_{\mathrm{II}}}(W) \right| \le d(\sigma_{\mathrm{I}}, \sigma_{\mathrm{I}}'), \quad \forall \sigma_{\mathrm{II}} \in \Sigma_{\mathrm{II}}(T, P).$$

We now define for each $\delta > 0$ a finite subset of the set $\Delta(\mathcal{A})$ of mixed moves, that is dense in $\Delta(\mathcal{A})$.

Definition 7.39 (δ-dense set). *Let $\mathcal{A}$ be a finite set, and let $\delta > 0$. A subset $\Delta' \subseteq \Delta(\mathcal{A})$ is* δ-dense *if for every $\mu \in \Delta(\mathcal{A})$ there is $\mu' \in \Delta'$ such that $\|\mu - \mu'\|_1 \le \delta$.*

Definition 7.40 (Δ_ε-discrete strategy). *Let $G = (T, W, P)$ be a game with imperfect monitoring, and let $\varepsilon > 0$. For every $n \in \mathbb{N}$, let $\Delta_{n,\varepsilon}$ be a finite $\frac{\varepsilon}{2^{n+1}}$-dense subset of $\Delta(\mathcal{A})$, and denote $\Delta_\varepsilon = (\Delta_{n,\varepsilon})_{n \in \mathbb{N}}$. A strategy $\sigma_{\mathrm{I}} \in \Sigma_{\mathrm{I}}(T, P)$ is* Δ_ε-discrete *if for every $n \in \mathbb{N}$ and every position $p \in T$ of length n, we have*

$$\sigma_{\mathrm{I}}(p) \in \Delta_{n,\varepsilon}.$$

Theorem 7.38 implies that every strategy can be approximated by a Δ_ε-discrete strategy. This is the content of the next result, whose proof is left to the reader (Exercise 7.10).

Theorem 7.41. *Let $G = (T, W, P)$ be a game with imperfect monitoring, and let $\varepsilon > 0$. For every $n \in \mathbb{N}$ let $\Delta_{n,\varepsilon}$ be a finite $\frac{\varepsilon}{2^{n+1}}$-dense subset of $\Delta(\mathcal{A})$, and denote $\Delta_\varepsilon = (\Delta_{n,\varepsilon})_{n \in \mathbb{N}}$. For every behavior strategy $\sigma_{\mathrm{I}} \in \Sigma_{\mathrm{I}}(T, P)$ there exists a Δ_ε-discrete strategy $\sigma_{\mathrm{I}}' \in \Sigma_{\mathrm{I}}(T, P)$ such that*

$$\left| \mathbf{P}_{\sigma_{\mathrm{I}}, \sigma_{\mathrm{II}}}(W) - \mathbf{P}_{\sigma_{\mathrm{I}}', \sigma_{\mathrm{II}}}(W) \right| \le \varepsilon, \quad \forall \sigma_{\mathrm{II}} \in \Sigma_{\mathrm{II}}(T, P).$$

7.8 RECAP: PROOF OF SHMAYA'S THEOREM

In this section we prove Theorem 7.27, which states that all games with imperfect monitoring where the information structure satisfies perfect recall and eventual perfect monitoring have a value. In the proof we will analyze games with a winning set different from W (but with the same tree T and the same information structure P). We therefore denote by $G(W) = (T, W, P)$ the game with the winning set W.

We will define for every $\varepsilon > 0$ an auxiliary game with perfect information $\widehat{G}_\varepsilon(W)$ which approximates the original game $G(W)$ in a proper sense. Three players will participate in the auxiliary game $\widehat{G}_\varepsilon(W)$: Player 1, who will represent Player I of $G(W)$, Player 2, who will represent Player II of $G(W)$, and a third player, who will follow a pre-specified behavior strategy, as we have seen in Exercise 5.15. The role of the third player will be to execute the lotteries selected by Players 1 and 2. Namely, Players 1 and 2 will select mixed moves for Players I and II in $G(W)$, and Player 3 will select the actual moves according to the mixed moves of Players 1 and 2.

7.9 NOTATIONS

Fix $\varepsilon > 0$. For every $\delta > 0$ let Δ_δ be a finite δ-dense subset of $\Delta(\mathcal{A})$.

Recall that a behavior strategy for Player I (resp., Player II) is a function from $\bigcup_{n\in\mathbb{N}} P_{2n}$ (resp., $\bigcup_{n\in\mathbb{N}} P_{2n+1}$) to $\Delta(\mathcal{A})$. The part of the behavior strategy that corresponds to stage n is a function from P_n to $\Delta(\mathcal{A})$. For every $n \in \mathbb{N}$, let B_n be the following set, which approximates the set of functions from P_n to $\Delta(\mathcal{A})$:

$$B_n := \{b : P_n \to \Delta_{\varepsilon/2^n}\}.$$

By Theorem 7.15, for every $m \in \mathbb{N}$ there exists a minimal integer $k(m) \in \mathbb{N}$ such that

- $k(m) > m$.

- $k(m)$ and m have different parity.

- At stage $k(m)$, the player who does *not* control stage m knows the move of her opponent at stage m.

Denote
$$K_n := \{m \in \mathbb{N} : k(m) = n\} \subseteq \{0, 1, \ldots, n-1\}, \quad \forall n \in \mathbb{N}.$$

We will devise an auxiliary game in which the move of a player at stage m is observed by her opponent at stage $k(m)$. As a result, in that auxiliary game, the player who controls stage n learns the moves of her opponent in the stages in K_n, that is, an element of the set
$$Z_n := \mathcal{A}^{K_n}.$$

For every $n \in \mathbb{N}$, denote by
$$\psi_n : \mathcal{A}^{\mathbb{N}} \to Z_n$$

the projection onto the coordinates in K_n. Let $F : \prod_{n \in \mathbb{N}} Z_n \to \mathcal{A}^{\mathbb{N}}$ be the function that satisfies

$$F(\psi_0(x), \psi_1(x), \dots) = x, \quad \forall x \in [T]. \tag{7.5}$$

When the play is x, at each stage $n \in \mathbb{N}$ the player who moves at stage n learns that her opponent selected the moves $\psi_n(x)$ in stages in K_n. The function F combines the information learnt by the players along the play into the actual play.

We note that F is well-defined. Indeed, since P satisfies eventual perfect monitoring, Theorem 7.15 implies that

$$\bigcup_{n \in \mathbb{N}} K_n = \mathbb{N}, \tag{7.6}$$

and hence $\prod_{n \in \mathbb{N}} Z_n$ is naturally mapped to $\mathcal{A}^{\mathbb{N}}$ by a proper permutation of the coordinates.

Since the information structure satisfies eventual perfect monitoring, the function F is continuous.

7.10 DEFINITION OF THE AUXILIARY GAME $\widehat{G}_\varepsilon(W)$

Let $\widehat{G}_\varepsilon(W) = (\widehat{T}_\varepsilon, \widehat{W}_\varepsilon)$ be the following auxiliary alternating-move win-lose game with perfect information that involves three players. Each stage $n \in \mathbb{N}$ is divided into two sub-stages: in the first sub-stage Player 3 selects a move, and in the second sub-stage Player 1 (if n is even) or Player 2 (if n is odd) selects a move. Formally, at each stage $n \in \mathbb{N}$,

- If n is even, Player 3 selects an element of Z_n, and then Player 1 selects an element of B_n.

- If n is odd, Player 3 selects an element of Z_n, and then Player 2 selects an element of B_n.

Thus, the tree of the game $\widehat{G}_\varepsilon(W)$ is

$$\widehat{T}_\varepsilon := \left(\bigcup_{n \in \mathbb{N}} \left(\prod_{k=0}^{n-1} (Z_k \times B_k) \right) \right) \cup \left(\bigcup_{n \in \mathbb{N}} \left(\prod_{k=0}^{n-1} (Z_k \times B_k) \right) \times Z_n \right).$$

The term $\bigcup_{n \in \mathbb{N}} \left(\prod_{k=0}^{n-1} (Z_k \times B_k) \right)$ is the set of positions in the first sub-stage and the term $\bigcup_{n \in \mathbb{N}} \left(\prod_{k=0}^{n-1} (Z_k \times B_k) \right) \times Z_n$ is the set of positions in the second sub-stage.

Remark 7.42 (The stages in $\widehat{G}_\varepsilon(W)$). *In our definition of $\widehat{G}_\varepsilon(W)$, two players move sequentially in every stage. This definition is made for convenience, so that stage n in the auxiliary game will correspond to stage n in the original game. The "correct" definition of $\widehat{G}_\varepsilon(W)$ would require Player 3 to move in all even stages, Player 1 to move in stages $n = 1 \bmod 4$, and Player 2 to move in stages $n = 3 \bmod 4$.*

The moves of Player 1 in $\widehat{G}_\varepsilon(W)$ define a behavior strategy of Player I in $G(W)$: at every even stage n, Player 1 selects a mixed move for every element of the information partition P_n. Similarly, the moves of Player 2 in $\widehat{G}_\varepsilon(W)$ define a behavior strategy of Player II in $G(W)$.

While Players 1 and 2 are "real" players who select mixed moves, Player 3 is a dummy player, who has no free will and follows a given behavior strategy, which we define now. To explain this behavior strategy, we introduce two fictitious persons, Dian and Ingrid. Suppose that Dian observes the moves $(b_0, b_1, \dots)$ made by Players 1 and 2 in $\widehat{G}_\varepsilon(W)$ (where $b_n \in B_n$ for every $n \in \mathbb{N}$), and randomly selects moves according to these distributions. That is, Dian selects a move $a_0 \in \mathcal{A}$ according to b_0, then for the element $\pi(\langle a_0 \rangle, P_1)$ of the partition P_1 she selects a move $a_1 \in \mathcal{A}$ according to $b_1(\pi(\langle a_0 \rangle), P_1)$, and so on. Ingrid observes the choices of Dian with delay: For each $m \in \mathbb{N}$, she observes the move a_m at stage $k(m)$. Thus, Dian plays the role of the *dice*, and Ingrid represents the *information* that is revealed to Players I and II along the game (albeit with delay: the move a_m is observed at stage $k(m)$, while according to the information structure P, $k(m)$ is an upper bound on the stage in which a_m is observed).

Since the information structure satisfies eventual perfect monitoring, the information Ingrid obtains along the game $\widehat{G}_\varepsilon(W)$ is a whole play $x \in [T]$.

Suppose that in the first n stages, Players 1 and 2 selected the mixed moves $(b_0, b_1, \dots, b_{n-1})$, Dian selected some sequence of n moves, and Ingrid has observed Dian's choices for all stages m with $k(m) < n$. At stage n, Ingrid is going to observe Dian's choices for all stages $m \in K_n$. Denote by $\mu_n = \mu_n(z_0, b_0, z_1, b_1, \dots, z_{n-1}, b_{n-1}) \in \Delta(Z_n)$ the conditional probability on Z_n given Ingrid's information at the beginning of stage n.

We are now ready to define Player 3's behavior strategy in the game $\widehat{G}_\varepsilon(W)$: Player 3 selects her move, which is an element of Z_n, according to the probability distribution $\mu_n(z_0, b_0, z_1, b_1, \dots, z_{n-1}, b_{n-1})$:

$$\sigma_3(\langle z_0, b_0, z_1, b_1, \dots, z_{n-1}, b_{n-1} \rangle) := \mu_n(z_0, b_0, z_1, b_1, \dots, z_{n-1}, b_{n-1}).$$

Note that this behavior strategy is independent of the behavior strategies σ_1 and σ_2 of Players 1 and 2 in $\widehat{G}_\varepsilon(W)$. The position that is reached at stage n, namely, $(z_0, b_0, z_1, b_1, \dots, z_{n-1}, b_{n-1})$, depends on the behavior strategies σ_1 and σ_2, but the mixed move selected by Player 3 at this position is independent of these behavior strategies.

To complete the definition of the auxiliary game $\widehat{G}_\varepsilon(W)$, we need to define a winning set $\widehat{W}_\varepsilon \subseteq [\widehat{T}_\varepsilon]$. We set:

$$\widehat{W}_\varepsilon := F^{-1}(W) := \{\langle z_0, b_0, z_1, b_1, \dots \rangle : F(z_0, z_1, \dots) \in W\} \subseteq [\widehat{T}_\varepsilon].$$

Since F is continuous, it is in particular Borel measurable, hence the set $\widehat{W}_\varepsilon$ is Borel.

7.11 THE VALUE OF $\widehat{G}_\varepsilon(W)$ AND $G(W)$

We next show that the auxiliary game $\widehat{G}_\varepsilon(W)$ has a value, and that its value can be approximated by the value of an auxiliary game $\widehat{G}_\varepsilon(W_0)$ where $W_0 \subseteq W$ is compact.

Lemma 7.43. *For every $\varepsilon > 0$ and every Borel set $W \subseteq [T]$, the game $\widehat{G}_\varepsilon(W)$ has a value. Moreover, there is a compact set $W_0 \subseteq W$ such that $v(\widehat{G}_\varepsilon(W_0)) \geq v(\widehat{G}_\varepsilon(W)) - \varepsilon$.*

Proof. The game $\widehat{G}_\varepsilon(W)$ is an alternating-move game with three players, where Player 3 follows a fixed strategy, the winning set of Player 1 is the Borel set $F^{-1}(W)$, which is the losing set of Player 2. By Exercise 5.15, the game has a value. By Exercise 5.25, there is a compact set $C \subseteq F^{-1}(W)$ such that

$$v(\widehat{T}_\varepsilon, C) \geq v(\widehat{G}_\varepsilon(W)) - \varepsilon.$$

Denote $W_0 := F(C)$. Since F is continuous, W_0 is compact.

The winning set in the game $\widehat{G}_\varepsilon(W_0)$ is $F^{-1}(W_0)$, and the winning set in the game $(\widehat{T}_\varepsilon, C)$ is C. Since $F^{-1}(W_0) = F^{-1}(F(C)) \supseteq C$,

$$v(\widehat{G}_\varepsilon(W_0)) \geq v(\widehat{T}_\varepsilon, C) \geq v(\widehat{G}_\varepsilon(W)) - \varepsilon,$$

as desired. □

Remark 7.44 (On the necessity of the approximation of behavior strategies). *In Section 7.7 we approximated strategies by functions from positions to finite subsets of $\Delta(\mathcal{A})$. This approximation is required in the proof of Lemma 7.43. In this proof, we used Exercise 5.15, which allowed us to deduce that the game $\widehat{G}_\varepsilon(W)$ has a value, and Exercise 5.25, which allowed us to approximate the winning set W with a compact subset C. Both exercises assume that the sets of moves of the players at all positions are finite.*

Since the game $\widehat{G}_\varepsilon(W)$ is an alternating-move game, its value exists as long as at each position, at most one player has an infinite set of moves, see Exercise 5.20. Hence, the approximation is not necessary to deduce the existence of the value of $\widehat{G}_\varepsilon(W)$.

To use Exercise 5.25, though, we do need to assume that the sets of moves are finite. Indeed, when the sets of moves are not necessarily finite, the approximating subset C is closed. When the sets of moves at all positions are finite, the set of plays is compact, and then C is compact. When the sets of moves at some positions are infinite, the set of plays is not necessarily compact, and C may fail to be compact.

Lemma 7.45. *For every $\varepsilon > 0$ and every Borel set $W \subseteq [T]$,*

$$v(\widehat{G}_\varepsilon(W)) - 2\varepsilon \leq \overline{v}(G(W)).$$

Proof. To prove the result, we will fix an ε-minmax strategy σ_{II} of Player II in $G(W)$, and define a strategy $\widehat{s}_2$ of Player 2 in $\widehat{G}_\varepsilon(W)$ such that $\mathrm{val}\,(\widehat{s}_2) \leq \mathrm{val}\,(\sigma_{\mathrm{II}}) + \varepsilon$. This will imply that

$$v(\widehat{G}_\varepsilon(W)) = \overline{v}(\widehat{G}_\varepsilon(W)) = \inf_{\widehat{s}_2 \in S_2(\widehat{T}_\varepsilon)} \mathrm{val}\,(\widehat{s}_2) \leq \mathrm{val}\,(\sigma_{\mathrm{II}}) + \varepsilon \leq \overline{v}(G(W)) + 2\varepsilon,$$

as the lemma claims.

Fix then an ε-minmax strategy σ_{II} of Player II in $G(W)$.

Step 1: Constructing a strategy $\widehat{s}_2$ in $\widehat{G}_\varepsilon(W)$.

Theorem 7.41 (with the roles of the players exchanged) implies that there is a strategy $\sigma'_{II} \in S_{II}(T)$ such that $\sigma'_{II}(p) \in \Delta_{\varepsilon/2^{\mathrm{len}(p)}}$, for every position $p \in T$, and

$$\left| \mathbf{P}_{\sigma_I,\sigma_{II}}(W) - \mathbf{P}_{\sigma_I,\sigma'_{II}}(W) \right| \le \varepsilon, \quad \forall \sigma_I \in \Sigma_I(T,P).$$

This implies that

$$\left| \mathrm{val}\,(\sigma_{II}) - \mathrm{val}\,(\sigma'_{II}) \right| \le \varepsilon,$$

and, in particular, σ'_{II} is a 2ε-minmax strategy of Player II in $G(W)$. For every odd number $n \in \mathbb{N}$, collecting the mixed moves $\sigma'_{II}(Q)$ for atoms $Q \in P_n$, we obtain a function $\sigma'_{II,n} : P_n \to \Delta_{\varepsilon/2^n}$.

Let $\widehat{s}_2$ be the following strategy of Player 2 in the auxiliary game $\widehat{G}_\varepsilon(W)$, which mimics σ'_{II} and is defined by:

$$\widehat{s}_2(\langle z_0, b_0, z_1, b_1, \dots, z_n \rangle) := \sigma'_{II,n}, \quad \forall \langle z_0, b_0, z_1, b_1, \dots, z_n \rangle \in \widehat{T}_\varepsilon,$$

so that indeed $\widehat{s}_2(\langle z_0, b_0, z_1, b_1, \dots, z_n \rangle) \in B_n$.

Step 2: $\mathrm{val}\,(\widehat{s}_2) \le \mathrm{val}\,(\sigma'_{II})$.

To prove the claim, we will show that for every strategy $\widehat{s}_1$ in $\widehat{G}_\varepsilon(W)$ there is a strategy σ_I in $G(W)$ such that $\mathbf{P}_{\widehat{s}_1,\widehat{s}_2}(\widehat{W}_\varepsilon) = \mathbf{P}_{\sigma_I,\sigma'_{II}}(W)$. This will imply that

$$\mathrm{val}\,(\widehat{s}_2) = \sup_{\widehat{s}_1 \in S_1(\widehat{T}_\varepsilon)} \mathbf{P}_{\widehat{s}_1,\widehat{s}_2}(\widehat{W}_\varepsilon) \le \sup_{\sigma_I \in S_I(T)} \mathbf{P}_{\sigma_I,\sigma'_{II}}(W) = \mathrm{val}\,(\sigma'_{II}),$$

and Step 2 will follow.

Fix then any strategy $\widehat{s}_1$ of Player 1 in the auxiliary game $\widehat{G}_\varepsilon(W)$. We would like to define a strategy σ_I in $G(W)$, such that the play under (σ_I, σ'_{II}) mimics the play under $(\widehat{s}_1, \widehat{s}_2)$. To this end, fix an odd number $n \in \mathbb{N}$ and an atom $Q \in P_n$. We would like to define

$$\sigma_I(Q) := \widehat{s}_1(\langle z_0, b_0, z_1, b_1, \dots, z_n \rangle)(Q), \tag{7.7}$$

for properly chosen $z_0, b_0, z_1, b_1, \dots, z_n$. The atom Q in P_n determines the moves selected Player I in past stages, as well as the moves selected by Player II in past stages that were revealed to Player I up to stage n. These moves include the moves $z_0, z_1, \dots, z_n$ selected by Dian in $\widehat{G}_\varepsilon$ up to stage n. We thus know how to define $z_0, z_1, \dots, z_n$ in Equation (7.7).

The entries $(b_k)_{k=0}^{n-1}$ in Equation (7.7) are determined by $\widehat{s}_1$ and $\widehat{s}_2$:

$$b_k := \begin{cases} \widehat{s}_1(\langle z_0, b_0, z_1, b_1, \dots, z_{k-1}, b_{k-1}, z_k \rangle), & \text{if } 0 \le k \le n, k \text{ is even,} \\ \widehat{s}_2(\langle z_0, b_0, z_1, b_1, \dots, z_k \rangle) = \sigma'_{II}(p_k), & \text{if } 0 \le k \le n, k \text{ is odd.} \end{cases} \tag{7.8}$$

In words, σ_I selects the mixed move that $\widehat{s}_1$ selects when it faces $\widehat{s}_2$. It follows that

$$\mathbf{P}_{\widehat{s}_1,\widehat{s}_2}(\widehat{W}_\varepsilon) = \mathbf{P}_{\sigma_I,\sigma'_{II}}(W),$$

as we wanted to show. $\qquad\square$

We are now ready to conclude the proof of Theorem 7.27.

Proof of Theorem 7.27. Fix for the moment $\varepsilon > 0$, and consider the auxiliary game $\widehat{G}_\varepsilon(W)$. By Lemma 7.43, there is a compact set $W_0 \subseteq W$ such that

$$v(\widehat{G}_\varepsilon(W_0)) \geq v(\widehat{G}_\varepsilon(W)) - \varepsilon. \tag{7.9}$$

Consequently,

$$
\begin{aligned}
\underline{v}(G(W)) &\geq \underline{v}(G(W_0)) & (7.10) \\
&= \overline{v}(G(W_0)) & (7.11) \\
&\geq v(\widehat{G}_\varepsilon(W_0)) - 2\varepsilon & (7.12) \\
&> v(\widehat{G}_\varepsilon(W)) - 3\varepsilon, & (7.13)
\end{aligned}
$$

where Equation (7.10) holds since $W \supseteq W_0$, Equation (7.11) holds by Theorem 7.34, Equation (7.12) holds by Lemma 7.45, and Equation (7.13) holds by the choice of W_0.

Exchanging the roles of the two players we obtain that

$$\overline{v}(G(W)) < v(\widehat{G}_\varepsilon(W)) + 3\varepsilon.$$

It follows that

$$\overline{v}(G(W)) < \underline{v}(G(W)) + 6\varepsilon \leq \overline{v}(G(W)) + 6\varepsilon.$$

Since ε is arbitrary, we deduce that $\overline{v}(G(W)) = \underline{v}(G(W))$, and so the game $G(W)$ has a value. □

7.12 DISCUSSION

The main result of this chapter, Theorem 7.27, was proved in Shmaya (2011). This result assumes that the information structure is deterministic. What happens when the information arrives at a random stage? That is, a player does not know in advance when she will learn the move of the other player in, say, stage n; all she knows is that she will eventually learn it. Arieli and Levy (2015) studied this extension, and proved that under appropriate conditions on the information revelation process, the game is still determined, see Exercise 7.13.

The game that we presented is a win-lose game. Does the value exist when the outcome is given by any bounded and Borel-measurable payoff function? To date this question is still open.

We assumed that the information structure satisfies eventual perfect monitoring — at the end of the game (at time ∞), the players know the realized play. Suppose we weaken this requirement, and assume that the information structure satisfies *eventual monitoring of the outcome*: each player can eventually distinguish between any two plays that yield a different outcome. To date it is not known whether such games have a value.

7.13 EXERCISES

Exercises marked with a star are more challenging than unmarked ones, while those marked with two stars are even more difficult.

1. Suppose $\mathcal{A} = \{0,1\}$. For each $n \in \mathbb{N}$, let P_n be the information partition that consists the two sets $\{p = \langle a_0, a_1, \ldots, a_{n-1}\rangle \in T: a_{n-1} = 0\}$ and $\{p = \langle a_0, a_1, \ldots, a_{n-1}\rangle \in T: a_{n-1} = 1\}$. What does the player who selects the move at stage n know at that stage?

2. Write down the information partitions $(P_n)_{n \in \mathbb{N}}$ that correspond to the following situation, where $\mathcal{A} = \{0,1\}$:

   - Both players have perfect recall.
   - The move that Player I selects in stage $2n$ is told to Player II in stage $2n^2 + 1$.
   - If Player II selects in stage 1 the move 0, then Player I learns the moves of Player II with delay 3; if Player II selects in stage 1 the move 1, then Player I learns the moves of Player II with delay 5.

3. Consider a situation where for every $n \in \mathbb{N}$, the move that a player selects at stage n is known by her opponent at stage $n+1$, forgotten by the opponent at stage $n+3$, and then known again in stages $n+5$, $n+7$, $n+9$, ... In addition, each player always remembers the moves she selected in earlier stages.

   (a) Describe the information structure.
   (b) Does the information structure satisfy perfect recall?
   (c) Does the information structure satisfy eventual perfect monitoring?

4. (*) Does every information structure that satisfies eventual perfect monitoring also satisfy perfect recall? Justify your answer.

5. Prove that in the setup described in Subsection 7.3.3, the maxmin value is $\underline{v} = \frac{1}{2}$.

   Hint: You can prove that the following behavior strategy σ_I guarantees $\frac{1}{2}$ for Player I :

   - In stage 0, selects each of the moves c and s with probability $\frac{1}{2}$.
   - For each $n \geq 1$, in stage $2n$ select the move that Player II selected in stage $2n - 1$.

6. (**) Consider the example of stopping games that is described in Section 7.3 with the following information structure P:

   - P satisfies perfect recall.
   - Player II never observes the move taken by Player I in stage 0.

- Player I never observes the move taken by Player II in stage 1.
- For all $n \geq 3$, the player who moves at stage n, observes at that stage the move taken by the other player in the previous stage.

Prove that $\underline{v} < \frac{3}{8} < \frac{1}{2} \leq \overline{v}$. In particular, the stopping game with information structure P does not have a value.

7. (*) Consider the following two-player zero-sum alternating move game:

- In stage 0, Player I selects "a" or "b".
- In stage 1, Player I selects again "a" or "b".
- In stage 2, Player II selects "not aa" or "not bb".
- The information structure is complete ignorance: in stage 1 Player I does not know what she selected in stage 0, and in stage 2 Player II does not know what Player I selected in stages 0 and 1.
- Player I wins in the following cases:
  - Player I selects "a" in both stages 0 and 1, and Player II selects "not aa".
  - Player I selects "b" in both stages 0 and 1, and Player II selects "not bb".
- Player II wins otherwise.

Thus, in this game Player II is trying to guess whether Player I did not select the moves "aa" in stages 0 and 1, or whether Player I did not select the moves "bb" in stages 0 and 1.

(a) Does the game have perfect recall?

(b) What is the minmax value of the game?

(c) What is the maxmin value of the game?

(d) What is the value of the game when the players use mixed strategies? Recall that a mixed strategy is a probability distribution on pure strategies, see Remark 7.28.

8. In this exercise we extend Exercise 3.10 to games with imperfect monitoring.

Equip $S_I(T, P)$ and $S_{II}(T, P)$ with the topology that is analogous to the topology defined for these spaces in Exercise 3.10.

Prove that the function $x : S_I(T, P) \times S_{II}(T, P) \rightarrow [T]$ that assigns to each pair of strategies $(s_I, s_{II}) \in S_I(T, P) \times S_{II}(T, P)$ the play that they induce is continuous.

9. Prove Theorem 7.38: Let $G = (T, W, P)$ be a game with imperfect monitoring, and let $\sigma_I, \sigma_I' \in \Sigma_I(T, P)$ be two behavior strategies. Prove that

$$\left| \mathbf{P}_{\sigma_I, \sigma_{II}}(W) - \mathbf{P}_{\sigma_I', \sigma_{II}}(W) \right| \leq d(\sigma_I, \sigma_{II}), \quad \forall \sigma_{II} \in \Sigma_{II}(T, P).$$

10. Prove Theorem 7.41.

11. (*) List all arguments in the proof of Theorem 7.27 where the assumption of eventual perfect monitoring was used, and all arguments where the assumption of perfect recall was used.

12. (*) Let $\mathcal{A} = \{0, 1\}$ and $T = \mathcal{A}^{<\mathbb{N}}$. Denote by $v_1(W)$ the value of the game with perfect monitoring and winning set W. Denote by $v_2(W)$ the value of the game with imperfect monitoring and winning set W, where Player I knows all previous moves of Player II, and Player II knows all moves of Player I except the last move. Is it true that for every tail-measurable set $W \subseteq [T]$ we have $v_1(W) = v_2(W)$?

13. (**) In this exercise we study games where the moves of the players are observed with a random delay.

    Let $\mathcal{A}$ be a finite set of moves, let $T = \mathcal{A}^{<\mathbb{N}}$ be a tree, let $W \subseteq [T]$ be Borel measurable, and let $(K_n)_{n \in \mathbb{N}}$ be a sequence of i.i.d. random variable with a geometric distribution with parameter $\frac{1}{2}$:

    $$\mathbf{P}(K_n = m) = \frac{1}{2^{m+1}}, \quad \forall m \in \mathbb{N}, \forall n \in \mathbb{N}.$$

    Consider a game with perfect recall, where the players alternately select moves, and, for each $n \in \mathbb{N}$, the move a_n selected in stage n is told to the other player in stage $n + 1 + 2K_n$. Prove that the game has a value.

14. (**) In the proof of Theorem 7.27 we defined an auxiliary game $\widehat{G}_\varepsilon(W)$, where Player 3 selects the move for stage n at stage $k(n)$. We then proved that $v(\widehat{G}_\varepsilon(W))$ exists and is close to $v(G(W))$.

    For each of the following three variations of $\widehat{G}_\varepsilon(W)$, determine whether the value of the auxiliary game exists and is close to $v(G(W))$.

    (a) Player 3 selects the move for stage n at stage $\max\{k(n) - 2, n\}$.

    (b) Player 3 selects the move for stage n at stage $k(n) + 1$.

    (c) Player 3 selects the move for position p at stage $\ell(p)$, where $\ell(p)$ is some function that satisfies $\ell(p) \geq k(\text{len}(p))$, for each $p \in T$.

CHAPTER 8

Open Problems

I end the book by listing some problems related to the material I presented that were not solved at the time the book was published.

1. Consider a two-player zero-sum game with a winning set. In Section 3 we assumed that the players observe the past moves of their opponent. Here we consider a variation of this model, where the players do not observe each other's moves. Rather, after every stage n, with probability α the move a_n of stage n is publicly announced (so both players know that it was announced), and with probability $1 - \alpha$ the move a_n is not announced (and both players know that it was not announced), where $\alpha \in (0, 1)$ is some fixed real number. Does this game have a value?

2. In Section 6.4 we assumed that a single player deviates, and our goal was to identify the deviator. What happens when two players deviate? Can we then identify at least one of the deviators? And what happens when a larger subset of players may deviate? In Exercise 6.11 we studied this problem, but did not answer it.

3. In Chapter 6.8 we proved that two-player Big Match games with tail-measurable payoffs have an ε-equilibrium for every $\varepsilon > 0$. Does the same hold without the tail-measurability assumption?

4. Given $\varepsilon \geq 0$, a vector of behavior strategies σ in a multiplayer repeated game is a *subgame perfect ε-equilibrium* if it induces an ε-equilibrium in all subgames. Does there exist a subgame perfect ε-equilibrium in multiplayer repeated games with tail-measurable payoffs? Flesch, Kuipers, Mashiach-Yaakovi, Schoenmakers, Shmaya, Solan, and Vrieze (2014) provided an example of a two-player game without a subgame-perfect ε-equilibrium for $\varepsilon > 0$ sufficiently small, but where the payoff function is not tail measurable.

5. Consider the model of two-player zero-sum alternating-move games with imperfect monitoring, when the outcome is given by any bounded and Borel-measurable payoff function. Suppose that the information structure satisfies eventual monitoring of the outcome. Does the game have a value?

DOI: 10.1201/9781003582106-8

Bibliography

[1] Alon N., Gunby B., He X., Shmaya E., and Solan E. (2024) Identifying the Deviator, *Annals of Applied Probability*, **34**(5), 4694–4708.

[2] Arieli I. and Levy J.Y. (2011) Infinite Sequential Games with Perfect but Incomplete Information, *International Journal of Game Theory*, **40**, 207–211.

[3] Arieli I. and Levy Y.J. (2015) Determinacy of Games with Stochastic Eventual Perfect Monitoring, *Games and Economic Behavior*, **91**, 166–185.

[4] Ashkenazi-Golan G., Flesch J., Predtetchinski A., and Solan E. (2022) Existence of Equilibria in Repeated Games with Long-Run Payoffs, *Proceedings of the National Academy of Sciences of the United States*, **119**(11).

[5] Ashkenazi-Golan G., Flesch J., Predtetchinski A., and Solan E. (2025) Regularity of the Minmax Value and Equilibria in Multiplayer Blackwell Games, *Israel Journal of Mathematics*, **266**, 25—67.

[6] Ashkenazi-Golan G., Flesch J., and Solan E. (2022) Absorbing Blackwell games, arXiv:2208.11425.

[7] Aumann R.J. (1964) Mixed and Behavior Strategies in Infinite Extensive Games, in *Advances in Game Theory*, Annals of Mathematics Study 52, edited by M. Dresher, L.S. Shapley, and A.W. Tucker, Princeton University Press, pp. 627–650.

[8] Billingsley P. (1995) *Probability and Measure*, John Wiley and Sons.

[9] Blackwell D. (1967) Infinite Games and Analytic Sets, *Proceedings of the National Academy of Sciences of the USA*, **58**(5), 1836–1837.

[10] Blackwell D. (1969) Infinite G_δ-Games with Imperfect Information, *Applicationes Mathematicae*, **10**, 99–101.

[11] Blackwell D. and Ferguson T.S. (1968) The Big Match, *Annals of Mathematical Statistics*, **39**, 159–163.

[12] Borwein J.M. and Zhuang D. (1986) On Fan's Minimax Theorem, *Mathematical Programming*, **34**, 232–234.

[13] Bruyére V. (2021) Synthesis of Equilibria in Infinite-Duration Games on Graphs, *ACM SIGLOG News*, **8**(2), 4–29.

[14] Buffard T., Levrel G., and Mayo S. (2024) Borel Determinacy: a Streamlined Proof, arXiv preprint arXiv:2401.09659.

[15] Bryant R. (2002) Borel Determinacy and Metamathematics, MA Thesis, University of North Texas.

[16] Chatterjee K. and Henzinger T.A. (2012) A Survey of Stochastic ω-Regular Games, *Journal of Computer and System Sciences*, **78**, 394–413.

[17] Das T., Fishman L., Simmons D., and Urbański M. (2017) A Variational Principle in the Parametric Geometry of Numbers, with Applications to Metric Diophantine Approximation, *Comptes Rendus Mathematique*, **355**(8), 835–846.

[18] Das T., Fishman L., Simmons D., and Urbański M. (2024). A Variational Principle in the Parametric Geometry of Numbers, *Advances in Mathematics*, **437**, 109435.

[19] Debs G. and Saint Raymond J. (1996) Some Applications of Game Determinacy, *Acta Universitatis Carolinae. Mathematica et Physica*, **37**(2), 7–23.

[20] Doob J.L. (1953) *Stochastic Processes*, Wiley.

[21] Durrett R. (2019). *Probability: Theory and Examples*, Cambridge university press.

[22] Dynkin E.B. (1967) Game Variant of a Problem on Optimal Stopping, *Soviet Mathematics Doklady*, **10**, 270–274.

[23] Fan K. (1952) Fixed-Point and Minimax Theorems in Locally Convex Topological Linear Spaces, *Proceedings of the National Academy of Sciences of the USA*, **38**(2), 121–126.

[24] Feller W. (1971) An Introduction to Probability Theory and its Applications, Volume I, Wiley series in probability and mathematical statistics, 3rd edn.(Wiley, New York, 1968).

[25] Filar J. and Vrieze K. (1997) *Competitive Markov Decision Processes*, Springer Science and Business Media.

[26] Flesch J., Kuipers J., Mashiach-Yaakovi A., Schoenmakers G., Shmaya E., Solan E., and Vrieze K. (2014) Equilibrium Refinements in Perfect Information Games with Infinite Horizon. *International Journal of Game Theory*, **43**, 945–951.

[27] Flesch J. and Solan E. (2023) Equilibrium in Two-Player Stochastic Games with Shift-Invariant Payoffs, *Journal des Mathémathiques Pures et Apliquées*, **179**, 68–122.

[28] Flesch J. and Solan E. (2024) Stochastic Games with General Payoff Functions, *Mathematics of Operations Research*, **49**(3), 1349–1371.

[29] Flesch J. and Solan E. (2025) Repeated Games with Tail-Measurable Payoffs, In *David Gale: Mathematical Economist: Essays in Appreciation on his 100th Birthday*, Monographs in Mathematical Economics (Springer).

[30] Gale D. and Stewart F.M. (1953) Infinite Games with Perfect Information, *Contributions to the Theory of Games*, **2**, 245–266.

[31] Gillette D. (1957) Stochastic Games with Zero Stop Probabilities, *Contributions to the Theory of Games*, **3**, Princeton University Press.

[32] Gimbert H., Renault J., Sorin S., Venel X., and Zielonka W. (2016) On Values of Repeated Games with Signals, *Annals of Applied Probability*, **26**(1), 402–424.

[33] Gowers T. (2013) Determinacy of Borel Games, `https://gowers.wordpress.com/2013/08/23/determinacy-of-borel-games-i/`.

[34] Grädel E. and Ummels M. (2008) Solution Concepts and Algorithms for Infinite Multiplayer Games, In *New Perspectives on Games and Interaction*, **4**, 151–178.

[35] Jaśkiewicz A. and Nowak A.S. (2018) Non-zero-sum Stochastic Games, In *Handbook of Dynamic Game Theory*, **1**, 281–344.

[36] Kneser H. (1952) Sur un Théoreme Fondamental de la Théorie des Jeux, *Comptes Rendus de l'Académie des Sciences de Paris*, Séries I, Mathématique, **234**, 2418–2420.

[37] Kuhn H.W. (1957) Extensive Games and the Problem of Information, In Kuhn H. and Tucker A.W., *Contributions to the Theorem of Games*, Annals of Studies, **28**. Princeton University Press, 193–216.

[38] Levy Y.J. and Solan E. (2020). Stochastic Games, In *Complex Social and Behavioral Systems: Game Theory and Agent-Based Models*, 229–250.

[39] Maitra A. and Sudderth W. (1998) Finitely Additive Stochastic Games with Borel Measurable Payoffs, *International Journal of Game Theory*, **27**, 257–267.

[40] Marks A.S. (2016) A Determinacy Approach to Borel Combinatorics, *Journal of the American Mathematical Society*, **29**(2), 579–600.

[41] Martin D.A. (1970) Measurable Cardinals and Analytic Games, *Fundamenta Mathematicae*, LXVI, 287–291.

[42] Martin D.A. (1975) Borel determinacy, *Annals of Mathematics*, **102**(2), 363–371.

[43] Martin D.A. (1985) A Purely Inductive Proof of Borel Determinacy, *Recursion Theory* (Ithaca, NY, 1982), **42**, 303–308.

[44] Martin D.A. (1990) An Extension of Borel Determinacy, *Annals of Pure and Applied Logic*, **49**(3), 279–293.

[45] Martin D.A. (1998) The Determinacy of Blackwell Games, *Journal of Symbolic Logic*, **63**(4), 1565–1581.

[46] Martin D.A. (2020) Determinacy of Infinitely Long Games, unpublished manuscript, `https://www.math.ucla.edu/~dam/booketc/D.A._Martin,_Determinacy_of_Infinitely_Long_Games.pdf`.

[47] Maschler M., Solan E., and Zamir S. (2020) *Game Theory*, Cambridge University Press.

[48] Mashiah-Yaakovi A. (2015) Correlated Equilibria in Stochastic Games with Borel Measurable Payoffs, *Dynamic Games and Applications*, **5**(1), 120–135.

[49] Mertens J.-F. (1987) Repeated Games, *Proceedings of the International Congress of Mathematicians*, **0**, (Berkeley, California, 0986), 0528–0577, American Mathematical Society, Providence, RI.

[50] Mertens J.-F. and Neyman A. (1981) Stochastic Games, *International Journal of Game Theory*, **10**, 53–66.

[51] Montalbán A. and Shore R.A. (2012) The Limits of Determinacy in Second-Order Arithmetic, *Proceedings of the London Mathematical Society*, **104**(2), 223–252.

[52] Moschovakis Y.N. (2009) *Descriptive Set Theory*, American Mathematical Society.

[53] Mycielski J. (1964) On the Axiom of Determinateness, *Fundamenta Mathematicae*, **53**(2), 205–224.

[54] Neeman I. (2010) Determinacy in $L(\mathbb{R})$, In *Handbook of Set Theory*, pp. 1877–1950. Springer, Dordrecht.

[55] Orkin M. (1972). Infinite Games with Imperfect Information, *Transactions of the American Mathematical Society*, **171**, 501–507.

[56] Shmaya E. (2011) The Determinacy of Infinite Games with Eventual Perfect Monitoring, *Proceedings of the American Mathematical Society*, **139**(10), 3665–3678.

[57] Shmaya E., Solan E., and Vieille N. (2003) An Application of Ramsey Theorem to Stopping Games, *Games and Economic Behavior*, **42**, 300–306.

[58] Solan E. (2022) *Stochastic Games*, Cambridge University Press.

[59] Solan E. and Vieille N. (2002) Uniform Value in Recursive Games. *The Annals of Applied Probability*, **12**, 1185–1201.

[60] Solan E. and Vieille N. (2002) Correlated Equilibrium in Stochastic Games, *Games and Economic Behavior*, **38**, 362–399.

[61] Solan E. and Vieille N. (2003) Deterministic Multi-Player Dynkin Games, *Journal of Mathematical Economics*, **39**(8), 911–929.

[62] Solan E. and Vieille N. (2015) Stochastic Games: A Perspective, *Proceedings of the National Academy of Sciences of the USA*, **112**(45), 13743–13746.

[63] Vervoort M.R. (1996) Blackwell games, *Lecture Notes-Monograph Series*, 369–390.

[64] Vieille N. (2000) Solvable States in N-Player Stochastic Games, *SIAM Journal on Control and Optimization*, **38**(6), 1794–1804.

[65] von Neumann J. (1928) Zur Theorie der Gesellschaftsspiele, *Mathematische Annalen*, **100**(1), 501–504.

[66] Voorneveld M. (2010) The Possibility of Impossible Stairways: Tail Events and Countable Player Sets, *Games and Economic Behavior*, **68**(1), 403–410.

[67] Williams D. (1991) *Probability with Martingales*, Cambridge University Press.

[68] Zermelo E. (1913) Über eine Anwendung der Mengenlehre auf die Theorie des Schachspiels, In *Proceedings of the Fifth International Congress of Mathematicians II*, 501–504.

Index

For Product Safety Concerns and Information please contact our EU
representative GPSR@taylorandfrancis.com
Taylor & Francis Verlag GmbH, Kaufingerstraße 24, 80331 München, Germany

www.ingramcontent.com/pod-product-compliance
Lightning Source LLC
Chambersburg PA
CBHW082006190326
41458CB00010B/3088